Engineering Reliability: Fundamentals and Applications

R. Ramakumar

**PSO/Albrecht Naeter Professor
and Director,
Engineering Energy Laboratory**

**Oklahoma State University
Stillwater, Oklahoma**

PRENTICE HALL
Englewood Cliffs, New Jersey 07632

Library of Congress Cataloging-in-Publication Data

Ramakumar, R. (Ramachandra)
   Engineering reliability : fundamentals and applications / R.
Ramakumar.

     p. cm.
   Includes bibliographical references and index.
   ISBN 0-13-276759-7
   1. Reliability (Engineering)   I. Title.
TA169.R36 1993   91-44604
620′.00452—dc20   CIP

Acquisitions editor: Alan Apt
Production editor: Irwin Zucker
Copy editor: Brian Baker
Prepress buyer: Linda Behrens
Manufacturing buyer: David Dickey
Supplements editor: Alice Dworkin
Cover design: Wanda Lubelska Design
Editorial assistant: Shirley McGuire

 © 1993 by Prentice-Hall, Inc.
A Simon & Schuster Company
Englewood Cliffs, New Jersey 07632

Printed in the United States of America

10 9 8 7 6 5 4 3 2 1

ISBN   0-13-276759-7

Prentice-Hall International (UK) Limited, *London*
Prentice-Hall of Australia Pty. Limited, *Sydney*
Prentice-Hall Canada Inc., *Toronto*
Prentice-Hall Hispanoamericana, S.A., *Mexico*
Prentice-Hall of India Private Limited, *New Delhi*
Prentice-Hall of Japan, Inc., *Tokyo*
Simon & Schuster Asia Pte. Ltd., *Singapore*
Editora Prentice-Hall do Brasil, Ltda., *Rio de Janeiro*

Contents

4 PROBABILITY DISTRIBUTION FUNCTIONS AND THEIR APPLICATION IN RELIABILITY EVALUATION 76

5 *COMBINATORIAL ASPECTS OF SYSTEM RELIABLITY* 148

Preface

The ambit of reliability is vast, and applications permeate all branches of science and engineering. Senior-level engineering students and professional engineers can all benefit from the use of reliability methods. All engineering systems, from the simplest to the most complex, can benefit from integrating reliability evaluation concepts into their planning, design, and operational phases. As technological advances produce more and more complex devices and systems that are massively expensive to build, and even more expensive if they fail to operate as designed, performance evaluation using reliability analysis techniques takes on an ever-increasing importance. The emphasis currently placed on the quality and reliability of products, especially in the high-tech arena, further stresses the need for studying, quantifying, innovating, and designing to improve the reliability of engineering systems.

In spite of the growing importance of the field, courses on reliability in engineering curricula have been slow in coming. Most universities have one course at the senior-elective/first-year graduate level in one of their engineering schools (electrical, mechanical, or industrial) which draws students from all branches of engineering. Some schools also offer specialized application courses, such as power system reliability evaluation. Recent trends indicate a climate of change, with the introduction of courses in reliability on a larger scale with increasing enrollments.

This book is designed to serve as a text for a senior-elective/first-year graduate course on reliability for students from all branches of engineering. The emphasis is on the fundamentals and applications of classical concepts in reliability engineering. The text is not meant to be a compendium on reliability. The material contained in it has evolved over a period of more than a decade of teaching a first course on the subject to seniors and first-year graduate students from all branches of engineering. The book is written in a manner that is suitable for self-study, even if the reader has no prior background in probability theory. Nearly 500 carefully selected examples and problems are included to illustrate the basic concepts and to assist in self-study; answers to all the problems are given at the end of the text. Thus, the book should also serve as a helpful reference for practicing engineers trying to enter the field of reliability. There is more than enough material for a one-semester course (15 weeks, 3 lectures per week of 50 minutes duration each); hence, the instructor has some flexibility in picking and choosing topics to fit the needs of the class. A complete solution manual is available as an aid, and a series of appendices at the end of the book enhances its value as a reference source.

After a very brief introduction and overview of the subject of reliability, the text begins with a chapter reviewing the basics of probability and random variables to assist those who have not been exposed to these topics, as well as those requiring a quick review to jog their memory. Reliability functions and their significance are introduced immediately afterwards. This arrangement of topics is intended to make the reader appreciate the reasons for studying the material that follows on mathematical models and associated details. Chapter 4 deals with probability distribution functions and their application in reliabilty evaluation. Combinatorial aspects of system reliability are discussed in Chapter 5 in some detail, including the consideration of three-state devices. Markov models are introduced in Chapter 6 as a prelude to the study of repairable components. The reliability evaluation of engineering systems using Markov models is considered in Chapter 7. A collection of approximate, but useful methods for system reliability evaluation is presented in Chapter 8. Included are considerations of the influence of weather, scheduled maintenance, nonexponential failure distributions, common-mode failures, and rare-event approximations. The next two chapters deal with topics selected from an immense array of possibilities to give the reader a glimpse of how material learned thus far can be applied: Chapter 9 discusses the close relationship between reliability and economics, and Chapter 10 introduces accelerated testing and models.

Limiting the scope of the text to a one-semester first-course on classical engineering reliability necessitated the omission of several worthwhile topics. However, a course based on the text should still provide the reader with a solid foundation on the basics and key concepts of classical reliability.

It is the author's fervent hope that the contents and organization of this text will stimulate many seniors, first-year graduate students, and practicing engineers in all branches of engineering to enter the exciting and important world of reliability.

ACKNOWLEDGMENTS

The material contained in this book is based on the work of many stalwarts. I am indebted to all of them, especially the ones listed in the reference section at the end of the book, for their contributions. Also, scores of students at Oklahoma State University have been subjected to the material in bits and pieces over a period or more than 15 years. My appreciation extends to each and every one of them for their role in the development of the book.

I wish to thank Professor Gautam Dasgupta of Columbia University and Professor David Soldan of Kansas State University for reviewing the complete manuscipt and for their comments and suggestions, which have contributed to the evolution of a much better book than otherwise would have resulted.

The confidence Professor William L. Hughes placed in me nearly a quarter of a century ago enabled me to serve at Oklahoma State University, where I learned, taught, and developed the material for this text. My sincere thanks to him for the role he continues to play as a friend and colleague.

The preparation of this book has been a long and arduous task, interrupted by the departures of several loved ones from this world. Throughout this period, my wife Gokula, son Sanjay, and daughter Malini patiently endured my ups and downs, and I am deeply grateful to them for their support, encouragement, and the time I wanted to but could not spend with them.

The author would very much appreciate receiving any corrections or constructive suggestions from the readers.

Finally, it has been a pleasure to work with Elizabeth Kaster, Jaime Zampino, Alan Apt, Shirley McGuire, Irwin Zucker, and their colleagues at Prentice Hall. I thank them for the meticulous care and professionalism they exhibited throughout this project. The final form of the book is due largely to their efforts, for which I am profoundly grateful.

R. Ramakumar

Stillwater, Oklahoma

Dedicated to
my wife, Gokula;
to my children, Sanjay and Malini;
and
to the memories of
my mother, father, father-in-law, and sister, Umabai.

Introduction
and Overview

1.1 INTRODUCTION

All of us have our own notions about what the term "reliability" means. In the broadest sense, it is a *measure of performance:* A reliable person is dependable, trustworthy, and consistent. However, it is hard to draw a definite line demarcating people who are reliable and people who are not. It is even more difficult to compare two individuals and conclude which of them is more reliable.

We can apply the term "reliability" not only to human activity but also to assess the performance of systems. As systems have grown more complex, the consequences of their unreliable behavior have become severe in terms of cost, effort, lives, etc., and the interest in assessing system reliability and the need for improving the reliability of products and systems have become very important.

The degree of interest one has in the reliability of a system and the standard of reliability to be achieved are closely coupled to the consequences of unreliable behavior. Improving reliability will, in general, cost more, but the achievement of reliability usually saves money and sometimes saves lives. Accordingly, the need to maintain an "economic balance" determines the level of reliability one should aspire for in the design of components and systems. Instead of asking of a system, "Is it reliable?" one should ask, "Is it reliable enough?" Finding an answer to this question obviously requires the quantification of reliability, which is achieved by using the theory of probability and statistics.

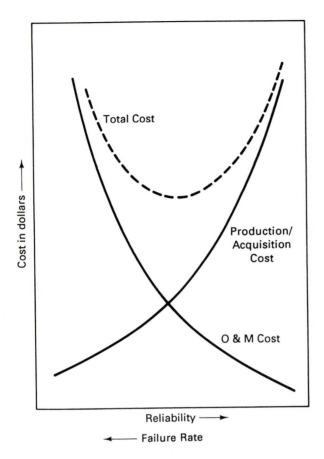

Figure 1.1 shows the relationship between reliability, failure rate, operation and maintenance cost, production and/or acquisition cost, and total cost. Total cost is at a minimum for a particular value of reliability, and prudence dictates that we aim for that value in our design choice, unless there are compelling countervailing reasons.

1.2 HISTORICAL PERSPECTIVE

During its expansion after World War I, the aircraft industry was the first industry to use reliability concepts. Initially, everything was qualitative—two or more engines are better than one, etc. As the number of aircraft grew during the 1930s, reliability was slowly being quantified as the mean failure rate and the average number of failures of an airship or airplane. By the 1940s, aircraft requirements were being expressed in terms of accident rates—one per 100,000 hours of flying time, etc. In Germany, the development of the V–1 and the V–2 missiles contributed much to the application of reliability concepts. During their development and testing, it was realized that a large number of fairly strong series of interconnections can be more unreliable than a single weak link if reliance is placed on all of them.

The unavailability of electronic equipment during the Korean war got the United States military interested in reliability assessment. The relationship between reliability, cost, and maintenance requirements became important, and military contracts started to carry bonus and penalty clauses tied to the degree of reliability realized during field testing.

During the 1950s, the nuclear industry started to develop, and reliability concepts were being increasingly used in the design of nuclear power plants and their control systems. By this time, the fundamentals of reliability theory were well worked out, and reliability technology was being applied to almost all major developments, from space missions and reconnaissance missions to electric power systems, computer systems, complex software, chemical processing plants, and military hardware. The great blackout of November 1965 in the northeastern states gave a strong impetus to the serious application of reliability concepts in the design and expansion of electric power systems.

Examples of high-reliability systems are all around us, from aircraft systems, electric power generating stations, chemical plants, and power systems to telephone and communication systems and computer systems and networks.

1.3 DEFINITION OF RELIABILITY

Although the term "reliability" can be applied to various types of human activities, as well as to the performance of physical systems or functional objects, our objective in this book is to focus only on the latter. Thus, we will use a variety of terms to denote the entity under study. The terms "unit," "component," "device," "system," and "equipment" will be used interchangeably as appropriate throughout this book.

System reliability is the probability that the system will perform its intended function for a specified interval of time under stated conditions. This statement looks rather simple and straightforward; however, a second look at it reveals that there are many terms that need definition or clarification.

First of all, a system need not fail completely in order for it not to perform its intended function successfully. An electric motor may work, but overheat; a beam may not break, but may deflect more than what can be tolerated; an amplifier may not operate at the required gain level or clarity; a gun may fire, but may be way off target; and so on. Consequently, we have to define rather clearly what is expected and what we mean by failure. Second, specification of a time interval need not always be in hours or months or years. Depending on the system under study, it could be time as measured by a clock, or the exact time of operation of the system, or the number of cycles of operation in the case of on-off and cyclically operated devices. Statements about the conditions of operations should include information on the nature and amount of load and the environmental conditions during operation. The term "load" is to be interpreted in a general sense; it could denote the electrical load on a transformer, generator, or a transmission line, the total weight on a bridge, the number of computations per second in the case of microprocessors, the volume of digitized information through a fiber optic cable, the number of barrels per day of crude oil being processed by a refinery, and the like. Environmental conditions refer to factors such as humidity, ambient temperature, visibility in the case of aircraft landings, amount of dust in the case of electric motors, and so forth.

It should be clear by now that we cannot talk intelligently about the reliability of a system, unless we specify a long list of terms and conditions in addition to defining failure in no uncertain terms.

If, during N trials of an experiment or mission, failure occurred n times, then the probability of occurrence of a failure can be estimated as

$$\hat{P}_f = \left(\frac{n}{N}\right) \qquad (1\text{--}1)$$

where the caret indicates that P is an estimator and not a true value. The exact probability of a failure is defined in the limiting sense as the number of trials becomes very large; thus,

$$P_f = \lim_{N \to \infty} \left(\frac{n}{N}\right) \qquad (1\text{--}2)$$

Obviously, in real life, we cannot repeat anything an infinite number of times. Hence, one has to surmise when the number of trials is large enough to assume that the estimator is close enough to the true value. Because a system's reliability is its probability of success, we have

$$R_s = 1 - P_f \qquad (1\text{--}3)$$

Thus, we are able to quantify system reliability in terms of a number that lies between zero and one. This number is absolutely meaningless, unless it is accompanied by all the conditions and clarifications we have discussed thus far.

1.4 THE ROLE OF RELIABILITY EVALUATION

Reliability considerations can be beneficial in almost all stages of engineering endeavors. The nature of the considerations may vary, however. Starting with the definition of the problem at hand or the system and its conceptual design, through the phases of detailed design and development, fabrication or manufacture, and operation and maintenance, attention to reliability can reduce the risk of failure, lower the associated costs, and improve the performance of the system for a given expenditure of resources.

In the conceptual design phase, system requirements are delineated and the definition of failure evolves. Several broad-brush types of questions are answered, including the characteristics of the operating environment. Attention to details such as ease of maintenance, design safety factors, the need to replace some of the components just before they wear out, avoidance of safety hazards, and simplicity of operation take place during the detailed design and development phase. It may even be necessary to build and test experimental prototypes to try out alternatives and fine-tune the design. Suitable reliability growth models are employed to analyze the data obtained from a repeated test-and-fix process to eventually achieve the desired reliability goals.

Reliability considerations during fabrication or manufacture come under the aegis of quality control. Adherence to reliability standards built in during the design phase must be tested by suitable sampling of the product and by employing statistical quality control techniques. With large one-of-a-kind systems, since testing is impossible, it becomes even more important to adhere to high standards as far as the individual components and subsystems are concerned. Finally, in the operation and maintenance phase, reliability considerations enable us to decide on the quality and frequency of maintenance that is needed. Also, in this phase, very good records of failure must be kept and the information fed back to the design group so that appropriate design modifications can be incorporated. Proper maintenance also includes the replacement of certain parts at predetermined time intervals and in accordance with accepted safety practices.

As this brief discussion indicates, the field of reliability has grown sufficiently large in the past few decades to include separate specialized subtopics, such as reliability analysis, failure modeling, reliability optimization, reliability growth and its modeling, reliability testing, reliability data analysis and plotting, quality control, maintenance engineering, system safety analysis, accelerated testing, and life cycle costing.

Classical reliability analysis techniques use parameters derived from actual test data in models to evaluate the performance of systems or components. As the systems become more and more complex and, consequently, expensive, sufficient data may not be available to extract the necessary parameters with a reasonable amount of confidence. Also, in some instances, an educated guess may have to be made based on similar equipment, data from operation of the system in different environments, or even theoretical or design considerations. In these cases, the assumed parameters and the associated distributions are labeled ''prior'' information. Bayesian reliability analysis techniques are then employed to combine whatever test data are available with such information to generate new, ''posterior'' distributions and other relevant information. As additional information becomes available, this procedure can repeatedly be used to develop improved estimates of parameters and improved distributions.

1.5 RELIABILITY ASSESSMENT

Irrespective of the type and complexity of a system under study, three major steps are essential for assessing the reliability of the system. First, we have to construct a reliability model. Then the model must be analyzed and the appropriate reliability indices calculated. Finally, the results obtained from the analysis should be evaluated and interpreted. The exact details and depth of these three activities may vary from case to case, but we can always delineate the steps for further scrutiny.

In many respects, modeling can be thought of as an art form. The model employed should be complicated enough to truly represent the major operational and other features of the system under study, but at the same time it should be simple enough to yield the desired results without an undue amount of effort. In addition, in selecting a model, the analyst should keep in mind the accuracy of the data available and the accuracy required of the results sought. Every model, whether it is simple or complicated, is based on a set of assumptions. The validity of these assumptions in the context of the ongoing analysis should be checked carefully. It is always a good idea to use the simplest possible model to get a feel for the problem at hand and then systematically relax the assumptions one by one, until the model becomes complicated enough. At the end of the model development phase, the system under study will have been divided into subunits, with models and block diagrams of their own, interconnected in a logical way to emulate the overall behavior of the system. As far as possible, it is desirable to split the system into blocks that are functionally independent of each other. If that is not possible, then any dependencies that exist should be clearly spelled out.

The second step, analysis of the model, involves the calculation of the various reliability performance indices from the block diagrams and the associated reliability functions and parameters. Combinatorial aspects of system reliability and approximate methods (in the case of large systems) are employed in this phase of reliability assessment.

Evaluation and interpretation of the results obtained from the analysis involves answering several key questions. First, we have to decide whether the results obtained are precise enough. If they have uncertainties, then their confidence limits must be established using well-known statistical techniques. Comparisons with data from field tests accumulated over a period of time for similar systems or the same system are always a welcome means of checking the validity of the theoretical results. The final decisions on possible design improvements or a determination of whether the system is reliable enough will be heavily based on a variety of economic factors and other considerations, some of which may even be intangible.

1.6 SCOPE AND ORGANIZATION OF THE TEXT

The primary purpose of this book is to serve as a text for a first course in engineering reliability. The structure, organization, and level of treatment are such that practicing engineers can also use the book, for self-study, even though they may have no prior background in reliability or may need a quick review of probability and random variables. The second chapter contains just the minimum amount of material on probability and random variables needed to follow the rest of the text. It may be skipped completely by those who already have a good background in these topics.

Catastrophic failure models, reliability functions, and their interrelationships are introduced early in the text, in Chapter 3. Probability distribution functions commonly used in reliability analyses are discussed in Chapter 4, which also contains applications in modeling the wearout region, ideal maintenance, and ideal repair. Combinatorial aspects of system reliability are considered in some detail in Chapter 5. Both binary (two-state) and ternary (three-state) devices are considered. The modeling of standby systems and of spares with instant replacement are also discussed in this chapter.

As a prelude to the study of repairable components, Markov models are introduced in Chapter 6. This is followed by a chapter on evaluating the reliability of engineering systems using Markov models. Frequency and duration techniques, cumulated states, the mixed product approach, the frequency balance approach, modeling normal repair and preventive maintenance, and standby systems are some of the topics considered in Chapter 7.

A collection of useful, but approximate methods for evaluating system reliability is presented in Chapter 8. Among these methods are a consideration of the influence of weather, scheduled maintenance, Markov representation of nonexponential failure distributions, common-mode failures, and rare-event approximations.

The topics discussed in the next two chapters are intended to arouse the curiosity of the reader as to how the basic material developed in Chapters 3 through 8 can be applied to investigate different problems in the field of reliability. Chapter 9 discusses the close relationship between economics and reliability. Four subtopics are examined: the economics of redundancy, the economics of repair and maintenance, availability and present-worth analysis, and the concept of acceptance number based on economics. The final chapter of the text deals with some of the models used in accelerated testing and the need for such procedures.

1.7 SOME IMPORTANT DEFINITIONS

Throughout this book, definitions of terms are given as they are introduced. However, definitions of some of the more important terms used in the reliability evaluation of engineering systems are collected in this section for easy access and reference. For the reader just being introduced to the field of reliability engineering, it may be worthwhile to scan these definitions before delving into the book.

Acceptance sampling is the procedure of taking a sample from a group (or lot) of typically similar items for testing to see whether to accept or reject the group (or lot) as a whole. Acceptance testing is employed when the number of items is too large to test all of them individually or if the testing is costly or destructive.

The *availability* $A(t)$ of a system is the probability that the system is operating successfully at time t. The steady-state availability of the system is

$$A = \lim_{t \to \infty} A(t)$$

A *bathtub curve* is the typical shape of the hazard (failure) rate function plotted against time for living entities, physical objects such as electronic and mechanical components, and even complex systems. This curve decreases at the beginning due to early failures (called *infant mortality*; see below), stays nearly constant over its useful lifetime, and then increases rapidly due to wearout (or aging).

Binary components are components that can exist in one of two states—successful, or the *up* state, and failure, or the *down* state.

The *catastrophic failure* of a component occurs when its performance level shifts to an extreme limit at which the component becomes useless. Catastrophic failures indicate that the component cannot be repaired or that repair is of no consequence to the success of the mission involving the component.

Common-mode failure is the simultaneous failure or outage of several units due to a common cause such as fire, flood, accidents, and the like.

Forced outage occurs when emergency conditions related to a component forces the component out of service immediately. The number of such happenings per unit of time is called the *forced outage rate*.

The *hazard rate* function $\lambda(t)$ is the same as the instantaneous failure rate. It is this function that exhibits the different life cycles of the component clearly and distinctly.

Infant mortality refers to the early failures experienced by a system, component, or species. As time progresses, the failure rate decreases due to what is known as *debugging* in the case of engineering systems and due to the loss of the weaker specimens in the case of living populations.

Maintainability is the ability of a component or unit to be retained in or restored to a state in which it can perform service under the conditions of use for which it is designed. Maintenance must be performed under prescribed conditions and using specified procedures and resources.

> *Corrective* or *unscheduled maintenance* is the action performed to restore an item to a specified condition following a failure.

> *Preventive* or *scheduled maintenance* is the action performed at periodic time intervals to maintain the component in a specified condition. These actions will, in general, include systematic inspection, diagnosis and detection of problems, and prevention of incipient failures.

The *mean time between failures* (*MTBF*) is the expected or mean value of the random variable called "time between failures." If the repair time is very short compared to the time between failures, then the mean time between failures is almost the same as the mean time to failure (see next entry), and the two terms are used interchangeably; otherwise, the mean time between failures is equal to the sum of the mean time to failure and the mean time to repair (see next two entries).

The *mean time to failure* (*MTTF*) is the expected or mean value of the random variable called "time to failure."

The *mean time to repair* (*MTTR*) is the expected or mean value of the random variable called "time to repair."

The *median time to failure* (T_{50}) is the time that splits the list of times to failure in half.

Parallel structure describes a system that can succeed when at least one of the components succeeds. Such a system is also known as a *fully or completely redundant* system.

An *r-out-of-n system* consists of *n* identical independent components of which at least $r < n$ of the components should succeed in order for the system to succeed. Such systems are also called *partly redundant* or *majority-vote* systems. For $r = 1$, they become truly parallel (fully redundant) systems, and for $r = n$, they become series (nonredundant) systems.

Redundancy is the existence of more than one means, identical or nonidentical, for accomplishing a given task or mission.

> *Active redundancy* is redundancy wherein all the components of a system are operating (are hot) all the time, rather than being brought into service only when needed.

> *Standby redundancy* is redundancy wherein a system's backup unit does not operate (is cold) until it is needed and is switched on only when the main unit fails to perform its task.

The *reliability R(t)* of a system is the probability that the system has operated successfully over the time interval from 0 to *t*. Successful operation is defined as the performance of the system's intended function for a specified interval of time under stated conditions. In the case of repairable components, the availability *A(t)* of the system gives no information on how many (if any) failure-repair cycles have occurred between 0 and *t*. If repair is impossible, then *A(t)* must be equal to *R(t)*. With repair, $R(t) \leq A(t)$; in general, *R(t)* is more stringent than *A(t)*.

The *security S(t)* of a system is the probability of a breach of system security at times in the near future (about a few hours), given a known operating condition at the time calculations are made. The security is also the probability of system insecurity at time *t* into the future. If *S(t)* is equal to 1 at $t = 0$, then there is no point in finding out what would happen in the future, since the system is insecure at $t = 0$. The basic difference between reliability and security is the time frame of interest: Reliability involves long-term considerations, whereas security typically involves short-term considerations.

Series (or chain) structure describes a system whose success depends on the success of all of its components. The components themselves need not be physically or topologically in series. Such a system is also known as a *nonredundant* system.

Ternary components are components that can exist in one of three states; one state corresponds to success, and the other two correspond to two different failure modes: open failure and short failure. Diodes and fluid valves are examples of ternary components.

The *unavailability U(t)* of a system is simply the complement of its availability *A(t)*. Or,

$$U(t) = 1 - A(t)$$

U(t) is the probability that the system is not operating successfully at time *t*. The steady-state unavailability of the system is

$$U = \lim_{t \to \infty} U(t)$$

The useful life of a component refers to that part of its life which lies between the early failure (infant mortality) phase and the wearout (aging) phase. Typically, during the useful life, failures are random and failure rates are approximately constant. Useful life corresponds to the nearly horizontal portion of the bathtub curve.

Wearout failures of components refer to failures that occur due to aging, after the end of the useful lives of the components.

1.8 FURTHER READING

Applications of reliability principles permeate all branches of engineering dealing with systems and devices whose complexity and character range from space rockets to social systems. As stated earlier, this book is intended to just crack open the door so that the reader can begin to appreciate and be able to plunge into various specialized fields and applications, depending on his or her interest. Many excellent books are available that deal with individual topics or branches of reliability engineering, applications, and related fields at varying levels of rigor. The list of references at the end of this text is a reasonable, although not exhaustive, collection of sources that readers can use to gain further knowledge about and insight into reliability.

Probability and Random Variables

2.1 INTRODUCTION

Interest in the study of uncertainties and random phenomena has grown concomitantly with an increase in the demands placed on system performance. This growth has led to the pervasive use of probability and statistics in almost every discipline imaginable—from biology to sociology and including all branches of engineering. Thus, engineers are often faced with problems involving the accumulation and analysis of data on random phenomena. Although random, these phenomena have certain statistical regularities and, therefore, are amenable to mathematical modeling using probability theory and statistics. Information on future outcomes of these phenomena can be deduced using probabilistic models. Since most of us are introduced to probability concepts via the tossing of a coin, we will consider a simple coin toss example first.

■■ **Example 2–1**

If a fair coin is tossed a number of times, the fraction of the times a head (or tail) appears will approach 0.5. This limiting value is the same whether we count every toss, every third toss, every fifth toss, etc. Alternatively, we can toss a number of coins simultaneously just one time and obtain the same result. ■■

If an event A occurs N_A times out of n trials, then the observed relative frequency $\hat{P}(A)$ is given by (N_A/n). The probability of occurrence of event A is then

$$P(A) = \lim_{n \to \infty} (N_A/n) \qquad (2-1)$$

In other words, as n becomes sufficiently large, $\hat{P}(A)$ approaches $P(A)$. Thus, probability is related to physical phenomena only in inexact terms. However, probability theory is an exact discipline developed logically from clearly defined axioms.

We should interpret the probability $P(A)$ of an event A as a number assigned to that event, just as mass is assigned to a body in mechanics and resistance is assigned to an element in circuit theory. We should not worry about the physical meaning of $P(A)$ in the development of the theory, just as we do not worry about the physical meaning of resistance in the development of circuit theory.

Statistics deals with the handling and processing of probabilistic data and with making physical predictions based on the data. We use statistics to arrive at models from actual data, use probability theory to derive certain required quantities, and then revert back to statistics to give physical meaning to the quantities derived.

■■   **Example 2–2**
The possible outcomes of a toss of two fair coins are

HH, HT, TH, TT

The probability of two heads showing is thus $\frac{1}{4}$. However, the probability of one head and one tail showing is $\frac{2}{4} = \frac{1}{2}$, since one head and one tail can occur in 2 out of 4 ways.   ■■

■■   **Example 2–3**
The probability of drawing, at random, a jack out of a well-shuffled deck of cards is $\frac{4}{52}$. But the probability of drawing a jack of diamonds is only $\frac{1}{52}$, since there is only one such card in the deck of 52 cards. The probability of drawing any diamond is $\frac{13}{52}$, or 0.25.   ■■

■■   **Example 2–4**
Suppose winning is defined as drawing a picture card at random out of a deck of cards. If the deck is well shuffled, then the probability of winning is $\frac{12}{52}$, or approximately 0.23. If the deck is doctored in some way, such as putting all the picture cards together, this probability value is no longer applicable.   ■■

■■   **Example 2–5**
Let us suppose that at the beginning of a month, 100 identical components are placed in service. If by the end of the month 6 of them fail, a good approximation to the probability of survival of a component over a period of one month is simply (100 –

6)/100, or 0.94. We consider this only as an approximation because, although 100 is a large number, it may not be sufficiently large to justify designating 0.94 as the correct value of the probability of survival over one month. We can repeat the experiment with a larger sample—say, 1,000 components—and if the number of failures at the end of one month of operation is close to 60, then we can surmise that the sample size of 100 is large enough to obtain an approximation to the desired probability.

If we let the experiment with 100 components continue for one more month, the number of failures over the next month may or may not be equal to 6. Our intuition tells us that if failures are due to purely random events, and not due to aging of components, then the number of failures over the next month will be quite close to 6. On the other hand, if the components have aged significantly, we would expect the number of failures to increase over the next month.

There is another possibility: If there is a problem in the way the components are made, some of them may fail fairly early in their life (equivalent to infant mortality), and if failures during the first month are due largely to this phenomenon, we can even expect a significant reduction in the number of failures during the next month. ■■

2.2 PERMUTATIONS AND COMBINATIONS

Permutations and combinations have fascinated thinkers since 350 B.C. However, their logical development is closely tied to the development of probability theory in the 1600s. Our interest in permutations and combinations is due to their use in systematically counting the number of ways certain events can occur and in solving problems in probability and statistics.

Ordered arrangements of a set of objects are called *permutations*. When the order in which they are arranged is disregarded, then the arrangements are called *combinations*. A combination of n objects can be arranged in $n!$ different ways. Therefore, we have many more possible permutations than combinations for any set of two or more objects.

Consider a set of n different objects. An arrangement of $r \leq n$ of these objects in every possible order is called a *permutation of the n objects taken r at a time*. The total number of possible permutations is designated as

$$_nP_r = \frac{n!}{(n-r)!} \tag{2-2}$$

For $r = n$, $_nP_n = n!$ since $0! = 1$.

■■ **Example 2–6**
Consider a room with r chairs. Suppose we want to seat n people, where $n \geq r$, in this room. How many different ways can this be done?

Note that the first chair can be filled n different ways, the second chair can be filled $(n - 1)$ ways, and so on. Following this line of thinking, we conclude that the rth chair can be filled $(n - r + 1)$ ways. Therefore, the total number of possible arrangements is equal to

$$n(n - 1)(n - 2) \cdot \ldots \cdot (n - r + 1) = \frac{n!}{(n - r)!}$$

If we have n chairs, the number of possible arrangements is equal to $n(n - 1)(n - 2) \ldots 1$, or $n!$

■■

Now let us consider the possibility of repetitions. Consider n objects of which r_1 are alike, r_2 are alike, ... , and r_k are alike, where $r_1 + r_2 + \ldots + r_k = n$. Since r_i identical objects can be arranged in $r_i!$ indistinguishable ways, for $i = 1,2, \ldots, k$, we can easily see that the number of distinguishable permutations possible is equal to $(n!)/(r_1!r_2! \ldots r_k!)$.

Next, consider a set of n different objects. An r-combination is any selection of r out of these n objects *without regard to their order or arrangement*. The number of such r-combinations is designated as $_nC_r$.

Since order does not matter, and since r objects can be arranged $r!$ different ways, we observe that

$$r! \, _nC_r = \, _nP_r = \frac{n!}{(n - r)!} \qquad (2\text{--}3)$$

so that

$$_nC_r = \frac{n!}{r!(n - r)!} \qquad (2\text{--}4)$$

Note that $_nC_r$ is also equal to the binomial coefficient $\binom{n}{r}$. For $r = n$, $_nC_n = 1$; that is, if order is disregarded, there is only one way to pick n items out of n items.

■■   **Example 2–7**

We want to select violin students to fill the first five chairs in a high school orchestra. If there are 17 students in the class, how many possible ways can the selection be made?

Obviously, the order of selection must be considered in the selection process. Therefore, we use the number of permutations and obtain the number of possible ways to fill the first five chairs:

$$_{17}P_5 = \frac{17!}{(17 - 5)!} = \frac{17!}{12!} = 742{,}560$$

Alternately, with 17 students, the first chair can be filled 17 different ways, the second chair can be filled 16 different ways, and so on. Therefore, the total number of ways of filling the first 5 chairs with 17 students is $(17 \times 16 \times 15 \times 14 \times 13)$, or 742,560. In arriving at this number, we have assumed that the chair assigned is not related to the talent of the student, a rather dubious assumption at best! Since our objective is to find the number of *possible* ways, without any concern about the students' capabilities, we are entitled to make this assumption.   ■■

■■   **Example 2–8**

A task force is to be assembled consisting of 2 men and 2 women. There are 6 men and 7 women qualified to fill the positions. How many different committees can be formed out of the finalists?

Since nothing is said about the positions in the committee, the order of selection is irrelevant. Out of 6 men, the number of ways of selecting 2 is $_6C_2$, or $[6!]/[2!(6 - 2)!] = 15$. Similarly, out of 7 women, the number of ways of selecting 2 is $_7C_2$ or $[7!]/[2!(7 - 2)!] = 21$. Therefore, the total number of possible committees that can be constituted, subject to the restriction that there should be 2 men and 2 women, is

$$15 \times 21 = 315$$ ■■

■■   **Example 2–9**

It is customary to use a few letters of the alphabet to designate states, airports, airlines, people, etc. Let us develop a table of the number of possibilities when 2, 3, 4, and 5 letters are used with and without repetition.

There are 26 letters in the English alphabet. Using only 2 letters, we have 26 × 26, or 676, possibilities if repetition is allowed. If not, the number of possibilities decreases to 26 × 25, or 650. Another way to arrive at the number of possibilities without repetition is to use the formula for the number of permutations:

$$_{26}P_2 = \frac{26!}{(26 - 2)!} = \frac{26!}{24!} = 26 \times 25 = 650$$

In a similar fashion, for the case of 3 letters, the number of possibilities with repetition is 26 × 26 × 26, or 17,576, and the number without repetition is $_{26}P_3 = 26 \times 25 \times 24$, or 15,600.

Generalizing this technique for the case of n letters, we have

$$\text{Number of possibilities with repetition} = 26^n$$

$$\text{Number of possibilities without repetition} = {_{26}P_n}$$

Table 2.1 lists the number of possibilities with and without repetition for various values of n.   ■■

TABLE 2.1   NUMBER OF POSSIBLE ACRONYMS WITH n LETTERS

n	Number of possibilities with repetition	Number of possibilities without repetition
2	676	650
3	17,576	15,600
4	456,976	358,800
5	11,881,376	7,893,600

The usefulness of combinations in probability evaluations is illustrated by the next two examples.

■■   **Example 2–10**

The State University car pool wants to purchase 4 cars. After all the models available were surveyed, 8 foreign-made and 10 U.S.-made cars were considered to satisfy all the requirements.

If the cars are chosen at random, what is the probability that 2 of the cars selected are foreign made?

$$\begin{array}{l}\text{Number of ways of selecting} \\ \text{4 out of 18}\end{array} = {}_{18}C_4 = 18!/(4!14!) = 3{,}060$$

$$\begin{array}{l}\text{Number of ways of selecting} \\ \text{2 out of 8}\end{array} = {}_{8}C_2 = 8!/(2!6!) = 28$$

$$\begin{array}{l}\text{Number of ways of selecting} \\ \text{2 out of 10}\end{array} = {}_{10}C_2 = 10!/(2!8!) = 45$$

$$\begin{array}{l}\text{Probability that the cars selected} \\ \text{are 2 foreign and 2 U.S. made}\end{array} = (28 \times 45)/3{,}060 = 0.4118$$

What is the probability that all of the four cars chosen are U.S. made? Following the approach of finding the number of ways of selecting cars, we have:

$$\text{Probability that all 4 are U.S. made} = [{}_{10}C_4\, {}_{8}C_0]/{}_{18}C_4 = 0.0686$$

What is the probability that all of the four cars chosen are foreign made?

$$\text{Probability that all 4 are foreign made} = [{}_{8}C_4\, {}_{10}C_0]/{}_{18}C_4 = 0.0229$$

What is the probability that at least two of the cars are U.S. made?

$$\begin{array}{l}\text{Probability that at least 2 of the cars are} \\ \text{U.S. made}\end{array} = P[2 \text{ or } 3 \text{ or } 4 \text{ are U.S. made}]$$

$$= \frac{{}_{8}C_2\, {}_{10}C_2 + {}_{8}C_1\, {}_{10}C_3 + {}_{8}C_0\, {}_{10}C_4}{{}_{18}C_4}$$

$$= \frac{(28 \times 45) + (8 \times 120) + (1 \times 210)}{3{,}060}$$

$$= 0.7941$$

Alternatively,

probability that at least 2 of the cars are U.S. made $= 1 - P[0 \text{ or } 1 \text{ is U.S. made}]$

$$= 1 - \frac{{}_{8}C_4\, {}_{10}C_0 + {}_{8}C_3\, {}_{10}C_1}{{}_{18}C_4}$$

$$= 1 - \frac{(70 \times 1 + 56 \times 10)}{3{,}060}$$

$$= 0.7941, \text{ as before} \qquad ■■$$

■■   **Example 2–11**

Five defective microchips were accidentally mixed with 45 good ones. After a thorough mixing, 5 microchips are picked simultaneously from the collection of 50. What is the probability that all 5 chips selected are good?

$$\text{Probability that all 5 are good} = \frac{_{45}C_5 \; _5C_0}{_{50}C_5}$$

$$= \frac{1{,}221{,}759 \times 1}{2{,}118{,}760}$$

$$= 0.57664$$

What is the probability that no more than one chip chosen is defective? The probability that no more than one chip is defective is the sum of the probability that no chips are defective and the probability that one chip is defective. So

$$\text{Probability of having 1 defective} = \frac{_5C_1 \; _{45}C_4}{_{50}C_5}$$

$$= \frac{(5)(148{,}995)}{2{,}118{,}760}$$

$$= 0.35161$$

and

$$P[\text{not more than 1 is defective}] = P[0 \text{ defective}] + P[1 \text{ defective}]$$

$$= 0.57664 + 0.35161$$

$$= 0.92825$$

What is the probability that all 5 chips chosen are defective?

$$\text{Probability that all 5 are defective} = \frac{_5C_5 \; _{45}C_0}{2{,}118{,}760}$$

$$= \frac{1}{2{,}118{,}760}$$

$$= 4.72 \times 10^{-7}$$

Although intuitively we feel that this last probability must be rather small, applying combinations enables us to quantify it exactly.   ■■

2.3 BINOMIAL THEOREM

The nth power of $(p + q)$ can be expressed in terms of binomial coefficients as

$$(p + q)^n = p^n + np^{n-1}q + \frac{n(n-1)}{2!} p^{n-2}q^2 + \dots$$

$$+ \frac{n!}{r!(n-r)!} p^{n-r}q^r + \dots + q^n = \sum_{r=0}^{n} \binom{n}{r} p^{n-r}q^r = \sum_{r=0}^{n} {}_nC_r \, p^{n-r}q^r \quad (2\text{-}5)$$

Since $(p + q) = (q + p)$, we can also write

$$(p + q)^n = \sum_{r=0}^{n} {}_nC_r p^r q^{n-r} \qquad (2\text{-}6)$$

2.4 ORDERED SAMPLES

Consider a box containing n different objects. Let us choose one object after another from the box, say, r times. The choice of r objects is called an *ordered sample of size r*.

The question we want to answer is how many different ordered samples of size r are there? There are two possibilities:

(*i*) *Sampling with replacement.* Choose one object at random, note what it is, put it back in the box, mix well, choose another object at random, note what it is, put it back, mix well, and so on. In this case,

$$\left.\begin{array}{l}\text{the number of different}\\ \text{ordered samples of size}\\ r \text{ with replacement from}\\ \text{a population of } n \text{ objects}\end{array}\right\} = \underbrace{n \times n \times n \times \ldots \times n}_{r \text{ times}} = n^r \qquad (2\text{-}7)$$

(*ii*) *Sampling without replacement.* Here the object is not replaced before choosing the next, and

$$\left.\begin{array}{l}\text{the number of different}\\ \text{ordered samples of size}\\ r \text{ without replacement}\\ \text{from a population of } n \text{ objects}\end{array}\right\} \begin{array}{l} = n(n-1)(n-2)\ldots(n-r+1)\\[2mm] = \dfrac{n!}{(n-r)!} = {}_nP_r \end{array} \qquad (2\text{-}8)$$

2.5 PROBABILITY SPACE

In an experiment, the union of all outcomes is the *certain event* or *sample space* S. An event is a subset of S. The *impossible event* or *empty* or *null* set is denoted by $\emptyset$ or $\overline{S}$.

A *probability space* is obtained by assigning, to ith element of S, a real number p_i, called its probability, such that $p_i \geq 0$ and $\sum_i p_i = 1.0$.

An *elementary event* consists of a single element. However, not all single elements need be events.

Two events, A and B, are *disjoint* (mutually exclusive) if they cannot occur simultaneously. Symbolically, $AB = \emptyset$.

Example: Let the events be drawing a king and drawing a 9 (both out of a fair deck of cards). These two events are disjoint, since one card cannot be both a king and a 9.

Example: Let the events be drawing a king and drawing a diamond (both out of a fair deck of cards). These two events are not disjoint, since one card can be both a king and a diamond at the same time.

Two events, A and B, are *independent* if the probability that B occurs is not influenced by whether A has occurred and if, likewise, the probability that A occurs is not influenced by whether B has occurred.

Example: Let the events be drawing a king and drawing another king without replacement. These two events are dependent, because their probabilities are $\frac{4}{52}$ and $\frac{3}{51}$, respectively.

Example: Let the events be drawing a king and drawing another king after replacement and shuffling. These two events are independent, because their probabilities are each $\frac{4}{52}$ separately.

2.6 STUDY APPROACHES

There are several approaches to defining and studying probability. The *classical* approach takes the ratio of the number of favorable cases to the total number of alternatives. Assuming that favorable implies success and unfavorable implies failure, we have:

$$P(\text{success}) = \frac{\text{number of successes}}{\text{total number of possible outcomes}} \qquad (2\text{–}9)$$

$$P(\text{failure}) = \frac{\text{number of failures}}{\text{total number of possible outcomes}} \qquad (2\text{–}10)$$

If each trial leads to success or failure, assuming s and f to be the number of successes and failures, we get:

$$P(\text{success}) = \frac{s}{s + f} \equiv p \qquad (2\text{–}11)$$

$$P(\text{failure}) = \frac{f}{s + f} \equiv q \qquad (2\text{–}12)$$

and

$$p + q = 1 \qquad (2\text{–}13)$$

This approach is too limited in its application because, in many cases, the experiment cannot be repeated. Even if it can be repeated, there is doubt about how many times it should be repeated to approach the exact probabilities of the events.

Other possible approaches are the *relative frequency* approach, *inductive reasoning* (as a measure of belief), and *axiomatic probability*.

Axiomatic probability is based on set theory. It is simple to use, and there are no ambiguities. It is based on three fundamental axioms. It is not based on any application or concept about the meaning of probability. The three axioms are:

(1) $0 \leq P(A) \leq 1$ (2–14)

(2) $P(S) = 1$ (2–15)

where S is the certain event (also called the *universal set* or the *entire sample space*)

(3)   If event A and event B are mutually exclusive (disjoint), then

$P(A + B) = P(A) + P(B)$ (2–16)

where $(A + B)$ means A *or* B, also written as $(A \cup B)$ where $\cup$ denotes union.

Axiom (3) can be extended to an infinite number of mutually exclusive events $A_1, A_2, \ldots$. This property is called *infinite additivity*. We have:

$P(A_1 + A_2 + A_3 + \ldots) = P(A_1) + P(A_2) + P(A_3) + \ldots$ (2–17)

if the events $A_1, A_2, A_3, \ldots$ are mutually exclusive.

2.7 BASIC PROPERTIES AND RULES OF PROBABILITY

We have already discussed disjoint events and independent events. Two events are *complementary* if, whenever one does not occur, the other does. The complement of event A is denoted by $\overline{A}$. The Venn diagram shown in Figure 2.1 illustrates complementation. From the figure, it is plain that

$$A + \overline{A} = S$$ (2–18)

$$A\overline{A} = \emptyset$$ (2–19)

2.7.1 Probability of the Null Event

$$P(\emptyset) = P(\overline{S}) = 0$$ (2–20)

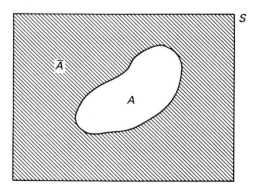

Figure 2.1   Complement of an event.

To prove equation (2–20), recall that S and $\overline{S}$ are disjoint since $S\overline{S} = \emptyset$. Therefore,

$$P(S + \overline{S}) = P(S) + P(\overline{S}) = 1 + P(\overline{S})$$

But the requirement that $[1 + P(\overline{S})]$ be ≤ 1 leads to the conclusion that $P(\overline{S}) = 0$.

2.7.2 Probability of the Union of an Event and Its Complement

$$P(A + \overline{A}) = P(S) = 1 \tag{2–21}$$

Alternatively,

$$P(A) = 1 - P(\overline{A}) \tag{2–22}$$

2.7.3 Conditional (or Dependent) Probability

The notation $P(A \mid B)$ denotes the conditional probability of event A occurring, given the knowledge that event B has occurred. Let us consider n trials of an experiment. In n trials, event B occurs $[nP(B)]$ times and event AB occurs $[n\,P(AB)]$ times. The notation AB means "event A *and* event B"; that is,

$$AB = A \cap B = A \text{ and } B$$

We can easily see that

$$nP(AB) = P(A \mid B)\, nP(B)$$

or

$$P(A \mid B) = \frac{P(AB)}{P(B)} \tag{2–23}$$

By a similar argument, we can show that

$$P(B \mid A) = \frac{P(AB)}{P(A)} \tag{2–24}$$

Note also that

$$P(A \mid B) = \frac{\text{number of ways } A \text{ and } B \text{ can occur}}{\text{number of ways } B \text{ can occur}} \tag{2–25}$$

2.7.4 Intersection of Two Events

The intersection of two events is their simultaneous occurrence (see Figure 2.2). We have

$$P(AB) = P(A)P(B \mid A) \tag{2–26a}$$

$$= P(B)P(A \mid B) \tag{2–26b}$$

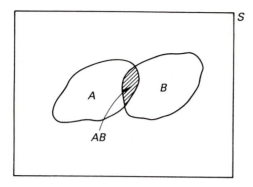

Figure 2.2   Intersection of two events.

If the events are disjoint, then

$$AB = \emptyset \qquad (2\text{--}27)$$

and

$$P(AB) = P(\emptyset) = 0 \qquad (2\text{--}28)$$

If $P(A) \neq 0$ and $P(B) \neq 0$, then we must have

$$P(B \mid A) = 0 \text{ and } P(A \mid B) = 0$$

If the events are independent, then

$$P(A \mid B) = P(A)$$
$$P(B \mid A) = P(B)$$

and

$$P(AB) = P(A)P(B) \qquad (2\text{--}29)$$

For the case of three events A_1, A_2, and A_3, let $A = A_1$ and $B = A_2A_3$. Then

$$P(AB) = P(A)P(B \mid A)$$
$$= P(A_1)P(A_2A_3 \mid A_1)$$
$$= P(A_1) \frac{P(A_1A_2A_3)}{P(A_1)} \frac{P(A_1A_2)}{P(A_1A_2)} \qquad (2\text{--}30)$$

$$P(A_1A_2A_3) = P(A_1) \frac{P(A_1A_2)}{P(A_1)} \frac{P(A_1A_2A_3)}{P(A_1A_2)}$$
$$= P(A_1)P(A_2 \mid A_1)P(A_3 \mid A_1A_2) \qquad (2\text{--}31)$$

Extending this reasoning to n events, we obtain

$$P(A_1A_2 \ldots A_n) = P(A_1)P(A_2 \mid A_1)P(A_3 \mid A_1A_2) \ldots P(A_n \mid A_1A_2 \ldots A_{n-1})$$

$$(2\text{--}32)$$

If the events are all independent of each other, then

$$P(A_1 A_2 \ldots A_n) = P(A_1)P(A_2)P(A_3) \ldots P(A_n)$$

$$= \prod_{i=1}^{n} P(A_i) \tag{2-33}$$

2.7.5 Probability of Union of Nondisjoint Events

Figure 2.3 shows two events, A and B, that are not disjoint. It is easily seen that

$$(A + B) = (A\overline{B} + AB) + (B\overline{A} + AB)$$

$$= A\overline{B} + B\overline{A} + AB$$

Since the three events on the right hand side are disjoint,

$$P(A + B) = P(A\overline{B}) + P(B\overline{A}) + P(AB) + [P(AB) - P(AB)]$$

$$= [P(A\overline{B}) + P(AB)] + [P(B\overline{A}) + P(AB)] - P(AB)$$

$$P(A + B) = P(A) + P(B) - P(AB) \tag{2-34}$$

If events A and B are disjoint, then $AB = \emptyset$, $P(AB) = 0$, and

$$P(A + B) = P(A) + P(B) \tag{2-35}$$

Equation (2–34) can be derived in an alternative manner as follows:

$$P(A \cup B) = P(A \text{ or } B \text{ or } (A \text{ and } B))$$

$$= 1 - P(\text{not } A \text{ and not } B)$$

$$= 1 - P(\overline{A} \cap \overline{B})$$

$$= 1 - P(\overline{A}) P(\overline{B}) \text{ (since } \overline{A} \text{ and } \overline{B} \text{ are independent)}$$

$$= 1 - (1 - P(A))(1 - P(B))$$

$$= P(A) + P(B) - P(AB)$$

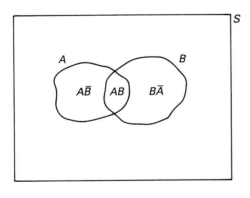

Figure 2.3 Union of two nondisjoint events.

Extending the foregoing reasoning to n nondisjoint events, we have

$$P(A_1 + A_2 + \ldots + A_n) = [P(A_1) + P(A_2) + \ldots + P(A_n)]$$
$$- [P(A_1A_2) + P(A_1A_3) + \ldots$$
$$+ P_{i \neq j}(A_iA_j)]$$
$$+ [P(A_1A_2A_3) + P(A_1A_2A_4) + \ldots$$
$$+ P_{i \neq j \neq k}(A_iA_jA_k)]$$
$$\vdots$$
$$(-1)^{n-1}[P(A_1A_2 \ldots A_n)] \qquad (2\text{--}36)$$

The number of terms on the right-hand side in Equation (2–36) is

$$\binom{n}{1} + \binom{n}{2} + \binom{n}{3} + \ldots + \binom{n}{n} = (2^n - 1)$$

Remember that

$$\binom{n}{k} = {}_nC_k = \frac{n!}{k!(n-k)!} \qquad (2\text{--}37)$$

2.7.6 Bayes' Theorem

Suppose the occurrence of event A is dependent on a number of events B_i that are mutually exclusive, as shown in the Venn diagram of Figure 2.4. Then, it is clear that

$$A = AB_1 + AB_2 + \ldots + AB_n$$

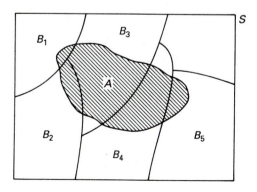

Figure 2.4 Dependence of an event on a set of mutually exclusive events.

Since the events on the right-hand side are disjoint,

$$P(A) = P(AB_1) + P(AB_2) + \ldots + P(AB_n)$$

$$= \sum_{i=1}^{n} P(AB_i)$$

$$= \sum_{i=1}^{n} P(B_i)P(A \mid B_i) \tag{2-38}$$

If $(B_1 + B_2 + \ldots + B_n) \neq S$, then the B_i's do not constitute a partition of S. On the other hand, if $(B_1 + B_2 + \ldots + B_n) = S$, then the B_i's are mutually exclusive and exhaustive and, therefore, constitute a partition of S. This result is expressed in Bayes' theorem as follows:

Theorem.   Let the events B_1, B_2, ..., B_n be mutually exclusive and exhaustive. Let A be any event. Then

$$A = \sum_{i=1}^{n} AB_i \tag{2-39}$$

$$P(A) = \sum_{i=1}^{n} P(B_i)P(A \mid B_i) \tag{2-40}$$

$$P(B_i \mid A) = \frac{P(AB_i)}{P(A)} = \frac{P(B_i)P(A \mid B_i)}{P(A)}$$

$$= \frac{P(B_i)P(A \mid B_i)}{\sum_{i=1}^{n} P(B_i)P(A \mid B_i)} \qquad \text{for all } i$$

Now consider a complex system whose reliability is to be evaluated. Let G be the event that the system is good. For the system to be good, a number of events, designated A_1, A_2, ... , A_n, must occur. These events are called *contingencies*. The Venn diagram shown in Figure 2.5 illustrates a contingency A_k with respect to event G. Clearly,

$$G = \overline{A}_kG + A_kG$$

and

$$P(G) = P(\overline{A}_kG) + P(A_kG)$$

$$= P(\overline{A}_k) P(G \mid \overline{A}_k) + P(A_k) P(G \mid A_k)$$

If the contingencies A_1, A_2, ... , A_n are mutually exclusive and exhaustive, then $G = GA_1 + GA_2 + \ldots + GA_n$, with the possibility that some of the intersections

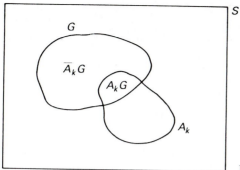

Figure 2.5   Event G and contingency A_k.

on the right-hand side are zero. Then

$$P(G) = \sum_{i=1}^{n} P(A_i)P(G \mid A_i)$$

Moreover,

$$P(A_i \mid G) = \frac{P(A_i)P(G \mid A_i)}{\displaystyle\sum_{i=1}^{n} P(A_i)P(G \mid A_i)} \qquad (2\text{--}41)$$

The basic properties and rules of probability can be applied to solve a wide variety of problems involving the calculation of probabilities of certain events or combinations of events using a set of known or assumed probabilities. The following examples illustrate this point.

■■   **Example 2–12**

Nonrepairable components such as diodes, transistors, microchips, and light bulbs are discarded when they fail. They can be assumed to be in one of two states—good and bad—for reliability modeling. These two states are complementary, so that

$$P[\text{component is good}] + P[\text{component is bad}] = 1$$

Thus, with one of the two probabilities known, the other can easily be calculated.

■■

■■   **Example 2–13**

The idea of complementary events can be applied effectively in many probability calculations. For example, if we have 10 identical devices in a system and we want the probability of two or more failing, instead of finding and summing the probabilities of 2, 3, 4, ... , 10 failing, we can use

$$P[2 \text{ or more failing}] = 1 - [P(0 \text{ failing}) + P(1 \text{ failing})] \qquad ■■$$

■■ **Example 2–14**
Suppose we have a collection of 85 bearings, of which the following are defective
as shown:

> 11 bearings with type I defects
> 8 bearings with type II defects
> 2 bearings with type III defects
> 4 bearings with type I and type II defects
> 2 bearings with all three types of defects

If a bearing picked out of this group has a type II defect, what is the probability
that it also has a type I defect?
Using the conditional probability relationship, we have

$$P(I \mid II) = \frac{P(I \text{ and } II)}{P(II)}$$

$$= \frac{4/85}{8/85} = 0.5$$

If the bearing picked has a type I defect, what is the probability that it also
has a type II defect? This is given as

$$P(II \mid I) = \frac{P(II \text{ and } I)}{P(I)} = \frac{4/85}{11/85} = \frac{4}{11} \qquad ■■$$

■■ **Example 2–15**
With thunderstorms present, the probability of a transmission line going down is
75%; the probability of it going down during calm weather conditions is only 10%.
The probability of thunderstorms appearing is 25%. What is the probability that
thunderstorms are present given that the line is down?
Let us define two events, A and B, as follows:

> A: Thunderstorms are present.
> B: The line is down.

These two events are not independent, since the presence of thunderstorms does
affect the probability of the line going down. We are given that

$$P(A) = 0.25; P(B \mid A) = 0.75; \text{ and } P(B \mid \overline{A}) = 0.1$$

We want $P(A \mid B)$. We have

$$P(A \mid B) = \frac{P(AB)}{P(B)} = \frac{P(B \mid A)P(A)}{P(B)}$$

But

$$P(B) = P(B \mid A)P(A) + P(B \mid \overline{A})P(\overline{A})$$

$$= (0.75)(0.25) + (0.1)(1 - 0.25)$$

$$= 0.2625$$

Therefore,

$$P(A \mid B) = \frac{(0.75)(0.25)}{0.2625} = 0.7143$$

■■

■■   **Example 2–16**

A rocket engine has been tested 1,610 times. On 290 occasions, the engine failed to perform as expected. A payload is fitted with three identical engines of this type and is launched. What is the probability of all three engines failing? Of at least one failing?

For each engine, the probability of failure $= 290/1{,}610 = 0.18$. So the probability of all three engines failing is $(0.18)^3 = 0.005832$, since the engines can be assumed independent. If this assumption is not valid, then we must have additional information on failures when two or three of these engines are tested in close proximity, duplicating the conditions expected under actual operations.

The probability of all three engines succeeding, again assuming independence, is $(1 - 0.18)^3$, or 0.5514. Therefore, the probability of at least one of the engines failing is $(1 - 0.5514)$, or 0.4486.   ■■

■■   **Example 2–17**

A family has two cars, one old and the other reasonably new. On any given Monday, the probabilities of finding the cars in operating condition are 0.95 and 0.85 for the new and old car, respectively. What is the probability that the new *or* the old car is operable?

Let event A be that the new car is in good working condition and event B be that the old car is in good working condition. These two events are not mutually exclusive (disjoint), since both cars can be in good condition simultaneously with a nonzero probability. Therefore, the probability of event *A or B* occurring is

$$P(A \cup B) = P(A) + P(B) - P(A)P(B) = 0.95 + 0.85 - (0.95)(0.85)$$
$$= 0.9925$$

The probability of finding both cars in working condition is

$$P(AB) = P(A)P(B) = 0.95 \times 0.85 = 0.8075$$

since these two events are obviously independent of each other. If both cars are needed, the probability of having to look for a ride on a Monday morning is $(1 - 0.8075) = 0.1925$.

If only one of the two cars is needed, the other car is considered to be a backup, we have a fully redundant situation, and the two cars are said to be logically in parallel. If both cars are needed, then the two cars are said to be logically in series. We will discuss such combinatorial aspects in detail later. For now, note that being logically in series and logically in parallel should not be confused with being physically in series and physically in parallel, as in the case of, say, electrical networks with resistors in series or parallel.   ■■

■■ **Example 2–18**

Three people, A, B, and C, are working independently to crack a secret code. Based on their previous performance, the probabilities that they will succeed are 0.15, 0.2, and 0.3, respectively. Calculate the probability that the code will be broken.

Since all we want is to break the code, either A or B or C should succeed. Therefore,

$$P(\text{code is broken}) = P(A + B + C)$$

where $P(n)$ denotes the probability of the event that person n breaks the code. Applying the expression for the probability of a nondisjoint union, we have

$$P(A + B + C) = [P(A) + P(B) + P(C)]$$
$$- [P(AB) + P(BC) + P(CA)]$$
$$+ [P(ABC)]$$

The three people are working independently and, as such, the probabilities of their successes are independent of each other. Thus,

$$P(A + B + C) = [0.15 + 0.2 + 0.3] - [0.15 \times 0.2 + 0.2 \times 0.3 + 0.3 \times 0.15]$$
$$+ [0.15 \times 0.2 \times 0.3]$$
$$= 0.524 \qquad\qquad ■■$$

■■ **Example 2–19**

A test setup has a 96% probability of correctly identifying a faulty item and a 5% probability of identifying a good item as faulty. A batch of 100 components, of which 2 are known to be defective, is subjected to testing. If the test identifies a component as defective, what is the probability that it is truly defective?

Let event D be that the component is truly defective and event C be that the component is classified by the test as defective. Then we have

$$P(C \mid D) = 0.96; P(C \mid \overline{D}) = 0.05; \text{ and } P(D) = 0.02$$

Applying Bayes' theorem, we obtain

$$P(D \mid C) = \frac{P(D)P(C \mid D)}{P(D)\,P(C \mid D) + P(\overline{D})\,P(C \mid \overline{D})}$$
$$= \frac{0.02 \times 0.96}{0.02 \times 0.96 + 0.98 \times 0.05}$$
$$= 0.2815, \text{ or } 28.15\% \,!$$

Plainly, it is very important that test equipment correctly identify the defective as well as the good components with a high degree of probability. ■■

■■ **Example 2–20**

A personal computer assembly plant purchases a certain component from three different vendors, A, B, and C, who supply 30%, 30%, and 40%, respectively, of the total number of such components required by the plant. On average, the percentage of defective components supplied by *A, B,* and *C* are 4%, 3%, and 2%, respectively. If a component is selected at random, find the probability that it is defective.

We have

$$P(\text{defective}) = P \text{ (it is supplied by } A) \, P(\text{defective} \mid \text{supplied by } A)$$

$$+ \, P(\text{it is supplied by } B) \, P(\text{defective} \mid \text{supplied by } B)$$

$$+ \, P(\text{it is supplied by } C) \, P(\text{defective} \mid \text{supplied by } C)$$

$$= (0.3)(0.04) + (0.03)(0.03) + (0.4)(0.02)$$

$$= 0.029, \text{ or } 2.9\%$$

If the selected component is defective, what is the probability that it was supplied by vendor *C*? By vendor *A*? By vendor *B*?

Applying Bayes' theorem, we obtain

$$P(\text{supplied by } C \mid \text{defective}) = P(C \text{ supplied}) \, P(\text{defective} \mid C \text{ supplied}) \div$$

$$[P(A \text{ supplied}) \, P(\text{defective} \mid A \text{ supplied})$$

$$+ \, P(B \text{ supplied}) \, P(\text{defective} \mid B \text{ supplied})$$

$$+ \, P(C \text{ supplied}) \, P(\text{defective} \mid C \text{ supplied})]$$

$$= \frac{0.4 \times 0.02}{0.029}$$

$$= 0.2759, \text{ or } 27.59\%$$

Thus, on average, 27.59% of the defective components are supplied by vendor *C*. Proceeding in a similar fashion, we have:

$$P(\text{supplied by } A \mid \text{defective}) = \frac{0.3 \times 0.04}{0.029}$$

$$= 0.4138, \text{ or } 41.38\%$$

$$P(\text{supplied by } B \mid \text{defective}) = \frac{0.3 \times 0.03}{0.029}$$

$$= 0.3103, \text{ or } 31.03\%$$

As we might have suspected, vendor *A* supplies the largest fraction (41.38%) of the defective parts. Hence, unless there is an overwhelming economic or other incentive, the overall product quality can be improved by dropping vendor *A* in favor of another vendor.   ■■

■■ **Example 2–21**

A system consists of 17 different components, of which one, component *K*, has been identified as the keystone component. The probability that the keystone component is good is 0.95. If the keystone component is good, then the system succeeds with a probability of 0.9. If not, then the probability of system success drops to 0.8. What is the probability of system success?

$$P(\text{system success}) = P(\text{system success} \mid K \text{ is good}) \, P(K \text{ is good})$$

$$+ \, P(\text{system success} \mid K \text{ is bad}) \, P(K \text{ is bad})$$

$$= (0.9)(0.95) + (0.8)(1 - 0.95)$$

$$= 0.895$$

Alternatively,

$$P(\text{system failure}) = P(\text{system failure} \mid K \text{ is good}) \, P(K \text{ is good})$$

$$+ \, P(\text{system failure} \mid K \text{ is bad}) \, P(K \text{ is bad})$$

$$= (0.1)(0.95) + (0.2)(0.05)$$

$$= 0.105$$

As expected, the sum of the probabilities of system success and system failure is equal to 1.   ■■

2.8 RANDOM VARIABLES

Let us consider an experiment, the outcomes of which have been defined and grouped into a collection called the *sample space*. Depending on the experiment, we can assign a probability value to each of the outcomes. Next, to every element of the sample space (in other words, to each outcome), we assign a specific real number. This assignment can be considered a mapping of every point (element) in the sample space onto the real axis. Such a mapping (or function) is called a *random variable*.

A random variable is a function whose domain is the sample space and whose range is a set of real numbers. These numbers could be the number of dots that show up on a throw of a pair of dice, the time taken to repair a failed component, the wind speed in meters per second, the voltage of a random source, the length of a manufactured part, etc.

■■ **Example 2–22**
Let the experiment just described in general terms be the tossing of a nickel and a dime. There are four possible outcomes: (NH,DH), (NH,DT), (NT,DH), and (NT,DT), where NT means the nickel lands tail up, DH means the dime lands head up, etc. To each of these four outcomes, let us assign a real number—the number of points scored in a hypothetical game. In effect, we have defined a random variable as illustrated in Figure 2.6.   ■■

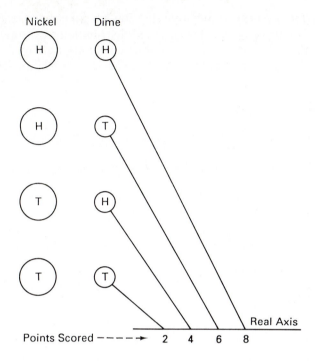

Figure 2.6   Definition of a random variable.

Random variables are numerical-valued quantities whose observed values are governed by the laws of probability. If the random variable X can take on only one of several (even countably infinite) discrete values $x_1, x_2, ..., x_n$ and no other value, it is called a *discrete random variable*. If it can take on a nondenumerably infinite number of values (not necessarily from $-\infty$ to $+\infty$), it is called a *continuous random variable*.

Some examples of discrete random variables are the following:

Toss of a coin or coins
Number of dots showing up on a throw of a die
Color of a ball drawn from a collection of balls

Some examples of continuous random variables are the following:

Length of a manufactured part
Time to failure of a component
Repair time
Peak load on a utility
Duration of a storm
Voltage of a random voltage source

Probability density functions or cumulative distribution functions are used to describe and model random variables. Density and distribution functions are denoted by f and F, respectively. These two functions constitute an integro-differential pair, and all we need is one of the two in order for the other to be easily obtained.

2.8.1 Discrete Random Variables

In the finite case, the random variable X takes on only certain discrete values x_1, x_2, ..., x_n and no other value.

The *probability density function* (or simply, density function) $f(x)$ is defined as follows:

$$f(x_i) = \text{probability that } X = x_i$$

$$= P(X = x_i) \tag{2-42}$$

$$\equiv P(x_i)$$

Since the sample space consists of only the elements x_1 through x_n, we have

$$P(S) = \sum_i f(x_i) = \sum_i P(x_i) = 1 \tag{2-43}$$

The *cumulative distribution function* (or simply, distribution function) $F(x)$ is defined as follows:

$$F(x_i) = P(X \le x_i) \tag{2-44}$$

Since this is a cumulative process, $F(x)$ is nondecreasing or monotonic in nature and eventually reaches the value 1. Graphs of the probability density and distribution functions are shown in Figure 2.7.

■■ **Example 2–23**

In a throw of a fair die, there are only six possible outcomes: 1, 2, 3, 4, 5, and 6. All are equally likely; that is,

$$P(1) = P(2) = P(3) = P(4) = P(5) = P(6) = \frac{1}{6} \qquad\qquad ■■$$

A quick look at the density and distribution functions shown for the throw of a fair die in Figure 2.8 leads us to the conclusion that $F(x)$ is related to $f(x)$ by

$$f(x) = F(x^+) - F(x^-) \tag{2-45}$$

where

$$F(x^+) = \lim_{\epsilon \to 0} F(x + \epsilon) \tag{2-46}$$

$$F(x^-) = \lim_{\epsilon \to 0} F(x - \epsilon) \tag{2-47}$$

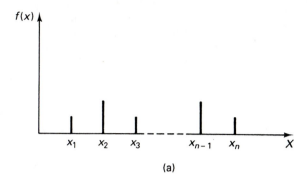

(a)

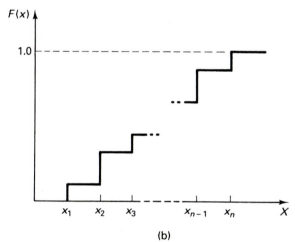

(b)

Figure 2.7 Typical (a) density and (b) distribution functions of a discrete random variable.

In other words, the density function is equal to the value of the discontinuity at x in the distribution function.

2.8.2 Continuous Random Variables

When the random variable X can take on any one of a nondenumerably infinite number of values, the cumulative distribution function $F(x)$ is defined as

$$F(x) = P(X \leq x)$$

Note that $F(x)$ can never exceed 1, can never be negative, and is a nondecreasing function because of its cumulative nature. Every value of x defines an event $X \leq x$ whose probability of occurrence is $F(x)$. Moreover,

$$F(x_1) \leq F(x_2) \quad \text{whenever } x_1 \leq x_2; \lim_{x \to \infty} F(x) = 1; \text{ and } \lim_{x \to -\infty} F(x) = 0$$

As x increases, the chance of X being less than or equal to x improves. The

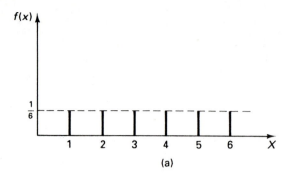

(a)

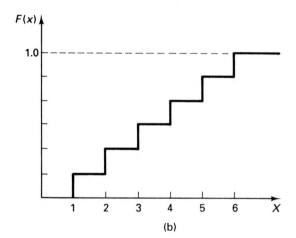

(b)

Figure 2.8 (a) Density and (b) distribution functions for the throw of a fair die.

probability density function $f(x)$ is defined as

$$f(x) \equiv \frac{d}{dx} F(x) \tag{2-48}$$

The function $f(x)$ is truly a density function, since it has a physical meaning only when integrated over a finite interval. For example,

$$P(a \le X \le b) = \int_a^b f(x) \, dx = F(b) - F(a) \tag{2-49}$$

The density function is nonnegative since the distribution function is monotonic (nondecreasing). Figure 2.9 illustrates the general nature of the distribution and density functions for a continuous random variable. The density and distribution functions form an integro-differential pair, and it is possible to treat discrete random variables as a special case of continuous random variables by employing delta functions. The area under the density function is always unity, since $F(x)$ approaches the value of unity as x increases.

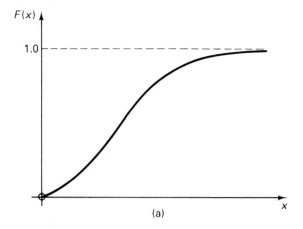

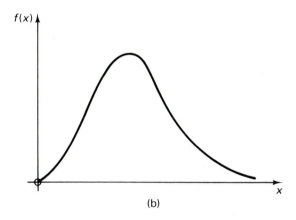

Figure 2.9 Typical (a) distribution and (b) density functions for a continuous random variable.

The probability of occurrence of any one particular value for the random variable is zero, since

$$P(X = a) = \int_{a}^{a} f(x) \, dx = 0 \qquad (2\text{--}50)$$

Modeling a random variable essentially boils down to the selection of an appropriate distribution (or density) function. The following factors enter into the selection:

(*i*) Physical nature of the problem
(*ii*) Assumptions associated with a particular distribution function
(*iii*) Plots of available data and curve fitting
(*iv*) Engineering judgment, simplicity, and convenience.

■■ **Example 2–24**
The density function for a continuous random variable X is given as

$$f(x) = 0.125(x - 2) \text{ for } 2 \leq x \leq 6$$

The corresponding distribution function is

$$F(x) = \int_{2}^{x} 0.125(\xi - 2) \, d\xi$$

$$= 0.0625x^2 - 0.25x + 0.25 \quad \text{for } 2 \leq x \leq 6$$

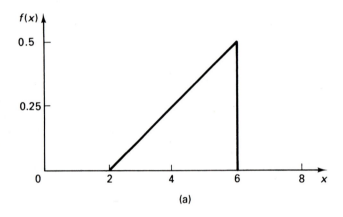

(a)

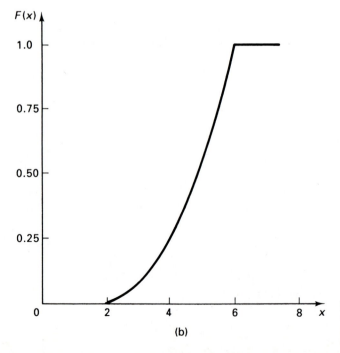

(b)

Figure 2.10   (a) Density and (b) distribution functions for Example 2–24.

These density and distribution functions are sketched in Figure 2.10. Also,

$$P[X > 6] = 0$$

$$P[X \leq 3] = F(3) = 0.0625$$

$$P[3 \leq X \leq 5] = F(5) - F(3) = 0.5$$ ■■

■■ **Example 2–25**

The time to failure of a component is a continuous random variable known to have the density function $0.5 \exp[-0.5t]$, where t is in years. The corresponding distribution function is

$$F(t) = \int_0^t f(\xi) \, d\xi = 0.5 \int_0^t e^{-0.5\xi} \, d\xi$$

$$= 1 - \exp[-0.5 \, t]$$

Note that as $t \to \infty$, $F(t) \to 1$ and the area under the density function is unity. Also, $F(t_1) \leq F(t_2)$ whenever $t_1 \leq t_2$. The probability that the time to failure is between 1 and 2 years is

$$\int_1^2 f(t) \, dt = \exp[-0.5] - \exp[-1]$$

$$= 0.23865$$

The probability that the component will fail within the first year of operation is

$$\int_0^1 f(t) \, dt = 1 - \exp[-0.5] = 0.39347$$

We will discuss this exponential distribution in detail later.   ■■

2.9 MOMENTS OF RANDOM VARIABLES

All the information about a random variable is contained in either the density or the distribution function. Often, the only information that is needed is some numerical value that characterizes the distribution sufficiently. Such a value (or values) is obtained by calculating the *moments of the distribution*. The rth moment of a random variable X computed about the origin is defined as

$$m_r \equiv \int_{-\infty}^{\infty} x^r f(x) \, dx \qquad (2\text{--}51)$$

If the random variable is defined only over the range $x_1 \leq x \leq x_2$, then

$$m_r = \int_{x_1}^{x_2} x^r f(x) \, dx \qquad (2\text{--}52)$$

Note that the zeroth-order moment is the area under the density function, so that

$m_0 = 1$ always. The first-order moment is called the *mean,* the *expected value,* or the *expectation E[X],* which is defined as

$$E[X] = \int_{-\infty}^{\infty} x f(x)\, dx \equiv \mu \tag{2–53}$$

For discrete random variables, we define the expected value as

$$E[X] = \sum_{i=1}^{n} x_i f(x_i) \tag{2–54}$$

If the density function is an even function about the origin, then the first, third, fifth, etc., moments are zero. In general, we shall denote

$$m_r = E[X^r] \tag{2–55}$$

In many practical situations, moments about the mean are often required instead of moments about the origin. The *r*th moment about the mean is defined as

$$\tilde{m}_r = E[(X - \mu)^r]$$

Alternatively,

$$\tilde{m}_r = \int_{-\infty}^{\infty} (x - \mu)^r f(x)\, dx \tag{2–56}$$

Then, for $r = 0$, we obtain $\tilde{m}_0 = 1$, as before. For $r = 1$,

$$\tilde{m}_1 = \int_{-\infty}^{\infty} (x - \mu) f(x)\, dx$$

$$= \int_{-\infty}^{\infty} x f(x)\, dx - \mu \int_{-\infty}^{\infty} f(x)\, dx$$

$$= \mu - \mu$$

$$= 0$$

The second moment about the mean, $\tilde{m}_2$, has a classical definition:

$$\tilde{m}_2 = \int_{-\infty}^{\infty} (x - \mu)^2 f(x)\, dx$$

$$= \text{variance of } X$$

$$= V(X)$$

$$= \text{var } X$$

The *variance* is a measure of the sum of the squares of the deviations from the mean value. A second classical measure, the *standard deviation,* is

$$\sigma \equiv \sqrt{\text{var } X} = + \sqrt{V(X)}$$

A convenient way of calculating the variance is as follows:

$$\text{var } X = \int_{-\infty}^{\infty} (x - \mu)^2 f(x) \, dx$$

$$= \int_{-\infty}^{\infty} x^2 f(x) \, dx - 2\mu \int_{-\infty}^{\infty} x f(x) \, dx + \mu^2 \int_{-\infty}^{\infty} f(x) \, dx$$

$$= E[X^2] - 2\mu \cdot \mu + \mu^2 \cdot 1$$

$$= E[X^2] - [E(X)]^2 \tag{2-57}$$

If the density function is symmetrical about the mean, then $\bar{m}_{2n+1} = 0$ for $n = 0, 1, 2, \ldots$. For most engineering problems, the mean and the variance are the only probabilistic quantities of interest.

The mean and the variance are respectively numerical measures of the central tendency and dispersion of a random variable. A measure of symmetry is provided by the *skewness*, the third moment of the random variable about its mean. The skewness is

$$\bar{m}_3 = E\left[(X - \mu)^3\right] \tag{2-58}$$

For symmetric density functions, the skewness is zero. Positive skewness implies an extended right tail, and negative skewness implies an extended left tail. The mean values of skewed distributions tend to lie on the side with the longer tail.

The fourth moment about the mean is called the *kurtosis* and is a measure of how peaked the data are around the mean. The kurtosis is

$$\bar{m}_4 = E[(X - \mu)^4] \tag{2-59}$$

Peaky distributions tend to have lighter tails, and flat-topped distributions tend to have heavier tails.

■■ **Example 2–26**
Consider the tossing of a fair die. The sample space consists of the six numbers 1, 2, ... , 6. Since the die is fair, we have an equiprobable space and $f(x_i) = \frac{1}{6}$ for all i. Let

$$x_1 = 1; \ x_2 = 2; \ \ldots, \ x_6 = 6$$

Suppose a player wins a number of dollars equal to the number shown on the tossed die. Then the number of dollars won is a random variable $X(x_i)$ with the following density function (given in tabular form):

x_i	1	2	3	4	5	6
$f(x_i)$	1/6	1/6	1/6	1/6	1/6	1/6

So

$$E[X] = \sum x_i f(x_i) = \frac{1}{6} + \frac{2}{6} + \frac{3}{6} + \frac{4}{6} + \frac{5}{6} + \frac{6}{6} = \frac{21}{6} = 3.5$$

Thus, on average, a player can expect to win \$3.50 per toss. $E[X]$ is the *weighted average* of the possible values of X, with each value weighted by its probability of occurrence. ■■

■■ **Example 2–27**

A discrete random variable X can take on any integer value between 1 and n. The probability density function for X is given as

$$f(x_i) = Ax_i \quad \text{for } x_i = 1, 2, 3, \ldots, n.$$

Find the variance of X.

We will first find the value of the constant A. Since the sum of the probabilities of all the possibilities must be equal to 1,

$$\sum_{i=1}^{n} Ai = A \sum_{i=1}^{n} i = 1$$

$$A \left[\frac{n(n+1)}{2} \right] = 1$$

$$A = \frac{2}{n(n+1)}$$

The expected value of the discrete random variable is

$$\sum_{i=1}^{n} Ai^2 = A \sum_{i=1}^{n} i^2 = \frac{An(n+1)(2n+1)}{6}$$

or

$$E[X] = \frac{2n+1}{3}$$

The second moment of the random variable is

$$E[X^2] = \sum_{i=1}^{n} Ai^3 = A \sum_{i=1}^{n} i^3$$

$$= A \frac{n^2(n+1)^2}{4}$$

$$= \frac{n(n+1)}{2}$$

Therefore, the variance of the random variable can be calculated as follows:

$$\text{var } X = E[X^2] - (E[X])^2$$

$$= \frac{n(n+1)}{2} - \frac{(2n+1)^2}{9}$$

$$= \frac{n^2 + n - 2}{18}$$

■■

■■ **Example 2–28**
The expected value of the random variable discussed in Example 2–24 is

$$\mu = \int_2^6 xf(x)\,dx = \int_2^6 (0.125)(x^2 - 2x)\,dx = \frac{14}{3}$$

■■

■■ **Example 2–29**
For the exponentially distributed random variable of Example 2–25, the expected value is

$$\mu = \int_0^\infty tf(t)\,dt = (0.5)\int_0^\infty te^{-0.5t}\,dt$$

Employing the method of integration by parts, we let

$$u = t \quad \text{and } dv = 0.5e^{-0.5t}\,dt$$

Then $v = -e^{-0.5t}$ and

$$\mu = \left[-te^{-0.5t} - \int -e^{-0.5t}\,dt \right]_0^\infty$$

$$= \frac{1}{0.5} = 2$$

■■

■■ **Example 2–30**
A continuous random variable can take on any value between 0 and 1, and its density function is

$$f(x) = kx^2, \quad 0 \le x \le 1$$

Find the mean, variance, and standard deviation of this random variable.
First of all, we should find the value of k that will make $f(x)$ a legitimate density function. We require that

$$\int_0^1 f(x)\,dx = \int_0^1 kx^2\,dx = 1$$

This gives the value $k = 3$.
The mean value is

$$\mu = \int_0^1 xf(x)\,dx = 3\int_0^1 x^3\,dx = \frac{3}{4}$$

The second moment about the origin is

$$E[X^2] = \int_0^1 x^2 f(x)\,dx = 3\int_0^1 x^4\,dx = \frac{3}{5}$$

Therefore, the variance of $X = (\frac{3}{5}) - (\frac{3}{4})^2 = \frac{3}{80} = 0.0375$, and the standard deviation $\sigma = \sqrt{\frac{3}{80}} = 0.19365$.

■■

2.10 EXTREME VALUES

In reliability work, we are often interested in an extreme value (maximum or minimum) of a set of random variables. The failure of a component or system is likely to be the result of a maximum load or stress or a minimum strength or capacity. These maximum and minimum values are also random variables, and in this section we will develop expressions for the distribution and density functions that describe them.

Let X_1, X_2, ..., X_n be a set of random variables with distribution functions F_1, F_2, ..., F_n, respectively. If Y is the random variable that is the maximum of these n values, its distribution function $F_Y(y)$ is called the *maximum extreme-value distribution*.

By the definition of a distribution function,

$$F_Y(y) = P(Y \leq y) \tag{2-60}$$

Since Y is the maximum of the set of n random variables, we can express $F_Y(y)$ as

$$F_Y(y) = P[(X_1 \leq y)(X_2 \leq y) \ldots (X_n \leq y)] \tag{2-61}$$

If the X_i's are independent, then we have

$$F_Y(y) = \prod_{i=1}^{n} P(X_i \leq y) \tag{2-62}$$

Next, we note that the distribution function $F_i(y)$ of the random variable X_i is the same as $P(X_i \leq y)$. Using this observation, we can develop the relationship between the individual distribution functions and the maximum extreme-value distribution:

$$F_Y(y) = \prod_{i=1}^{n} F_i(y) \tag{2-63}$$

If, in addition, all the n random variables are identically distributed, with the common distribution function denoted $F(y)$, we get

$$F_Y(y) = [F(y)]^n \tag{2-64}$$

The corresponding density function $f_Y(y)$ is obtained by differentiating $F_Y(y)$ with respect to y:

$$f_Y(y) = \frac{d}{dy} F_Y(y) = n[F(y)]^{n-1}f(y) \tag{2-65}$$

where

$$f(y) = \frac{d}{dy} F(y) \tag{2-66}$$

The individual distribution $F(y)$ is called the *parent distribution* of the maximum extreme-value distribution $F_Y(y)$.

A similar approach can be employed to develop the relationship between a parent distribution and the minimum extreme-value distribution. Let Z be the smallest of the n random variables under consideration. Then it is also a random variable. Let $F_Z(z)$ be its distribution function. Then

$$F_Z(z) = P(Z \leq z) \qquad (2\text{--}67)$$

or

$$F_Z(z) = 1 - P(Z > z) \qquad (2\text{--}68)$$

Since Z is the minimum of the set of n random variables,

$$P(Z > z) = P[(X_1 > z)(X_2 > z) \ldots (X_n > z)] \qquad (2\text{--}69)$$

If the X_i's are independent of each other,

$$P(Z > z) = \prod_{i=1}^{n} P(X_i > z) \qquad (2\text{--}70)$$

Moreover,

$$F_i(z) = P(X_i \leq z) = 1 - P(X_i > z) \qquad (2\text{--}71)$$

from which

$$P(X_i > z) = 1 - F_i(z) \qquad (2\text{--}72)$$

Therefore,

$$P(Z > z) = \prod_{i=1}^{n} [1 - F_i(z)] \qquad (2\text{--}73)$$

and

$$F_Z(z) = 1 - \prod_{i=1}^{n} [1 - F_i(z)] \qquad (2\text{--}74)$$

If, in addition, all the n random variables have the identical distribution function $F(z)$, the minimum extreme-value distribution $F_Z(z)$ can be expressed as

$$F_Z(z) = 1 - [1 - F(z)]^n \qquad (2\text{--}75)$$

The corresponding density function is

$$f_Z(z) = n[1 - F(z)]^{n-1} f(z) \qquad (2\text{--}76)$$

where

$$f(z) = \frac{d}{dz} F(z) \qquad (2\text{--}77)$$

■■ **Example 2–31**
Let us consider n random variables with identical distribution given by

$$F(t) = 1 - e^{-\lambda t} \tag{2-78}$$

The corresponding density function is

$$f(t) = \lambda e^{-\lambda t} \tag{2-79}$$

These are the expressions for the exponential distribution, one of the most commonly used models in reliability studies.

The maximum extreme-value distribution is, using Equation (2–64),

$$F_Y(y) = [1 - e^{-\lambda y}]^n \tag{2-80}$$

and the density function is, using Equation (2–65),

$$f_Y(y) = n\lambda[1 - e^{-\lambda y}]^{n-1}e^{-\lambda y} \tag{2-81}$$

The minimum extreme-value distribution and density functions are obtained from Equations (2–75) and (2–76) and are respectively

$$F_Z(z) = 1 - e^{-n\lambda z} \tag{2-82}$$

and

$$f_Z(z) = n\lambda e^{-n\lambda z} \tag{2-83}$$

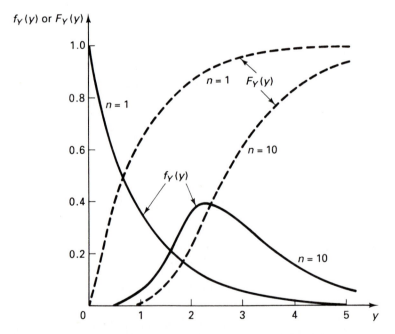

Figure 2.11 Maximum Extreme-Value Density and Distribution Functions for an Exponential Parent Distribution ($\lambda = 1$).

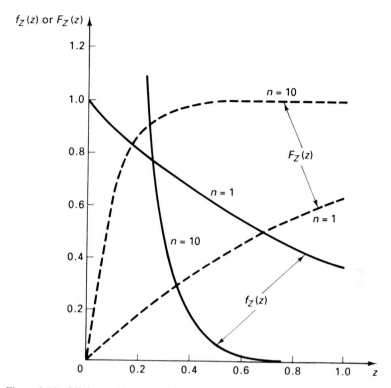

Figure 2.12   Minimum Extreme-Value Density and Distribution Functions for an Exponential Parent Distribution ($\lambda = 1$).

Plots of $F_Y(y)$ and $f_Y(y)$ for $\lambda = 1$ and $n = 1$ and 10 are shown in Figure 2.11. Plots of $F_Z(z)$ and $f_Z(z)$ for $\lambda = 1$ and $n = 1$ and 10 are shown in Figure 2.12. As n increases, the maximum extreme-value density function $f_Y(y)$ shifts more and more to the right, and the maximum extreme-value distribution function $F_Y(y)$ approaches its ultimate value of unity more and more reluctantly. As far as the minimum extreme-value distribution function $F_Z(z)$ is concerned, it reaches the value of unity rather rapidly as n increases. Correspondingly, the minimum extreme-value density function $f_Z(z)$ starts at higher and higher values and decreases towards zero more and more rapidly as n increases. Obviously, $n = 1$ corresponds to the standard exponential distribution in both cases.   ■■

2.11 SUMMARY

This chapter has presented a brief review of the fundamental concepts associated with probability and random variables that are necessary to follow the rest of the book. Readers with a good background in these topics may find the material rather elementary. However, the objective was not to present a rigorous mathematical treatise on these topics, but to provide an opportunity for a quick review of the basics for those who have never been exposed to the requisite concepts. It is hoped that those who were exposed to these ideas before but have not had the opportunity to use them recently will find a reading of the chapter refreshing and useful.

The depth and scope of probability and random variables are vast. The numerous examples presented should at least pique the reader's curiosity enough to refer to the rich collection of classical textbooks on these topics for additional information. The reader will find repeated application of the basic ideas presented in this chapter throughout the rest of the book.

PROBLEMS

2.1 Three identical fair coins are tossed simultaneously. What is the sample space? Using this sample space, find the probability of (a) two heads, (b) two tails, (c) one head, (d) one or more tails, and (e) two or three heads appearing.

2.2 A fair coin is tossed five times. Let X denote the longest string of tails occurring. Define the possible values of X, and find the probability of occurrence of each of these values.

2.3 A random sample of companies in two adjacent states (A and B) showed that 33 companies preferred to do business with a particular foreign country while the others did not. The sample consisted of 50 companies from state A and 35 companies from state B. Draw a Venn diagram, and
   (*i*) Find how many companies did not favor doing business with that particular foreign country.
   (*ii*) If 23 companies from A did not like to do business with that foreign country, how many from B preferred to do business with that country?

2.4 An ancient game of Indian origin employs two short metal sticks of square (8 mm × 8 mm) cross section and 6 cm in length as a pair of dice. The sticks are tossed together. Three of the four rectangular surfaces of each stick are marked with 1, 2, and 3 dots, respectively, and the fourth rectangular surface is left blank. If two blanks appear in the toss, the value of the toss is counted as 12. Otherwise, only the number of dots that show up are counted as the value of the toss.
   (a) Enumerate all the possible outcomes of a toss.
   (b) What is the probability of at least one blank showing up?
   (c) What is the probability of an even number of dots appearing?
   (d) Given that an even number of dots show up, what is the probability of a blank?
   (e) What is the probability of tossing a pair (both faces identical)?
   (f) What is the probability that the value of a toss is an even number?

2.5 How many different ways can five consonants and one vowel be selected out of the English alphabet?

2.6 In a word game, the player is asked to select five consonants and one vowel. How many different four-letter words can be formed out of these if each word must contain the vowel?

2.7 How many odd numbers consisting of six different digits can be formed out of 1, 2, 3, ... , 9?

2.8 Repeat Problem 2.7 if, in addition, the first digit must be odd.

2.9 Ignoring leap years and assuming that a person's birthday can fall on any day with equal likelihood, find the probability of two or more people having the same birthday among a group of five people.

2.10 The Southwest Youth Orchestra consists of 95 musicians. They are planning a concert tour of the Orient, and 38 parents and friends of the orchestra have volunteered to act as chaperones. The plan is to divide the musicians into groups of five and assign two chaperones to each group. In how many different ways can such assignments be made?

2.11 There are two boxes containing different colored cubes of the same size. Box A contains 4 red cubes and 1 brown cube. Box B contains 2 blue cubes. Three cubes are drawn at random from box A and transferred to box B. Then 4 cubes are drawn at random from box B and placed in box A. What is the probability that the brown cube is in box A after all these transfers are completed?

2.12 A product can have a critical defect in 4 different ways and a minor defect in 9 different ways. In how many different ways may it have
  (*i*) two minor defects and one critical defect,
  (*ii*) four minor defects, and
  (*iii*) three minor defects and two critical defects?

2.13 A solid-fuel booster engine has been test-fired 2,860 times. On 429 occasions, the engine failed to ignite. If a projectile is fitted with three identical and independent booster engines of this type, what is the probability on launching the projectile that
  (*i*) all the engines fail to ignite, and
  (*ii*) at least one of the engines fails to ignite?

2.14 A sensitive installation is to be supplied with extremely reliable electric power. The utility nearby can provide electric power with a probability of failure of 0.03. Calculate the allowable maximum probability of failure of an independent power source if the reliability of the overall power supply is to be at least 0.999.

2.15 Show that $_nC_1 + {}_nC_2 + \ldots + {}_nC_n = (2^n - 1)$.

2.16 Consider two events, A and B, with probabilities of occurrence of 0.31 and 0.45, respectively. Assuming that the events are independent, find (*i*) $P(A + B)$, (*ii*) $P(AB)$, (*iii*) $P(\bar{A} + B)$, and (*iv*) $P(\overline{AB})$.

2.17 The following two events are defined over the sample space of the experiment described in Problem 2.4:

  Event A: value of the toss is even
  Event B: at least one blank shows up

Find the probabilities of (*i*) AB, (*ii*) $A + B$, (*iii*) $A\bar{B}$, (*iv*) $\overline{AB}$, and (*v*) $\bar{A}\,\bar{B}$.

2.18 There is an 80% probability that a project will be completed on time. The probability that there will be no strikes is 0.7, and the probability that the project will be finished on time given that there will be no strikes is 0.9.

(*i*) What is the probability that the job is finished on time and there are no strikes?

(*ii*) What is the probability that there are no strikes given that the project is completed on time?

2.19 A pumping station has two identical pumps, each with a probability of failure of 0.1. The probability of both failing is 0.02.

(*i*) Are the failures disjoint?

(*ii*) What is the probability of at least one of the two pumps failing?

(*iii*) Are the pump failures independent?

(*iv*) What is the probability of a pump failing given that the other pump has failed?

(*v*) What is the probability of both pumps working?

2.20 There are 10 identical-looking transformers in a storeroom, of which 2 are defective. Three are chosen at random. Find the probability that (*i*) all chosen are good, (*ii*) only one of those chosen is defective, and (*iii*) one or more chosen are defective. If a transformer is chosen and tested until one good one is obtained, what is the expected number of transformers chosen?

2.21 In a high school, 5% of the boys and 10% of the girls perform in the school orchestra. The girls constitute 55% of the student body. If an orchestra player is selected at random, what is the probability that the student is a girl?

2.22 A graduate course in reliability has 50% master's students, 20% doctoral students, and 30% advanced seniors. Based on previous history, the probabilities of receiving an A in the course are 0.4, 0.6, and 0.2 for the masters, doctors, and senior students, respectively.

(*i*) What is the probability of a randomly selected student receiving an A?

(*ii*) If the student selected does not receive an A, what is the probability of that student being a senior?

(*iii*) If a student selected at random receives an A, find the probabilities of that student being (a) a doctoral student, (b) a master's student, and (c) a senior.

2.23 A satchel contains two 1-dollar bills, two 5-dollar bills, and one 10-dollar bill. Two bills are selected at random from the satchel. Find the distribution of the money drawn.

2.24 The density function of a discrete random variable is

$$f(x_i) = Kx_i^2 \quad \text{for } x_i = 1, 2, 3, 4$$

Find K, $F(x_i^+)$, the expected value, the variance, the standard deviation, the skewness, and the kurtosis of the variable.

2.25 A discrete random variable has the density function

$$f(x_i) = \frac{1}{n} \quad \text{for } x_i = 1, 2, 3, \ldots, n$$

Calculate the mean and variance of the variable.

2.26 A box has 10 microchips, of which 2 are defective. A chip is selected from the box and tested until a good one is chosen. What is the expected number of chips chosen?

2.27 It costs $50 million each time a satellite company attempts to launch a communications satellite. If the launch is successful, the company earns $100 million. If not, the launch cost must be absorbed by the company. A maximum of three launches are attempted, one after the other, until a successful launch results. There is a 90% probability of success each time a launch is attempted.

(*i*) If N denotes the number of launches before a successful launch is made, find the probability density function for N.

(*ii*) What is the expected value of the net profit?

2.28 Find the distribution function for a discrete random variable X with Poisson density

$$f(x) = \sum_{k=0}^{\infty} \frac{a^k e^{-a}}{k!} \delta (x - k) \quad \text{for any } a > 0$$

2.29 A fair coin is tossed five times. Let X denote the number of tails showing up. Find the distribution, mean, variance, standard deviation, skewness, and kurtosis of the discrete random variable X.

2.30 A continuous random variable has the density function

$$f(x) = a + 0.25a(x - 2) \quad \text{for } 2 \leq x \leq 6$$

(*i*) What is the value of a?

(*ii*) Find $P(3 < x < 5)$ and $P(x < 4)$.

(*iii*) Sketch the density and distribution functions of the random variable.

2.31 The probability density function for a continuous random variable X is

$$f(t) = \frac{d}{dt}F(t) = \frac{d}{dt} [P(X \leq t)] = 3t^2 \quad \text{for } 0 < t < 1.$$

(*i*) Find the cumulative distribution function for X.

(*ii*) If $P(X \leq a) = 0.25$, what is a?

(*iii*) Find the mean and the variance of the random variable.

2.32 The density function of a continuous random variable has the shape of the positive half of a sinusoid, as shown in Figure P 2.32.

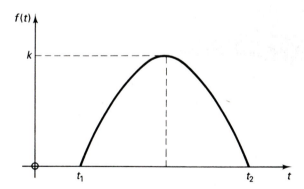

Figure P 2.32

a. Find the value of k.

b. Develop an expression, and sketch the corresponding distribution function.

2.33 For the density function shown in Figure P 2.33,

a. Find the relationship between K and t_1.

b. Develop an expression for the corresponding distribution function.

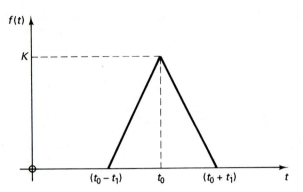

Figure P 2.33

2.34 Cauchy random variables have the probability density function

$$f(t) = \frac{a/\pi}{a^2 + (t - b)^2}$$

for $a > 0$ and $-\infty < b < +\infty$. Show that the corresponding distribution function is

$$F(t) = 0.5 + \left(\frac{1}{\pi}\right) \tan^{-1}\left[\frac{(t - b)}{a}\right]$$

2.35 The complementary cumulative distribution function $\bar{F}$ is defined as follows:

For discrete random variables,

$$\bar{F}(x_i) = P(X > x_i) = 1 - F(x_i)$$

For continuous random variables,

$$\bar{F}(t) = P(X > t) = 1 - F(t)$$

Sketch this function for

(*i*) $f(x_i) = 0.1i$ for $i = 0, 1, 2, 3,$ and 4.
(*ii*) $f(t) = \lambda \exp(-\lambda t)$ for $t > 0$.

2.36 Repeat the plots shown in Figures 2.11 and 2.12 for $n = 5$ and $n = 20$. Assume $\lambda = 1$.

3

Catastrophic Failure Models
and Reliability Functions

3.1 INTRODUCTION

A failure of a component is called *catastrophic* if repair of the component is not possible, not available, or of no relevance to the success of the mission utilizing the component. Life-support, surveillance, and safety systems fall into this category. Failure models for such components are typically based on life test results and failure rate data using probability theory. Although it is possible in some cases to develop a failure model based on the physics of failure, the procedure is usually difficult and requires a considerable amount of study and analysis.

Suppose a set of N_0 identical items are placed in operation at time $t = 0$. As time progresses, some of the items will fail. Let $N_s(t)$ be the number of survivors at time t. Typically, the unit of time used is hours. Then the number N_f that have failed at time t is

$$N_f(t) = N_0 - N_s(t) \qquad (3\text{--}1)$$

Next, we construct two functions of time, the piecewise-continuous failure data density function

$$f_d(t) = \frac{[N_s(t_i) - N_s(t_i + \Delta t_i)]/N_0}{\Delta t_i}, \quad t_i < t \leq (t_i + \Delta t_i) \qquad (3\text{--}2)$$

and the piecewise-continuous hazard rate function

$$\lambda_d(t) = \frac{[N_s(t_i) - N_s(t_i + \Delta t_i)]/N_s(t_i)}{\Delta t_i}, \quad t_i < t \le (t_i + \Delta t_i) \tag{3-3}$$

Notice that $f_d(t)$ is the ratio of the number of failures occurring in the time interval Δt_i to the *size of the original population,* divided by the length of the time interval. The hazard rate function $\lambda_d(t)$ is the ratio of the number of failures in the time interval Δt_i to the *number of survivors at the beginning of the time interval,* divided by the length of the time interval. Intuitively, we can see that $f_d(t)$ is a measure of the overall speed at which failures are occurring and $\lambda_d(t)$ is a measure of the *instantaneous speed* of failure. Both functions have the unit time^{-1}. The choice of t_i and Δt_i is unspecified, and Δt_i need not be the same for all i. Usually, Δt_i is chosen as the times between failures. In such an event, the failure is assumed to have occurred just before (or could be just after) the end of the interval.

The data failure distribution function $Q_d(t)$ is obtained by integrating $f_d(t)$. It is piecewise continuous and consists of a sum of ramp functions. Thus,

$$Q_d(t) \equiv \int_0^t f_d(\xi)\, d\xi \tag{3-4}$$

The data success distribution function is defined as

$$R_d(t) = 1 - Q_d(t) \tag{3-5}$$

Since eventually all the components must fail, the area under $f_d(t)$ is unity, and consequently, as t increases, $Q_d(t) \to 1$ and $R_d(t) \to 0$. Figure 3.1 illustrates the nature of $Q_d(t)$ and $R_d(t)$. In general,

$f_d(t)$ is a decreasing function
$\lambda_d(t)$ decreases initially and later increases rapidly with time
$Q_d(t)$ is a monotonically increasing function
$R_d(t)$ is a monotonically decreasing function

If a large number of items are tested (equivalent to the availability of a large amount of failure data), the time scale is usually divided into a number K of equally spaced intervals and the functions are plotted. A rule of thumb for finding K is

$$K = 1 + 3.3 \log N \tag{3-6}$$

where N is the number of failures. For example, if $N = 10$, $K = 4.3$, and if $N = 1,000$, $K = 10.9$. The choice of K is not very critical; $K = 5$ to 10 is sufficient in most cases.

It can be shown that as the amount of data becomes large and as the time

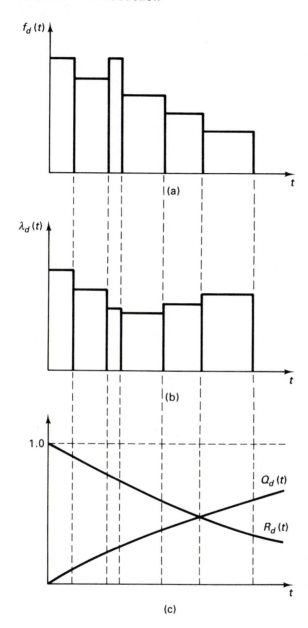

Figure 3.1 Nature of the functions derived from failure data. (a) Failure data density function. (b) Hazard rate function. (c) Data failure and success distribution functions.

interval between failures approaches zero, all the discrete functions defined thus far approach the continuous functions denoted by $f(t)$, $\lambda(t)$, $Q(t)$, and $R(t)$.

The hazard rate function $\lambda(t)$ is the function most commonly used to fit or find a model from available data, because it exhibits most clearly the various stages in the life of components. This topic will be discussed in detail later.

3.2 RELATIONSHIPS BETWEEN THE DIFFERENT RELIABILITY FUNCTIONS

In this section, we will discuss the relationships between the four functions $f(t)$, $\lambda(t)$, $Q(t)$, and $R(t)$ used in the study of the reliability of components.

First of all, note that

$$R(t) = \frac{N_s(t)}{N_0}$$

$$= \frac{N_0 - N_f(t)}{N_0}$$

$$= 1 - \frac{N_f(t)}{N_0}$$

or

$$R(t) = 1 - Q(t) \tag{3-7}$$

Since

$$f(t) = \lim_{\Delta t \to 0} \left[\frac{1}{N_0} \frac{N_s(t) - N_s(t + \Delta t)}{\Delta t} \right]$$

$$= -\frac{1}{N_0} \frac{d}{dt} N_s(t)$$

and

$$N_s(t) = N_0 R(t)$$

we have

$$f(t) = -\frac{d}{dt} R(t) \tag{3-8a}$$

$$= \frac{d}{dt} Q(t) \tag{3-8b}$$

Moreover,

$$\lambda(t) = \lim_{\Delta t \to 0} \left[\frac{1}{N_s(t)} \frac{N_s(t) - N_s(t + \Delta t)}{\Delta t} \right]$$

$$= -\frac{1}{N_s(t)} \frac{d}{dt} N_s(t)$$

$$= \frac{N_0 f(t)}{N_s(t)} = \frac{f(t)}{[N_s(t)/N_0]}$$

or

$$\lambda(t) = \frac{f(t)}{R(t)} \tag{3-9}$$

To obtain the relationship between the reliability function $R(t)$ and the hazard function $\lambda(t)$, we observe that

$$\lambda(t) = \frac{f(t)}{R(t)} = \frac{-\dfrac{d}{dt} R(t)}{R(t)}$$

$$= -\frac{d}{dt} [\ln R(t)]$$

or

$$\ln R(t) = -\int_0^t \lambda(\xi)\, d\xi + c$$

in which c is the constant of integration. Raising both sides to a power of e, we get

$$R(t) = e^c \exp\left[-\int_0^t \lambda(\xi)\, d\xi \right]$$

Since, at $t = 0$, $R(t) = 1$, we conclude that $e^c = 1$ and

$$R(t) = \exp\left[-\int_0^t \lambda(\xi)\, d\xi \right] \tag{3–10}$$

Because $f(t) = \lambda(t)R(t)$, we also have

$$f(t) = \lambda(t) \exp\left[-\int_0^t \lambda(\xi)\, d\xi \right] \tag{3–11}$$

and

$$Q(t) = 1 - R(t) = 1 - \exp\left[-\int_0^t \lambda(\xi)\, d\xi \right] \tag{3–12}$$

Moreover,

$$\lambda(t) = \frac{f(t)}{R(t)} = \frac{f(t)}{1 - Q(t)} = \left[\frac{\dfrac{d}{dt} Q(t)}{1 - Q(t)} \right] \tag{3–13}$$

or

$$\lambda(t) = \frac{f(t)}{1 - Q(t)} = \frac{f(t)}{1 - \int_0^t f(\xi)\, d\xi} \tag{3–14}$$

Since $\lim_{t \to \infty} R(t) = 0$ and $R(t) = \exp\left[-\int_0^t \lambda(\xi)\, d\xi \right]$, we see that the area under $\lambda(t)$ should go to ∞ as $t \to \infty$.

TABLE 3.1   RELATIONSHIPS BETWEEN DIFFERENT RELIABILITY FUNCTIONS

	$f(t)$	$\lambda(t)$	$Q(t)$	$R(t)$
$f(t) =$	$f(t)$	$\lambda(t) \exp\left[-\int_0^t \lambda(\xi)\, d\xi \right]$	$\dfrac{d}{dt} Q(t)$	$-\dfrac{d}{dt} R(t)$
$\lambda(t) =$	$\dfrac{f(t)}{1 - \int_0^t f(\xi)\, d\xi}$	$\lambda(t)$	$\dfrac{1}{1 - Q(t)} \dfrac{d}{dt}(Q(t))$	$-\dfrac{d}{dt}[\ln R(t)]$
$Q(t) =$	$\int_0^t f(\xi)\, d\xi$	$1 - \exp\left[-\int_0^t \lambda(\xi)\, d\xi \right]$	$Q(t)$	$1 - R(t)$
$R(t) =$	$1 - \int_0^t f(\xi)\, d\xi$	$\exp\left[-\int_0^t \lambda(\xi)\, d\xi \right]$	$1 - Q(t)$	$R(t)$

The relationships between the four reliability functions are summarized in Table 3.1.

3.3 ALTERNATIVE APPROACH TO RELIABILITY FUNCTIONS

Let T be the continuous random variable representing the time to failure or, simply, failure time of a component (or unit or system). Then the probability of the component surviving up to time t is

$$P(T > t) = R(t) \tag{3-15}$$

where $R(t)$ is the time-dependent reliability (or probability of success) of the component. As t increases, the chance of the component failing increases, and therefore, as $t \rightarrow \infty$, $R(t) \rightarrow 0$. The probability of the component's failure as a function of time is

$$P(T \leq t) \equiv Q(t) \tag{3-16}$$

where $Q(t)$ is the distribution function for the random variable T. Therefore, the failure density function is

$$f(t) = \frac{d}{dt} Q(t) \tag{3-17}$$

The probability of failure in the time interval $(t, t + \Delta t]$ can now be expressed in terms of $F(t)$. We have

$$P(t < T \leq t + \Delta t) = Q(t + \Delta t) - Q(t) \tag{3-18}$$

The conditional probability of failure in the interval $(t, t + \Delta t]$, given survival of the component up to time t is

$$P(t < T \leq t + \Delta t \mid T > t) = \frac{P(t < T \leq t + \Delta t)}{P(T > t)}$$

Therefore,

$$P(t < T \leq t + \Delta t \mid T > t) = \frac{Q(t + \Delta t) - Q(t)}{R(t)} \tag{3-19}$$

Dividing both sides by Δt and taking the limit as $\Delta t \to 0$ leads to

$$\lim_{\Delta t \to 0} \frac{P(t < T \leq t + \Delta t \mid T > t)}{\Delta t} = \frac{\dfrac{d}{dt} Q(t)}{R(t)} = \frac{f(t)}{R(t)}$$

Now,

$$\lambda(t) = \lim_{\Delta t \to 0} \left[\frac{1}{N_s(t)} \frac{N_s(t) - N_s(t + \Delta t)}{\Delta t} \right]$$

$$= \lim_{\Delta t \to 0} \left(\frac{1}{\Delta t} \right) \left[\frac{\text{number of failures in } (t, t + \Delta t]}{\text{number of survivors at } t} \right]$$

$$= \lim_{\Delta t \to 0} \frac{1}{\Delta t} \left[\begin{array}{c} \text{probability of failure in } (t, t + \Delta t] \\ \text{given survival up to } t \end{array} \right]$$

$$= \lim_{\Delta t \to 0} \frac{P(t < T \leq t + \Delta t \mid T > t)}{\Delta t}$$

Therefore,

$$\lambda(t) = \frac{f(t)}{R(t)} \tag{3-20}$$

Note that the event $t < T \leq t + \Delta t$ is more stringent than the event $t < T \leq (t + \Delta t)$ given $T > t$. In other words,

$$P(t < T \leq t + \Delta t) < P(t < T \leq t + \Delta t \mid T > t)$$

3.4 TYPICAL HAZARD FUNCTION

Examination of the failure data for a variety of components over a number of years indicates that the general form of $\lambda(t)$ has a bathtub shape with three distinct regions, as shown in Figure 3.2. Region I corresponds to early failures (infant mortality) during debugging. As the debugging process continues, the hazard rate goes down. Region II corresponds to the useful lifetime of the component. During this period, failures are very nearly random and the hazard rate is constant, corresponding to an exponentially decreasing density function (see later). Region III corresponds to the wearout or fatigue phase, during which the hazard rate increases rapidly with time. These three regions are easily distinguishable in the plot of the hazard function.

Most manufacturers of high-reliability components subject their products to an initial ''burn-in'' period of t_1 to eliminate the initial failure region in the actual

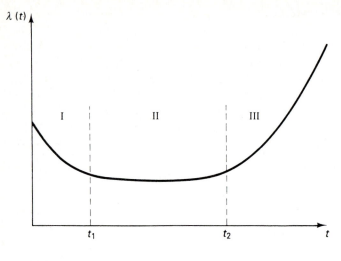

Figure 3.2 Bathtub-shaped hazard function.

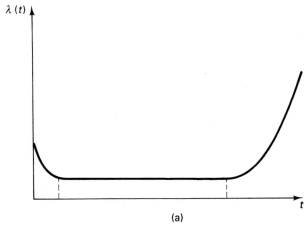

(a)

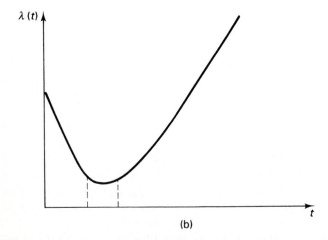

(b)

Figure 3.3 Two extreme shapes for the hazard function. (a) Components with long useful life. (b) Components dominated by wearout.

operation of the component. It is also wise to replace a component after $(t_2 - t_1)$ hours of operation, even though it has not failed. To achieve the highest reliability, all the components must be operating in their period of useful life.

Electronic components have a fairly long period of useful life, exhibited by a large value of $(t_2 - t_1)$. In the case of mechanical components, the wearout region tends to dominate. These two extreme cases are illustrated in Figure 3.3. Most of the large manufacturers collect life test and field failure data for their components and publish them in handbooks of failure rate data.

■■ **Example 3–1**

Figure 3.4 shows a piecewise-linear approximation to the bathtub hazard function. The expression for this function is

$$\lambda(t) = \begin{cases} k_1 - [(k_1 - k)/t_1]t & \text{for } 0 < t \leq t_1 \\ k & \text{for } t_1 < t \leq t_2 \\ k + k_2 (t - t_2) & \text{for } t > t_2 \end{cases}$$

Our objective is to find all the other reliability functions using the relationships given in Table 3.1.

We will first compute $\int_0^t \lambda(\xi)\, d\xi$. The resulting expressions are:

$$\Lambda(t) = \begin{cases} k_1 t - \left(\dfrac{k_1 - k}{t_1}\right)\dfrac{t^2}{2} & \text{for } 0 < t \leq t_1 \\ \dfrac{(k_1 + k)t_1}{2} + k(t - t_1) & \text{for } t_1 < t \leq t_2 \\ \dfrac{(k_1 + k)t_1}{2} + k(t_2 - t_1) + (k - k_2 t_2)(t - t_2) + \left(\dfrac{k_2}{2}\right)(t^2 - t_2^2) & \text{for } t > t_2 \end{cases}$$

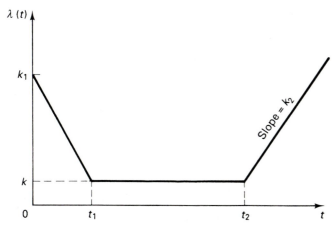

Figure 3.4 Piecewise-linear approximation to the bathtub hazard function.

For convenience, let us label these three functions as $\Lambda_1(t)$, $\Lambda_2(t)$, and $\Lambda_3(t)$, respectively, for the three time regions shown. The failure density function for this model is then

$$
f(t) = \begin{cases}
\{k_1 - [(k_1 - k)/t_1]t\}\ \exp\ -\ \Lambda_1(t) & \text{for } 0 < t \le t_1 \\
k & \exp\ -\ \Lambda_2(t) & \text{for } t_1 < t \le t_2 \\
\{k + k_2(t - t_2)\} & \exp\ -\ \Lambda_3(t) & \text{for } t > t_2
\end{cases}
$$

The reliability function is

$$
R(t) = \begin{cases}
\exp\ -\ \Lambda_1(t) & \text{for } 0 < t \le t_1 \\
\exp\ -\ \Lambda_2(t) & \text{for } t_1 < t \le t_2 \\
\exp\ -\ \Lambda_3(t) & \text{for } t > t_2
\end{cases}
$$

The failure distribution function $Q(t)$ is, of course, equal to $1 - R(t)$.   ■■

■■ **Example 3–2**

The failure distribution function for a certain class of items is

$$
Q(t) = 1 - \exp[-\lambda_1 t]; \quad t, \lambda_1 > 0
$$

Find the distribution function for the random variable τ, the time to failure of the same items under a different set of operating conditions. Assume that

$$
\tau = \left(\frac{\lambda_1}{\lambda_2}\right) t
$$

If X and Y are two random variables related by a linear function

$$
y = g(x)
$$

then the density functions of Y and X are related by

$$
f(y) = \frac{f(x_1)}{|g'(x_1)|}
$$

where x_1 is a real root of $y = g(x)$ and

$$
g'(x) = \frac{d}{dx} g(x)
$$

In the present case,

$$
f(t) = \frac{d}{dt} Q(t) = \lambda_1 \exp[-\lambda_1 t]
$$

and

$$
\tau = g(t) = \left(\frac{\lambda_1}{\lambda_2}\right) t
$$

Therefore,

$$f(\tau) = \frac{f(t_1)}{|g'(t_1)|}$$

where

$$t_1 = \left(\frac{\lambda_2}{\lambda_1}\right)\tau$$

$$f(\tau) = \frac{\lambda_1 \exp[-\lambda_1(\lambda_2/\lambda_1)\tau]}{(\lambda_1/\lambda_2)}$$

$$= \lambda_2 \exp[-\lambda_2\tau]$$

The corresponding distribution function is obtained by integrating $f(\tau)$ and using the initial condition $Q(\tau) = 0$ at $\tau = 0$. We obtain

$$Q(\tau) = 1 - \exp[-\lambda_2\tau]; \quad \tau, \lambda_2 > 0 \qquad \blacksquare\blacksquare$$

■■   **Example 3–3**

The failure density function for a class of components is

$$f(t) = 0.25 - \left(\frac{0.25}{8}\right)t$$

where t is in years. Find the remaining three functions associated with reliability evaluation, and sketch all four functions.

The failure distribution function is

$$Q(t) = \int_0^t f(\xi)\, d\xi$$

$$= 0.25t - \left(\frac{0.25}{16}\right)t^2$$

The reliability function is

$$R(t) = 1 - Q(t)$$

$$= 1 - 0.25t + \left(\frac{0.25}{16}\right)t^2$$

The hazard function is

$$\lambda(t) = \frac{f(t)}{R(t)}$$

$$= \frac{2 - 0.25t}{8 - 2t + 0.125t^2}$$

All four functions are sketched in Figure 3.5.

The given density function implies that the probability of a component sur-

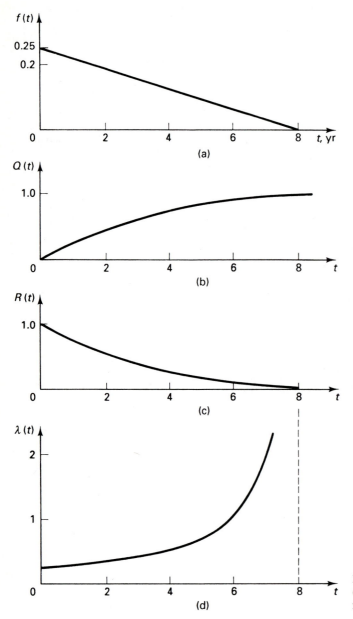

Figure 3.5 (a) Density, (b) distribution (c) reliability, and (d) hazard function for Example 3–3.

viving longer than eight years is zero. In other words, if N components are put into service at $t = 0$, then all of them will fail within eight years.

The probability of a component lasting more than four years is the same as the probability of its not failing within four years. The latter is equal to $R(4)$ or $[1 - Q(4)]$ or the area under the density function between $t = 4$ and $t = 8$. We have

$$R(4) = 1 - Q(4) = 0.25$$

The probability of the component failing within the first year of operation is

$$Q(1) = 1 - R(1) = 0.2344$$

A host of similar questions can be answered using these four functions.

■■

■■ **Example 3–4**

Based on experimental evidence, the ability of a thyristor to withstand the nth voltage spike can be expressed as $(1 - p)$, a constant, if it has withstood all the previous

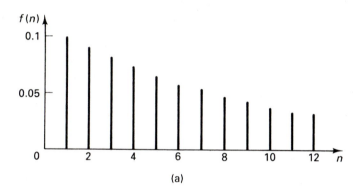

(a)

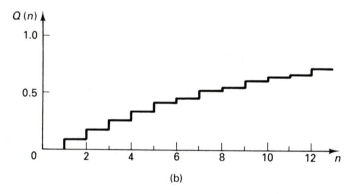

(b)

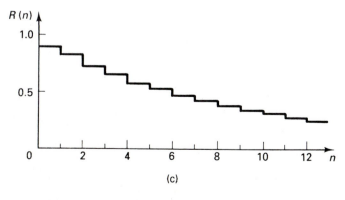

(c)

Figure 3.6 (a) Density, (b) distribution, and (c) reliability functions for the geometric distribution (Example 3–4).

$(n - 1)$ spikes. Then the probability of failure in the nth trial is

$$f(n) = p(1 - p)^{n-1} \quad \text{for } n = 1, 2, 3...$$

This is the failure density function (discrete in this case), and it is of the general form shown in Figure 3.6 for $p = 0.1$. The corresponding failure distribution function $Q(n)$ and reliability function $R(n) = 1 - Q(n)$ are also shown in the figure. The function $Q(n)$ is sometimes referred to as the *geometric distribution*.   ■ ■

3.5 MEAN TIME TO FAILURE (MTTF)

The expected value of the continuous random variable called time to failure is known as the *mean time to failure,* or simply, MTTF. In many practical situations, a knowledge of the MTTF is enough to assess the quality and usefulness of a certain component. By definition,

$$\text{MTTF} = \int_0^\infty tf(t) \, dt \qquad (3\text{–}21)$$

Since

$$R(t) = 1 - Q(t) \qquad (3\text{–}22)$$

we have

$$\frac{d}{dt} R(t) = -\frac{d}{dt} Q(t) = -f(t) \qquad (3\text{–}23)$$

and

$$\text{MTTF} = -\int_0^\infty t \frac{d}{dt} R(t) \, dt \qquad (3\text{–}24)$$

$$= -\left[tR(t) \, |_0^\infty - \int_0^\infty R(t) \, dt \right]$$

The component is assumed to be good and working at time $t = 0$. Therefore, $R(0) = 1$ and $\lim_{t \to 0} tR(t) = 0$. Moreover, the fact that the area under the hazard function goes to ∞ as $t \to \infty$ yields

$$R(t) = \exp\left[-\int_0^t \lambda(\xi) \, d\xi \right] \qquad (3\text{–}25)$$

and

$$\lim_{x \to \infty} xe^{-x} = 0 \qquad (3\text{–}26)$$

assures that

$$\lim_{t \to \infty} tR(t) = 0 \qquad (3\text{–}27)$$

Therefore,

$$\text{MTTF} = \int_0^\infty R(t)\,dt \qquad (3\text{--}28)$$

Using the final value theorem of Laplace transforms, the MTTF can also be expressed as

$$\text{MTTF} = \lim_{s \to 0} \mathcal{R}(s) \qquad (3\text{--}29)$$

where

$$\mathcal{R}(s) = \mathcal{L}[R(t)] \qquad (3\text{--}30)$$

■■ **Example 3–5**

For the components discussed in Example 3–3, the mean time to failure is

$$\text{MTTF} = \int_0^\infty R(t)\,dt$$

$$= \int_0^8 \left[1 - 0.25t + \left(\frac{0.25}{16} \right) t^2 \right] dt$$

$$= \left[t - \frac{0.25 t^2}{2} + \frac{0.25 t^3}{48} \right]_0^8$$

$$= 2.667 \text{ years}$$

The MTTF is simply equal to the area under the reliability function.   ■■

■■ **Example 3–6**

Given that $R(t) = \exp[-\lambda t]$ we have

$$\text{MTTF} = \int_0^\infty \exp[-\lambda t]\,dt$$

$$= \left[-\left(\frac{1}{\lambda} \right) \exp(-\lambda t) \right]_0^\infty$$

$$= \frac{1}{\lambda}$$

Also,

$$\mathcal{R}(s) = \mathcal{L}[R(t)]$$

$$= \mathcal{L}[\exp(-\lambda t)]$$

$$= \frac{1}{s + \lambda}$$

$$\text{MTTF} = \lim_{s \to 0} \mathcal{R}(s)$$

$$= \frac{1}{\lambda}$$ ■■

The *median time to failure* is sometimes employed to characterize the lifetimes of components. Let us consider an experiment involving the testing of n identical components or units. At time $t = 0$, all the n units are put in service. The median time to failure, denoted T_{50}, is the time taken for one-half of the components to fail. Put another way, if we have data on the time to failure of n items, then the median time to failure effectively divides the data in half.

Since $Q(t)$ is the cumulative distribution function of the random variable called time to failure, T_{50} is also the value of t for which $Q(t)$ reaches the value one-half; that is, $Q(T_{50}) = 0.5$.

■■ **Example 3–7**

The median time to failure for the components of Example 3–3 is found by equating $Q(T_{50})$ to 0.5. Thus,

$$0.5 = 0.25 T_{50} - \left(\frac{0.25}{16}\right) T_{50}^2$$

Solving for T_{50}, we get $T_{50} = 13.657$ or $T_{50} = 2.343$. Since $t > 8$ has no physical relevance in this case, we conclude that $T_{50} = 2.343$ years.   ■■

■■ **Example 3–8**

For components with linearly increasing hazard,

$$\lambda(t) = kt$$

and

$$Q(t) = 1 - \exp\left[-\int_0^t k\xi \, d\xi\right]$$

$$= 1 - \exp\left[-\frac{kt^2}{2}\right]$$

The median time to failure can be found from

$$0.5 = 1 - \exp\left[-\frac{kT_{50}^2}{2}\right]$$

or

$$T_{50} = \sqrt{\frac{2 \ln 2}{k}}$$ ■■

3.6 CUMULATIVE HAZARD FUNCTION AND AVERAGE FAILURE RATE

The cumulative hazard function $\Lambda(t)$ is related to the hazard function $\lambda(t)$ in a manner similar to the way the cumulative distribution function $F(t)$ and the probability density function $f(t)$ are related. Therefore,

$$\Lambda(t) = \int_0^t \lambda(\xi) \, d\xi \tag{3-31}$$

Since

$$\lambda(t) = -\frac{d}{dt} [\ln R(t)] \tag{3-32}$$

we have

$$\Lambda(t) = -\int_0^t \frac{d}{d\xi} [\ln R(\xi)] \, d\xi$$

or

$$\Lambda(t) = -\ln R(t) \tag{3-33}$$

The average failure rate (AFR) over a time interval (t_1, t_2) is defined as

$$\text{AFR}(t_1, t_2) = \frac{1}{(t_2 - t_1)} \int_{t_1}^{t_2} \lambda(\xi) \, d\xi \tag{3-34}$$

AFR is a single number that can be used to characterize the failure rate of a component over a specified time interval, typically its useful lifetime. Using the definition of $\Lambda(t)$, we have

$$\text{AFR}(t_1, t_2) = \frac{\Lambda(t_2) - \Lambda(t_1)}{t_2 - t_1} \tag{3-35}$$

Alternatively,

$$\text{AFR}(t_1, t_2) = \frac{\ln R(t_1) - \ln R(t_2)}{t_2 - t_1} \tag{3-36}$$

If the time interval of interest is 0 to T, then $t_1 = 0$, $t_2 = T$, and

$$\text{AFR}(0, T) \equiv \text{AFR}(T) = \frac{\Lambda(T)}{T} = -\frac{\ln R(T)}{T} \tag{3-37}$$

For well-designed components, failure rates and average failure rates expressed in number of failures per hour will be very small. Therefore, certain special units are commonly used to express these rates. If $\lambda(t)$ is the failure rate in numbers per hour, then

%/K $\equiv$ failure rate in percent per thousand hours

$$= 10^5 \, \lambda(t) \tag{3-38}$$

PPM/K $\equiv$ failure rate in parts per million per thousand hours

(also known as FIT, for "fails in time")

$$= 10^9 \lambda(t) \tag{3-39}$$

Therefore,

$$\text{FIT} = 10^4 (\%/\text{K}) \qquad (3\text{--}40)$$

Similarly, if AFR is the average failure rate in numbers per hour over a specified time interval, then over the same interval,

$$\text{AFR in } \%/\text{K} = 10^5 \text{AFR} \qquad (3\text{--}41)$$

and

$$\text{AFR in PPM/K} = 10^9 \text{AFR} \qquad (3\text{--}42)$$

The relationships between the various units used to express failure rates are summarized in Table 3.2.

TABLE 3.2   RELATIONSHIPS BETWEEN DIFFERENT FAILURE RATE UNITS

	λ (#/hr)	%/K	PPM/K (FIT)
$\lambda =$	λ	10^{-5} (%/K)	10^{-9} (PPM/K)
%/K $=$	$10^5 \lambda$	%/K	10^{-4} (PPM/K)
PPM/K (FIT) $=$	$10^9 \lambda$	10^4 (%/K)	PPM/K

■■ **Example 3–9**
For constant hazard components,

$$\lambda(t) = \lambda, \text{ a constant}$$

and

$$\Lambda(t) = \lambda t$$

As we would expect, the average failure rate over a time interval (t_1, t_2) is

$$\text{AFR}(t_1, t_2) = \frac{1}{(t_2 - t_1)} (t_2 - t_1) \lambda$$

$$= \lambda \qquad \qquad ■■$$

■■ **Example 3–10**
Using Equation (3–31), for the components considered in Example 3–3, we find that the cumulative hazard function is

$$\Lambda(t) = \int_0^t \left(\frac{2 - 0.25\xi}{8 - 2\xi + 0.125\xi^2} \right) d\xi$$

$$= -\ln(1 - \frac{1}{4}t + \frac{1}{64}t^2)$$

which is equal to $-\ln R(t)$, as expected.

Also,

$$AFR(T) = -\frac{1}{T} \ln(1 - \frac{1}{4}T + \frac{1}{64}T^2)$$

and

$$AFR(2, 6) = \frac{1}{(6 - 2)} \left[\ln\left(1 - \frac{2}{4} + \frac{2^2}{64}\right) - \ln\left(1 - \frac{6}{64} + \frac{6^2}{64}\right) \right]$$

$$= 0.5493 \text{ failure/year} \qquad \blacksquare\blacksquare$$

3.7 A POSTERIORI FAILURE PROBABILITY AND WEARIN PERIOD

The reliability function $R(t)$ is the probability of no failure during the time interval $(0, t)$. Thus,

$$R(t) = 1 - \int_0^t f(\xi) \, d\xi \qquad (3\text{–}43)$$

Since the area under the density function must be unity, we have

$$R(t) = \int_t^\infty f(\xi) \, d\xi \qquad (3\text{–}44)$$

Next, we ask the following question: Given that a component (or system or item or device) has survived during $(0, T)$, what is the probability of its failing during $(T, T + t)$? This probability is called the *a posteriori* failure probability $Q_c(t)$.

The probability of the event that the component survives up to T *and* fails during $(T, T + t)$ is simply equal to the area under the failure density function calculated from T to $(T + t)$. Using the conditional probability relationships, we can easily see that

$$Q_c(t) = P[\text{failure during } (T, T + t) \mid \text{survival during } (0, T)] \qquad (3\text{–}45)$$
$$= \frac{P[\text{survival during } (0, T) \text{ and failure during } (T, T + t)]}{P[\text{survival during } (0, T)]}$$

or

$$Q_c(t) = \frac{\int_T^{T+t} f(\xi) \, d\xi}{\int_T^\infty f(\xi) \, d\xi} \qquad (3\text{–}46)$$

Given survival during $(0, T)$, the probability of surviving the next t, or during $(T, T + t)$, is

$$R(t \mid T) = 1 - Q_c(t) \qquad (3\text{–}47)$$

Using Equation (3–46) in Equation (3–47), we get

$$R(t \mid T) = 1 - \frac{\int_{T}^{T+t} f(\xi)\, d\xi}{\int_{T}^{\infty} f(\xi)\, d\xi}$$

$$= \frac{\int_{T+t}^{\infty} f(\xi)\, d\xi}{\int_{T}^{\infty} f(\xi)\, d\xi} \qquad (3\text{--}48)$$

or

$$R(t \mid T) = \frac{R(T + t)}{R(T)}$$

$$= \exp\left[-\int_{T}^{T+t} \lambda(\xi)\, d\xi \right] \qquad (3\text{--}49)$$

The quantity $R(t \mid T)$ includes the effect of the accumulated operating time T on the survival of the device during an additional time t. Next, let us consider $R(t \mid T)$ as a function of T and take the partial derivative with respect to T. This gives

$$\frac{\partial}{\partial T} R(t \mid T) = [\lambda(T) - \lambda(T + t)] R(t \mid T) \qquad (3\text{--}50)$$

If the hazard function is decreasing with time, $\lambda(T)$ is greater than $\lambda(T + t)$ and the initial wearin period $(0, T)$ improves the reliability of the device during $(T, T + t)$. If the hazard function is increasing with time, initial wearin only adds to the decline of the device with age, and the reliability during $(T, T + t)$ decreases.

■■ **Example 3–11**
If the hazard function is constant,

$$R(t \mid T) = \exp[-\lambda t]$$

implying that the probability of survival during $(T, T + t)$ does not depend on T, the duration of prior operation. In other words, a constant hazard model implies no aging, and there is no need to subject such constant hazard components to an initial wearin period to improve their reliability of operation at future periods. We will discuss these aspects of the constant hazard model further in the next chapter.   ■■

3.8 SUMMARY

The four reliability functions—namely, the failure density function $f(t)$, hazard function $\lambda(t)$, failure distribution function $Q(t)$, and reliability function $R(t)$—were introduced in this chapter from two different viewpoints. First, they were developed as limiting cases of the corresponding piecewise-linear functions constructed using catastrophic failure data. Then, the time to failure was considered as a continuous random variable. Both viewpoints led to the same relationships between the four functions (tabulated in Table 3.1).

The bathtub-shaped hazard function was introduced next, and its various regions and their implications were discussed. Important reliability parameters, such as the mean time to failure (MTTF), median time to failure (T_{50}), and average failure rate over a time interval (AFR(t_1, t_2)), were introduced, and their determination by means of the reliability functions was discussed. Finally, the cumulative hazard function and the *a posteriori* failure probability were introduced.

PROBLEMS

3.1 Is $\lambda(t) = kt$ a valid hazard model? Find all the other reliability functions and sketch them.

3.2 For the cumulative failure distribution function given in Figure P 3.2, find and sketch all the other reliability functions.

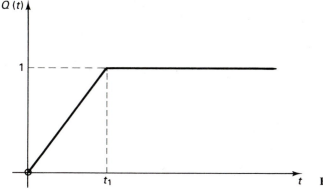

Figure P 3.2

3.3 Find the hazard function for the failure distribution

$$Q(t) = 1 - \exp\left[- \frac{Kt^{m+1}}{(m+1)} \right]$$

3.4 The time to failure (in hours) of a system has the density function

$$f(t) = 0.001 \exp[-0.001t]$$

What is the probability that this system will fail in (*i*) less than 100 hours and (*ii*) less than 1,000 hours?

3.5 For the failure density function

$$f(t) = \frac{a}{(b + t)^n}, \quad t > 0, n > 1$$

find (*i*) the relationship between *a*, *b*, and *n* and (*ii*) the corresponding reliability function.

3.6 One useful empirical expression for the failure rates of some mechanical components is

$$\lambda(t) = a[C_1 + C_2 e^{-bt} + C_3 e^{-ct}]$$

By adjusting the parameters *a*, *b*, *c*, C_1, C_2, and C_3, we can fit a variety of hazard function shapes. Find the corresponding failure density and reliability functions.

3.7 A small fraction (α) of a component population is susceptible to a certain infant mortality (early failures) syndrome. The rest of the population is not susceptible to this syndrome and fails in a normal manner. Let the failure distribution functions that apply to these two mechanisms be $Q_e(t)$ and $Q_n(t)$, respectively.
   (*i*) Construct the distribution function $Q_T(t)$ for the total population by taking a weighted average of $Q_e(t)$ and $Q_n(t)$.
   (*ii*) Find the failure rate $\lambda_T(t)$ applicable to the total population.

3.8 Suppose we are concerned only with initial failures. Suppose also that a component, once it survives past a time t_c, is considered failproof for all practical purposes. Such components can be considered to have a failure rate given by

$$\lambda(t) = \begin{cases} k[1 - (t/t_c)] & \text{for } 0 < t < t_c \\ 0 & \text{otherwise} \end{cases}$$

   (a) Find expressions for the failure density and distribution functions.
   (b) What is the probability of an initial failure?

3.9 An electronic amplifier is required to maintain its output voltage between 39 and 41.5 volts. The output voltage has random variations with a density function

$$f(V) = \begin{cases} 0 & \text{for } 0 \le V \le 38 \\ 0.25V - 9.5 & \text{for } 38 < V < 40 \\ 10.5 - 0.25V & \text{for } 40 \le V \le 42 \\ 0 & \text{for } 42 < V < \infty \end{cases}$$

The amplifier has a probability of 0.02 of being in the catastrophically failed state with zero output. Find the reliability of the amplifier under these conditions.

3.10 The failure density function for a machine tool component can be approximated as

$$f(t) = 2b(1 - bt) \quad \text{for } 0 \le t < (1/b)$$

   (*i*) Is this a valid density function?
   (*ii*) Find the corresponding failure rate.
   (*iii*) What is the MTTF?

3.11 Given that the failure rate for a certain type of capacitor is 2 PPM/K, find $R(t)$ for t in hours. Calculate the one-year (8,760 hours) reliability, MTTF in hours, and allowable operating time if the reliability is to be no lower than 0.9999.

3.12 Repeat Problem 3.11 for components with a constant failure rate of 0.4%/K.

3.13 The reliability of a class of components can be approximated by

$$R(t) = (1 - at)^n \quad \text{for } 0 \le t < (1/a); \quad a > 0, n > 0$$

Find the corresponding failure density function, distribution function, hazard function, MTTF, and median time to failure.

3.14 For constant hazard components (λ is constant), show that the median time to failure is less than the MTTF.

3.15 Derive expressions for $f(t)$, $\lambda(t)$, $Q(t)$, and $R(t)$ in terms of the cumulative hazard function $\Lambda(t)$.

3.16 For the linearly increasing hazard $\lambda(t) = kt$, $k > 0$, find the cumulative hazard function, AFR(t_1, t_2), and AFR(0, T).

3.17 The lifetimes of a certain group of components have the distribution given by

$$Q(t) = 1 - \frac{1}{(1 + kt)}$$

(a) What is the probability of n of these components operating successfully over the time interval 0 to T? Assume the failures to be independent.

(b) Find the corresponding hazard function.

(c) Find the average failure rate applicable to the time interval 0 to T.

3.18 For constant hazard components (λ is constant), what is the median time to failure (T_{50})? Show that the reliability function can be expressed as

$$R(t) = 0.5^{t/T_{50}}$$

3.19 Show that

$$R(t) = 0.5 + \int_{t}^{T_{50}} f(\xi)\, d\xi$$

3.20 An important electronic component for use in a computer has a decreasing hazard rate given by

$$\lambda(t) = 0.048/\sqrt{t} \ \text{year}^{-1}$$

with the time t measured in years. Calculate the one-year design life reliability of the component (i) without any initial wearin and (ii) with an initial wearin period of five weeks.

4

Probability Distribution
Functions and Their
Application
in Reliability Evaluation

4.1 THE BINOMIAL DISTRIBUTION

The binomial distribution, also known as the Bernoulli distribution, has applications in many combinational-type reliability problems. Let us consider an experiment, repeated several times, that satisfies the following conditions:

(*i*) The number of trials, n, is fixed.

(*ii*) Each trial results in success or failure, with probabilities of p and q, respectively, where $(p + q) = 1$.

(*iii*) All trials have identical probabilities of success (and therefore failure); in other words, p and q are constant.

(*iv*) All trials are independent.

We now ask the following question: What is the probability of exactly r successes in n trials? Clearly, exactly r successes imply exactly $(n - r)$ failures in n trials. The order of occurrence of failures and successes is not important; what is important is that the number of successes should be equal to r.

The number of possible sequences of exactly r successes and $(n - r)$ failures in n trials is the same as the number of ways we can pick r out of n without regard to order. This number is

$$_nC_r = \binom{n}{r} = \frac{n!}{r!(n-r)!} \qquad (4\text{--}1)$$

Since each trial is independent of all the others, the probability of *any one sequence* of r successes and $(n-r)$ failures is $p^r q^{n-r}$. However, all the possible sequences are also independent events. Therefore, the probability P_r of exactly r successes and $(n-r)$ failures in n trials should be

$$P_r = \sum_{\substack{\text{all possible sequences of} \\ r \text{ successes and } (n-r) \text{ failures}}} (\text{probability of any one sequence})$$

$$= \frac{n!}{r!(n-r)!} p^r q^{n-r} \qquad (4\text{--}2)$$

$$= {_nC_r} p^r (1-p)^{n-r}$$

Sometimes, the notation $B(r; n, p)$ is used instead of P_r. Expanding $(q + p)^n$ using the binomial theorem, we get

$$(q + p)^n = \sum_{r=0}^{n} {_nC_r} p^r q^{n-r} = \sum_{r=0}^{n} P_r = 1 \qquad (4\text{--}3)$$

The name "binomial distribution" comes from the fact that the rth term in the expansion of $(q + p)^n$ is indeed P_r. The binomial coefficients $\binom{n}{r}$ can be evaluated using Pascal's triangle:

n	r												
0	0						1						
1	0,1						1	1					
2	0,1,2					1	2	1					
3	0,1,2,3				1	3	3	1					
4	0,1,2,3,4			1	4	6	4	1					
5	0,1,2,3,4,5		1	5	10	10	5	1					
6	0,1,2,3,4,5,6	1	6	15	20	15	6	1					
$\vdots$					$\vdots$								

Factorials of large numbers can be approximated by Sterling's formula,

$$n! \cong e^{-n} n^n \sqrt{2\pi n} \qquad (4\text{--}4)$$

Another commonly used approximation is

$$\ln n! \cong n(\ln n - 1) \qquad (4\text{--}5)$$

4.1.1 Expected Value and Standard Deviation

Consider a random variable X with a binomial distribution. For our purposes, X denotes the number of successes in n trials, and it can have any value between 0 and n. The probability of exactly r successes and $(n - r)$ failures in n trials is

$$P_r = \frac{n!}{r!(n - r)!} p^r q^{n-r}$$

The expected value of X is

$$E[X] = \sum_{r=0}^{n} rP_r = \sum_{r=0}^{n} \frac{rn!}{r!(n - r)!} p^r q^{n-r}$$

$$= np \sum_{r=1}^{n} \frac{(n - 1)!}{(r - 1)!(n - r)!} p^{r-1} q^{n-r}$$

Let $i = r - 1$. Then r going from 1 to n is the same as i going from 0 to $(n - 1)$; thus,

$$E[X] = np \sum_{i=0}^{n-1} \frac{(n - 1)!}{i!(n - 1 - i)!} p^i q^{n-1-i}$$

Note that the probability of i successes and $(n - 1 - i)$ failures in $(n - 1)$ trials is

$$\frac{(n - 1)!}{i!(n - 1 - i)!} p^i q^{n-1-i}$$

As i goes from 0 to $(n - 1)$, the foregoing summation must be unity. Therefore, we conclude that

$$E[X] = np \tag{4-6}$$

The second moment of X is

$$E[X^2] = \sum_{r=0}^{n} r^2 P_r = \sum_{r=0}^{n} r^2 \frac{n!}{r!(n - r)!} p^r q^{n-r}$$

$$= np \sum_{r=1}^{n} r \frac{(n - 1)!}{(r - 1)!(n - r)!} p^{r-1} q^{n-r}$$

As before, we let $i = (r - 1)$ and proceed with the summation from $i = 0$ to $i = (n - 1)$ to obtain

$$E[X^2] = np \sum_{i=0}^{n-1} (i + 1) \frac{(n - 1)!}{i!(n - i - 1)!} p^i q^{n-i-1}$$

$$= np \left[\sum_{i=0}^{n-1} i \frac{(n - 1)!}{i!(n - i - 1)!} p^i q^{n-i-1} + \sum_{i=0}^{n-1} \frac{(n - 1)!}{i!(n - i - 1)!} p^i q^{n-i-1} \right]$$

$$= np[(n - 1)p + 1]$$

$$= (np)^2 - p^2 n + np$$

Using Equation (2–57), the variance is found to be

$$\sigma^2 = \text{var } X = E[X^2] - (E[X])^2$$
$$= (np)^2 - p^2n + np - (np)^2$$
$$= npq$$

The standard deviation is the positive square root of the variance. Thus,

$$\sigma = \sqrt{npq} \tag{4–7}$$

4.1.2 Applications

1. Consider a collection of n identical components. If p is the probability that a component is defective, then the expected number of defective components out of a total of n is equal to np.

   a. What is the probability of finding r defects out of n?

$$P(r \text{ defective}) = {}_nC_r p^r (1 - p)^{n-r}$$

   b. What is the probability of finding r or fewer defects out of n?

$$P(r \text{ or fewer defects}) = \sum_{i=0}^{r} {}_nC_i p^i (1 - p)^{n-i}$$

   c. Assuming that np is an integer, what is the probability of finding np defects out of n?

$$P(np \text{ defects}) = \frac{n!}{(np)!(n - np)!} p^{np}(1 - p)^{n - np}$$

2. Consider a system with n identical components, out of which at least r of them must be good for the system to succeed. Examples of such systems are a cable with n strands, out of which at least r of them must be good; a power plant with n identical generators, out of which at least r of them must be working to meet the demand; and a pumping station with n identical pumps, out of which at least r of them must be good. The system will succeed if r or $(r + 1)$ or $(r + 2)$ or ... or n components are good. That is,

$$P(\text{system success}) = R = \sum_{k=r}^{n} {}_nC_k p^k (1 - p)^{n-k} \tag{4–8}$$

where p is the probability of success of a component. Systems for which $r < n$ are said to have *redundancy*.

3. Consider a system with three components, each of which could be good or bad. One way to enumerate all the possible states of the system is to start with the binary number 000 and keep on adding 1 in binary arithmetic until the binary number 111 is reached. The result is tabulated as follows, along with the corresponding state probabilities. (For $i = 1, 2, 3$, we assume that

$$p_i = \text{probability that the } i\text{th component is good}$$

$$q_i = \text{probability that the } i\text{th component is bad})$$

Component Number 1	2	3		State Probability		
1	1	1		p_1	p_2	p_3
1	1	0		p_1	p_2	q_3
1	0	1		p_1	q_2	p_3
1	0	0		p_1	q_2	q_3
0	1	1		q_1	p_2	p_3
0	1	0		q_1	p_2	q_3
0	0	1		q_1	q_2	p_3
0	0	0		q_1	q_2	q_3

1 means good
0 means bad

$\Sigma = 1$

We note that the state probabilities can also be obtained by expanding the product $(p_1 + q_1)(p_2 + q_2)(p_3 + q_3)$. We obtain

$$(p_1 + q_1)(p_2 + q_2)(p_3 + q_3) = p_1p_2p_3 + p_1q_2p_3 + q_1p_2p_3$$
$$+ q_1q_2p_3 + q_3p_1p_2 + q_3p_1q_2 + q_3q_1p_2 + q_1q_2q_3$$

If the components are identical, then we can use the binomial expression for $(p + q)^3$ to enumerate the different states. We get

$$(p + q)^3 = p^3 + 3p^2q + 3pq^2 + q^3$$

The states and their probabilities are as follows:

All components are good p^3
Only one component failed $3p^2q$
Only two components failed $3pq^2$
All components failed q^3

In general, in the case of n identical components,

$$(p + q)^n = \sum_{r=0}^{n} {}_nC_r p^r (1 - p)^{n-r} = 1 \qquad (4\text{--}9)$$

and the probability of the state with r good components and $(n - r)$ bad components is given by

$${}_nC_r p^r (1 - p)^{n-r}$$

where the factor $_nC_r$ indicates that there are $_nC_r$ states with r good and $(n - r)$ bad components.

■■   **Example 4–1**

Let us consider a power plant with three generators supplying a 250-MW load. Data on the individual units, the possible states and their probabilities, the MW capacities in and out, and the corresponding losses of load are listed in the following table:

UNIT # CAPACITY, MW P(failure)			1 100 0.01	2 150 0.02	3 200 0.03	
Unit #			MW CAP IN	MW CAP OUT		Load Loss, MW
1	2	3			Probability	
G	G	G	450	0	0.941094	0
B	G	G	350	100	0.009506	0
G	B	G	300	150	0.019206	0
G	G	B	250	200	0.029106	0
B	B	G	200	250	0.000194	50
B	G	B	150	300	0.000294	100
G	B	B	100	350	0.000594	150
B	B	B	0	450	0.000006	250
					Σ = 1.000000	

Calculate the expected value of the load loss.

We have

$$E[\text{load loss}] = (50 \times 0.000194) + (100 \times 0.000294)$$

$$+ (150 \times 0.000594) + (250 \times 0.000006)$$

$$= 0.1297 \text{ MW}$$

If the load is 200 MW,

$$E[\text{load loss}] = (50 \times 0.000294) + (100 \times 0.000594)$$

$$+ (200 \times 0.000006)$$

$$= 0.0753 \text{ MW} \qquad ■■$$

■■   **Example 4–2**

Bolts manufactured in a factory have a 5% probability of defects. They are sold in lots of 1,000 to a quality-conscious customer who picks 20 bolts at random from each lot and tests them. If there are two or more defects out of this random sample of 20, the entire lot is rejected. Calculate the probability that a lot will be rejected.

First,

$$P[\text{a bolt is good}] = p = 0.95$$

$$P[\text{a bolt is defective}] = q = (1 - p) = 0.05$$

Using the binomial distribution, we obtain

$$P[0 \text{ defects out of } 20] = {}_{20}C_0(0.05)^0(0.95)^{20} = 0.3585$$

$$P[1 \text{ defect out of } 20] = {}_{20}C_1(0.05)^1(0.95)^{19} = 0.3774$$

Therefore,

$$P[\text{lot is accepted}] = P[0 \text{ defects}] + P[1 \text{ defect}]$$

$$= 0.7359$$

Thus, there is a 73.59% probability that the lot will be accepted; or, alternatively, there is a 26.41% probability that the lot will be rejected.

In order to improve the probability of acceptance, some improvements are made in the manufacturing process that cut the defects down to 4%. In that case, the new acceptance probability is

$$P[\text{lot is accepted}] = {}_{20}C_0(0.04)^0(0.96)^{20}$$

$$+ \; {}_{20}C_1(0.04)^1(0.96)^{19}$$

$$= 0.4420 + 0.3683 = 0.8103$$

The probability of acceptance has been increased from 73.59% to 81.03%.

If the manufacturer convinces the customer to pick a sample of 10 instead of 20 and continues with the same manufacturing process, then the probability of acceptance (still requiring less than two defects) becomes

$$P[\text{lot is accepted}] = {}_{10}C_0(0.05)^0(0.95)^{10} + {}_{10}C_1(0.05)^1(0.95)^9$$

$$= 0.5987 + 0.3151 = 0.9138$$

The probability of acceptance is improved dramatically, from 73.59% to 91.38%!

■■

■■   **Example 4–3**

It is known that the probability of a certain kind of power transistor surviving a thermal test is 0.9. Calculate, from among a group of 17 such devices, the probability that (a) exactly 15 will survive; (b) at least 15 will survive; and (c) at least 2 will fail.

$$P[\text{exactly 15 will survive}] = {}_{17}C_{15}(0.9)^{15}(0.1)^2$$

$$= \frac{17!}{15!2!}\,(0.9)^{15}(0.1)^2$$

$$= 0.28$$

$$P[\text{at least 15 will survive}] = P[\text{15 or more will survive}]$$

$$= \sum_{k=15}^{17} {}_{17}C_k(0.9)^k(0.1)^{17-k}$$

$$= 0.28 + {}_{17}C_{16}(0.9)^{16}(0.1) + (0.9)^{17}$$

$$= 0.280 + 0.315 + 0.167$$

$$= 0.762$$

$$P[\text{at least 2 will fail}] = 1 - P[\text{16 surviving}] - P[\text{17 surviving}]$$

$$= 1 - 0.315 - 0.167$$

$$= 0.518 \qquad \blacksquare\blacksquare$$

■■ **Example 4–4**

A small aircraft landing gear has three tires. It can make a safe landing if no more than one tire bursts. An examination of past records reveals that, on average, tire bursts occur once in every 1,000 landings. What is the probability that a given aircraft will make a safe landing?

$$P[\text{tire bursting on landing}] = 0.001$$

$$P[\text{safe landing}] = P[\text{no tire bursting}] + P[\text{1 tire bursting}]$$

$$= {}_3C_0(0.001)^0(0.999)^3$$

$$+ {}_3C_1(0.001)^1(0.999)^2$$

$$= 0.99700 + 0.00299$$

$$= 0.99999$$

Alternatively,

$$P[\text{unsafe landing}] = 0.00001 \text{ or } 0.001\% \qquad \blacksquare\blacksquare$$

4.2 THE POISSON DISTRIBUTION

In the case of events that occur at a certain average rate, say, λ occurrences per unit time, we would like to know the probability $P_x(t)$ of the event occurring exactly x times in the time interval $(0, t)$. Here we count only the occurrences, and not the nonoccurrences, as we did in the case of the binomial distribution.

Consider a small time interval dt, an interval so small that the probability of the event occurring more than once during this time interval is zero. If λ is the average rate of occurrences, then

λdt = probability of the event occurring once in the time interval $(t, t + dt)$

Let

$P_x(t)$ = probability of the event occurring x times during the time interval $(0, t)$

and

$P_x(t + dt)$ = probability of the event occurring x times during the time interval $(0, t + dt)$

= $P_x(t)$ [probability of zero occurrences in $(t, t + dt)$] + $P_{x-1}(t)$ [probability of one occurrence in $(t, t + dt)$] + $P_{x-2}(t)$ [probability of two occurrences in $(t, t + dt)$] + ... + $P_0(t)$ [probability of x occurrences in $(t, t + dt)$]

Since the probability of more than one occurrence in an interval of length dt is zero, the foregoing equation becomes

$$P_x(t + dt) = P_x(t)[1 - \lambda dt] + P_{x-1}(t)[\lambda dt]$$

$$= P_x(t) - \lambda dt[P_x(t) - P_{x-1}(t)]$$

For zero occurrences in the time interval $(0, t)$, we set $x = 0$ and get

$$P_0(t + dt) = P_0(t)[1 - \lambda dt]$$

Rearranging terms yields

$$\frac{P_0(t + dt) - P_0(t)}{dt} = -\lambda P_0(t)$$

which, in the limit as $dt \to 0$, becomes

$$\frac{d}{dt} P_0(t) + \lambda P_0(t) = 0$$

the general solution of which is

$$P_0(t) = ke^{-\lambda t}$$

Since at $t = 0$, zero occurrences are assured, $P_0(0) = 1$, which leads to $k = 1$. Therefore, the time-dependent probability $P_0(t)$ is

$$P_0(t) = e^{-\lambda t} \qquad (4-10)$$

This is the probability of zero occurrences of the event. If the event is failure, then Equation (4–10) is the no-failure or reliability expression.

For one occurrence during $(0, t)$, we set $x = 1$ and obtain

$$P_1(t + dt) = P_1(t) - \lambda dt[P_1(t) - P_0(t)]$$

or

$$\frac{P_1(t + dt) - P_1(t)}{dt} = \lambda[P_0(t) - P_1(t)] = \lambda[e^{-\lambda t} - P_1(t)]$$

In the limit as $dt \to 0$, the preceding becomes

$$\frac{d}{dt} P_1(t) + \lambda P_1(t) = \lambda e^{-\lambda t}$$

The solution of this differential equation is

$$P_1(t) = ke^{-\lambda t} + \lambda t e^{-\lambda t}$$

At $t = 0$, $P_1(t) = 0$. Therefore, k must be zero and

$$P_1(t) = \lambda t e^{-\lambda t} \tag{4–11}$$

Proceeding in a similar fashion for $x = 2, 3$, and so on, we can show that

$$P_x(t) = \frac{(\lambda t)^x e^{-\lambda t}}{x!} \tag{4–12}$$

If the event is failure of a component, then, in discussing multiple failures, we assume instant replacement after each failure. In practice, repair or replacement times are much shorter than the time to failure, and it is justifiable to assume instant replacement. If this assumption is not valid, and if we want to include repair or replacement times, we have to use different techniques, such as Markov processes. Even if the repair or replacement time is not negligible, the Poisson distribution is valid for zero failures and for one failure.

4.2.1 Expected Value and Standard Deviation

The expected value of the Poisson distribution is

$$\mu = E(X) = \sum_{x=0}^{\infty} xP_x$$

$$= \sum_{x=0}^{\infty} \frac{x(\lambda t)^x e^{-\lambda t}}{x!}$$

$$= (\lambda t) \sum_{x=1}^{\infty} \frac{(\lambda t)^{x-1} e^{-\lambda t}}{(x-1)!} \tag{4–13}$$

$$= (\lambda t) \sum_{y=0}^{\infty} \frac{(\lambda t)^y e^{-\lambda t}}{y!}$$

$$= (\lambda t)$$

The summation is simply the sum of 0, 1, 2, ... , ∞ occurrences of the event and must be unity. We can also see that if λ is the average rate of occurrence of the event, then the average (or expected) number of occurrences during a time interval of t is simply λt. In terms of μ, the Poisson distribution can be written as

$$P_x(t) = \frac{\mu^x e^{-\mu}}{x!} \tag{4–14}$$

The second moment, variance, and standard deviation are calculated for the Poisson distribution as follows:

$$E[X^2] = \sum_{x=0}^{\infty} x^2 P_x = \sum_{x=0}^{\infty} \frac{x^2(\lambda t)^x e^{-\lambda t}}{x!}$$

$$= (\lambda t) \sum_{x=1}^{\infty} \frac{x(\lambda t)^{x-1}}{(x-1)!} e^{-\lambda t}$$

$$= \lambda t \sum_{x=1}^{\infty} \frac{[(x-1)+1](\lambda t)^{x-1} e^{-\lambda t}}{(x-1)!}$$

$$= \lambda t \left[\sum_{x=1}^{\infty} \frac{(x-1)(\lambda t)^{x-1}}{(x-1)!} e^{-\lambda t} + \sum_{x=1}^{\infty} \frac{(\lambda t)^{x-1} e^{-\lambda t}}{(x-1)!} \right]$$

$$= \lambda t \left(\sum_{y=0}^{\infty} \frac{y(\lambda t)^y}{y!} e^{-\lambda t} + \sum_{y=0}^{\infty} \frac{(\lambda t)^y}{y!} e^{-\lambda t} \right)$$

$$= \lambda t[(\lambda t) + 1]$$

$$\text{Var } X = E[X^2] - [E(X)]^2$$
$$= (\lambda t)^2 + (\lambda t) - (\lambda t)^2 \qquad (4\text{--}15)$$
$$= \lambda t$$

$$\text{Standard deviation} = \sigma = \sqrt{\text{Var } x} = \sqrt{\lambda t} \qquad (4\text{--}16)$$

4.2.2 Alternative Derivation of the Poisson Distribution

Although the Poisson distribution is a distribution in its own right, it can also be derived as a limiting form of the binomial distribution. Under certain conditions, the binomial distribution can be approximated by the Poisson distribution.

According to the binomial distribution, the probability of exactly r successes and $(n - r)$ failures in n trials is

$$P_r = \frac{n!}{r!(n-r)!} p^r q^{n-r}$$

If $n \gg r$, then

$$\frac{n!}{(n-r)!} = n(n-1)(n-2) \dots (n-r+1)$$

$$\simeq n^r$$

Now,

$$q^{n-r} = (1-p)^{n-r}$$

Moreover, if p is very small and if $n \gg r$, then

$$q^{n-r} \cong (1 - p)^n$$

$$= 1 - np + \frac{n(n-1)}{2!} p^2 - \cdots$$

$$\simeq 1 - np + \frac{(np)^2}{2!} - \cdots$$

$$= e^{-np}$$

Therefore, the limiting form of the binomial distribution is

$$P_r = \frac{n^r}{r!} p^r e^{-np}$$

$$= \frac{(np)^r}{r!} e^{-np} = \frac{\mu^r e^{-\mu}}{r!}$$

which is of the same form as Equation (4–14), in which P_x is the probability of x successes in n trials when the expected number of successes is equal to μ (or np).

■■ **Example 4–5**

Suppose that events occur at an average rate of λ per unit of time and that the time between occurrences is a continuous random variable. Find the cumulative distribution function of the variable.

We assume that the event occurred once at time $t = 0$ and ask the following question: What is the probability that the event will occur again within the time interval $(0, t]$? In other words, we seek the distribution function $F(t)$ of the random variable called time between occurrences. We have

$$F(t) = P[\text{time between occurrences} \leq t] = 1 - P[\text{no occurrence in } (0, t)]$$

$$= 1 - P_0(t) = 1 - e^{-\lambda t}$$

The corresponding density function is

$$f(t) = \frac{d}{dt} F(t) = \lambda e^{-\lambda t}$$

As we will see in the next section, this function is known as the *exponential distribution*, and it is employed widely in reliability analyses. ■■

■■ **Example 4–6**

Failures in a large web-handling system occur at an average rate of one every three months. For a period of one year,

$$P[\text{more than 5 failures}] = 1 - P\begin{bmatrix}\text{less than or equal to}\\ \text{5 failures}\end{bmatrix}$$

$$\text{Average failure rate} = (1/3)\ \text{month}^{-1}$$

$$P[0\ \text{failure}] = e^{-4} \qquad = 0.01832$$

$$P[1\ \text{failure}] = 4e^{-4} \qquad = 0.07326$$

$$P[2\ \text{failures}] = [4^2 e^{-4}]/2 = 0.14653$$

$$P[3\ \text{failures}] = [4^3 e^{-4}]/3! = 0.19537$$

$$P[4\ \text{failures}] = [4^4 e^{-4}]/4! = 0.19537$$

$$P[5\ \text{failures}] = [4^5 e^{-4}]/5! = \underline{0.15629}$$

$$\Sigma = 0.78514$$

Therefore,

$$P[\text{more than 5 failures in one year}] = 1 - 0.78514$$

$$= 0.21486$$

The probability of two consecutive failures occurring less than one month apart is equal to $P[T < 1]$, where T is the continuous random variable called time between consecutive failures. Since T is exponentially distributed,

$$P[T < 1] = 1 - e^{-1/3} = 1 - 0.71653 = 0.28347 \qquad ∎∎$$

■■ **Example 4–7**

We have seen that the Poisson distribution can be obtained as the limiting case of the binomial distribution when the number of trials is large compared to the number of successes sought and when the probability of success in each trial is small. The following table shows the results obtained using these two distributions for $n = 50$ and $r = 1, 2, 3$ for three different values of p, the probability of success in each trial. Close agreement between the values are obtained for small values of p and when r is small compared to the number of trials.

			BINOMIAL	POISSON
n	r	p	$_nC_r p^r (1 - p)^{n-r}$	$\dfrac{(np)^r e^{-np}}{r!}$
50	1	0.1	0.02863	0.03369
50	2	0.1	0.07794	0.08422
50	3	0.1	0.13857	0.14037
50	1	0.01	0.30558	0.30327
50	2	0.01	0.07562	0.07582
50	3	0.01	0.01222	0.01264
50	1	0.001	0.04761	0.04756
50	2	0.001	0.00117	0.00119
50	3	0.001	0.00002	0.00002

■■

4.3 CONSTANT-HAZARD MODEL AND THE EXPONENTIAL DISTRIBUTION

During their useful lifetimes, many components exhibit a constant hazard rate. Such a rate implies that the occurrence of failures is purely random and that there is no deterioration of the strength or soundness of the components with time. In that case, the instantaneous rate of failure is the same in the 100th hour as it is in the 1,000th hour. Although this analysis is not realistic for all time, it is a good approximation during the useful lifetime (the horizontal portion of the bathtub curve) of the component. A constant hazard rate leads to the exponential distribution, a model that is simple, requires only one parameter to be defined, and is probably the most widely used distribution in reliability analysis. We let

$$\lambda(t) = \lambda, \text{ a constant} \tag{4-17}$$

Then

$$f(t) = \lambda \exp\left[- \int_0^t \lambda \, d\xi \right] = \lambda e^{-\lambda t}, \quad t > 0 \tag{4-18}$$

$$R(t) = \exp\left[- \int_0^t \lambda \, d\xi \right] = e^{-\lambda t} \tag{4-19}$$

$$Q(t) = 1 - e^{-\lambda t} \tag{4-20}$$

These four functions representing the exponential distribution are shown in Figure 4.1. Expressions for $Q(t)$ and $R(t)$ can be derived using the failure rate λ as follows:

$$Q(t) = \int_0^t \lambda e^{-\lambda \xi} \, d\xi = \left[\frac{\lambda e^{-\lambda \xi}}{(-\lambda)} \right]_0^t = (1 - e^{-\lambda t}) \tag{4-21a}$$

$$R(t) = \int_t^\infty \lambda e^{-\lambda \xi} \, d\xi = \left[\frac{\lambda e^{-\lambda \xi}}{(-\lambda)} \right]_t^\infty = (e^{-\lambda t}) \tag{4-21b}$$

The sum of the two expressions given in Equations (4–21) is always equal to unity, as illustrated in Figure 4.2.

The parameter λ is called the *failure rate*, and it is equal to the number of failures per unit time. Using this information, we can derive the following expressions for reliability and unreliability, respectively:

$$P(\text{no failure in the time interval 0 to } T) = R(T) = e^{-\lambda T} \tag{4-22}$$

$$P(\text{failure in the time interval 0 to } T) = Q(T) = 1 - e^{-\lambda T} \tag{4-23}$$

Next, let us consider an interval of time $(T, T + t)$ as illustrated in Figure 4.3. Let

$$\text{event } A = \text{failure during time } t$$

$$\text{event } B = \text{survival up to time } T$$

$$= \text{no failure in the time interval } (0, T)$$

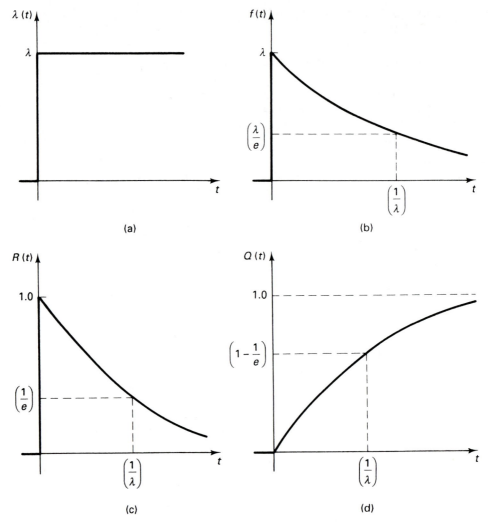

Figure 4.1 The exponential distribution model. (a) Hazard function. (b) Failure density function. (c) Reliability function. (d) Failure distribution function.

Then

$$A \cap B = \text{survival up to } T \textit{ and } \text{failure during } (T, T + t)$$

$$P(A \cap B) = \int_T^{T+t} \lambda e^{-\lambda \xi} \, d\xi = \left(\frac{\lambda e^{-\lambda \xi}}{-\lambda} \right)_T^{T+t} = e^{-\lambda T} - e^{-\lambda(T+t)}$$

Also,

$$P(B) = \int_T^{\infty} \lambda e^{-\lambda \xi} \, d\xi = e^{-\lambda T}$$

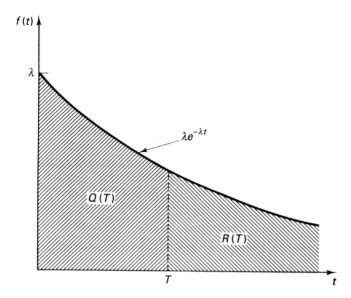

Figure 4.2   Exponential failure density function.

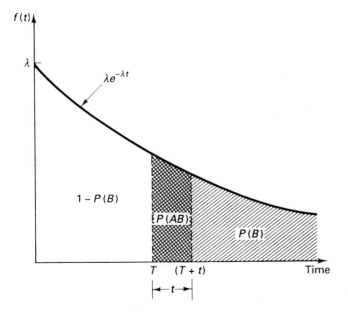

Figure 4.3   *A posteriori* probability of failure, $Q_c(t) = [P(AB)]/[P(B)]$.

The *a posteriori* failure probability $Q_c(t)$ is defined as the probability of failure during t, *given* survival up to T. Thus,

$$Q_c(t) \equiv P(A \mid B) = \frac{P(A \cap B)}{P(B)}$$

or

$$Q_c(t) = \frac{e^{-\lambda T} - e^{-\lambda(T+t)}}{e^{-\lambda T}} = 1 - e^{-\lambda t} \tag{4-24}$$

Note that $Q_c(t)$ is independent of the prior operating time T, and it depends only on the duration t of the time interval. In other words, no matter how long the component has been in operation, the probability of failure during the next t is the same. This is the same as saying that the component does not degrade in quality with the time of operation. It is also simply a restatement of our original assumption that the hazard rate λ is a constant, and it is valid only during the normal or useful lifetime of a component.

We notice that the *a priori* failure probability of the component, which is the probability of failure between 0 and t, is

$$Q(t) = 1 - e^{-\lambda t} \tag{4-25}$$

Therefore, for the exponential distribution,

$$Q(t) = Q_c(t) = 1 - e^{-\lambda t} \tag{4-26}$$

The probability of failure depends only on the duration of time considered, and not on previous history. In other words, the exponential distribution is a memoryless distribution. It is applicable only to the normal operating period during which λ is constant.

If

$$\lambda t \ll 1, \quad Q(t) = Q_c(t) \simeq \lambda t$$

and

$$R(t) \simeq (1 - \lambda t)$$

If λ is not constant, we can always find $Q_c(t)$ from Equation (3–46):

$$Q_c(t) = \frac{\displaystyle\int_T^{T+t} f(\xi)\, d\xi}{\displaystyle\int_T^{\infty} f(\xi)\, d\xi} \tag{4-27}$$

4.3.1 Expected Value and Standard Deviation

For an exponentially distributed continuous random variable, the expected value μ can be obtained from

$$\mu = \int_0^{\infty} t\lambda e^{-\lambda t}\, dt = \int_0^{\infty} t\, d(-e^{-\lambda t})$$

Integrating by parts, we get

$$\mu = [-te^{-\lambda t}]_0^\infty - \int_0^\infty (-e^{-\lambda t})\, dt$$

$$= 0 - \left[\frac{e^{-\lambda t}}{\lambda}\right]_0^\infty \qquad (4\text{--}28)$$

$$= \frac{1}{\lambda}$$

For the case of a failure density function, μ is called the *mean time to failure* (MTTF). The variance is

$$\text{var } t = \sigma^2 = E[t^2] - (E[t])^2$$

Since

$$E[t^2] = \int_0^\infty t^2 \lambda e^{-\lambda t}\, dt$$

$$= \int_0^\infty t^2\, d(-e^{-\lambda t})$$

$$= (-t^2 e^{-\lambda t})_0^\infty - \int_0^\infty - e^{-\lambda t} 2t\, dt$$

$$= 0 + \frac{2}{\lambda} \int_0^\infty t\lambda e^{-\lambda t}\, dt = \frac{2}{\lambda}\frac{1}{\lambda} = \frac{2}{\lambda^2}$$

we get

$$\sigma^2 = \frac{2}{\lambda^2} - \frac{1}{\lambda^2} = \frac{1}{\lambda^2}$$

and the standard deviation becomes

$$\sigma = \frac{1}{\lambda} \qquad (4\text{--}29)$$

For components with exponentially distributed failure times, the MTTF is equal to the reciprocal of the failure rate. It is possible for the MTTF to be greater than the useful lifetime of the component. This only means that if the constant-λ portion of the hazard rate curve continues to be valid, then the MTTF is what it is computed to be. Otherwise, it simply means that the failure rate is equal to the reciprocal of the MTTF.

In the case of repairable components, the term MTBF is used to indicate the mean time between failures:

MTBF = cycle time between failures

= mean time to failure + mean time to repair

= MTTF + MTTR

If MTTR ≪ MTTF, then MTBF ≅ MTTF.

When a failed component is replaced by another, new component,

$$\text{MTBF} = \text{MTTF} + \text{mean time to install a replacement}$$

$$= \text{MTTF} + \text{MTTI}$$

Again, if $\text{MTTI} \ll \text{MTTF}$, $\text{MTBF} \cong \text{MTTF}$.

■■ **Example 4–8**

We can show that if the ratio of the number of failures in a small time interval to the time interval involved is proportional to the number of survivors at the beginning of this time interval, then the time to failure is exponentially distributed.

Let $N(t)$ be the number of survivors at time t, and consider a small time interval Δt. Then

$$\frac{N(t + \Delta t) - N(t)}{\Delta t} = -\lambda N(t)$$

The negative sign is necessary, since N is decreasing with time; λ is the proportionality constant. In the limit as $\Delta t \to 0$, we obtain

$$\frac{dN}{dt} = -\lambda N$$

which, upon integration, yields

$$\ln N = -\lambda t + k$$

or

$$N = N_0 e^{-\lambda t}$$

where N_0 is the number of items at time $t = 0$. The reliability function is

$$R(t) = \frac{N(t)}{N_0} = e^{-\lambda t}$$

This reliability function corresponds to the exponential distribution of the random variable called time to failure.   ■■

■■ **Example 4–9**

The exponential distribution of the time to failure is equivalent to the Poisson distribution of the probability of x failures. If the MTTF is 200 hours, the probability of x failures in 1,000 hours is

$$P_x = \frac{(1{,}000/200)^x \exp[-1{,}000/200]}{x!} = \frac{5^x e^{-5}}{x!}$$

Note that $1{,}000/200 = 5$ is the expected number of failures in 1,000 hours. The following tabulation gives the probability of x failures for various values of x.

x:	0	1	2	3	4	5	6
P_x:	0.00674	0.03369	0.08422	0.14037	0.17547	0.17547	0.14622

■■

■■   **Example 4–10**

The lifetime of a certain device is exponentially distributed with a mean value of 500 hours. What is the probability of this device operating successfully for at least 600 hours?

Since the MTTF is 500 hours, the hazard rate $\lambda = (1/500)$ hour^{-1} and

$$P[\text{lifetime} \geq 600 \text{ hours}] = \exp[-600/500] = 0.3012$$

If we have three such devices, the probability of at least one failing during the first 400 hours is equal to $[1 - P(\text{all 3 surviving for 400 h}]$. So

$$P[\text{at least 1 fails within 400 hours}] = 1 - [\exp(-400/500)]^3$$

$$= 0.90928$$

Another approach to arriving at this result is to employ the exponential distribution to get the probabilities of failure and success and then to use these values in the binomial distribution. For each device,

$$p = P[\text{success in 400 hours}] = \exp(-400/500) = 0.44933$$

$$q = P[\text{failure in 400 hours}] = 1 - 0.44933 = 0.55067$$

$$P[\text{at least 1 fails wthin 400 hours}] = {_3}C_1 q p^2 + {_3}C_2 q^2 p + {_3}C_3 q^3$$

$$= 0.33354 + 0.40876 + 0.16698$$

$$= 0.90928, \text{ as before}$$

If we have four such devices, the probability of exactly two of them failing during the first 300 hours is calculated as follows. For each device,

$$p = P[\text{success in 300 hours}] = \exp(-300/500) = 0.5488$$

$$P\left[\begin{array}{l}\text{exactly 2 out of 4 failing} \\ \text{during the first 300 hours}\end{array}\right] = {_4}C_2 (1 - 0.5488)^2 (0.5488)^2$$

$$= 0.36789$$

Finally, the probability of the device failing between the Tth and the $(T + 100)$th hours of operation is the *a posteriori* probability

$$Q_c(100) = 1 - \exp(-100/500) = 0.18127$$

For the exponential distribution, this value does not depend on T; it depends only on the duration of time considered, which, for this example, is 100 hours.   ■■

4.4 THE NORMAL DISTRIBUTION

The *normal distribution* is one of the best known and most widely used two-parameter distributions. It is also known as the *Gaussian distribution* and, in France, as the *Laplacian distribution*. It was discovered by De Moivre in 1733 as the limiting form of the binomial distribution for discrete random variables. The normal distribution is often a good fit for the sizes of manufactured parts, populations of living organisms, magnitudes of certain electrical signals, and other natural phenomena. Its use in reliability evaluation is rather limited, except in the wearout region.

The Central Limit theorem states that if the sample size is very large, the distribution of the sample means approximates a normal distribution. In other words, a superposition of independent random variables always tends towards normality, regardless of the distribution of the individual random variables contributing to the sum. How large the sample size should be for assuming normality depends on the specific problem at hand. In any case, normal random variables occur quite commonly in nature.

In the context of reliability, suppose a phenomenon (the wearing out of a component is a good example) depends on several factors, each of which is random with a different distribution. If no one of the factors is dominant and if, say, the wearout time can be expressed as a sum of several random variables, then the wearout time itself can be described by a normal distribution, even though the individual random variables contributing to wearout are not normally distributed.

Another frequently occurring situation is when a random variable can be expressed as a product of several random variables with different distributions. If

$$t = t_1 t_2 \ldots t_n$$

then

$$\ln t = \ln t_1 + \ln t_2 + \ldots + \ln t_n$$

and $\ln t$ will tend to be normally distributed, even though the individual $\ln t_i$ are not. This analysis leads us to the lognormal distribution, in which the natural logarithm of the random variable of interest is normally distributed. The lognormal distribution is discussed in Section 4.9.

The two parameters defining the normal distribution are the mean μ and the standard deviation σ. We should remember that, regardless of the distribution, a particular random variable can be characterized by its mean and standard deviation. However, just because we know only μ and σ for a random variable, we are not justified in assuming that the distribution is normal, although normal random variables occur quite commonly in nature.

The density function for the normal distribution is

$$f(t) = \frac{1}{\sigma\sqrt{2\pi}} \exp\left[-\frac{(t - \mu)^2}{2\sigma^2} \right]; \quad -\infty < t < +\infty \qquad (4\text{--}30)$$

The corresponding distribution function, reliability function, and hazard function are, respectively,

$$F(t) = \frac{1}{\sigma\sqrt{2\pi}} \int_{-\infty}^{t} \exp\left\{ -\frac{(\xi - \mu)^2}{2\sigma^2} \right\} d\xi$$

$$= Q(t)$$

$$(4\text{--}31)$$

$$R(t) = 1 - Q(t) = 1 - F(t) = \int_t^\infty f(\xi) \, d\xi \qquad (4\text{--}32)$$

and

$$\lambda(t) = \frac{f(t)}{R(t)} \qquad (4\text{--}33)$$

Typical density and distribution curves for the normal distribution are shown in Figure 4.4. Larger values of σ result in low, broad curves, and smaller values

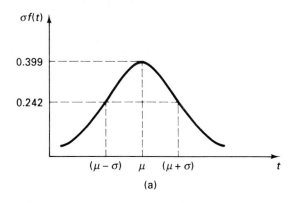

(a)

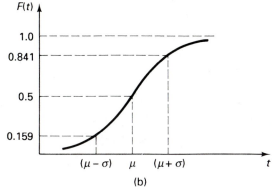

(b)

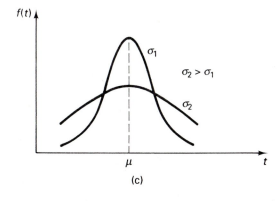

(c)

Figure 4.4   The normal distribution. (a) Probability density function multiplied by σ. (b) Distribution function. (c) Density functions for different σ values.

of σ result in narrow, high curves. Changes in μ merely slide the curve along the t-axis.

By the definition of the density function, we have $P(t_1 < t < t_2) = \int_{t_1}^{t_2} f(t)\, dt$ which is the shaded area shown in Figure 4.5.

Standard tables are available to determine the area under a normal curve for different values of z, a new random variable defined as

$$z \equiv \frac{t - \mu}{\sigma}$$

Density and distribution functions for the newly defined random variable are

$$f(z) = \frac{1}{\sqrt{2\pi}} \exp\left(-\frac{z^2}{2}\right) \tag{4–34a}$$

and

$$Q(z) = \frac{1}{\sqrt{2\pi}} \int_{-\infty}^{(t-\mu)/\sigma} \exp\left(-\frac{z^2}{2}\right) dz \tag{4–34b}$$

The random variable z has zero mean and a standard deviation of unity.

The total areas under the density curve for one, two, and three standard deviations above and below the mean value are

$$\pm 1\sigma:\ 0.6826$$

$$\pm 2\sigma:\ 0.9544$$

$$\pm 3\sigma:\ 0.9972$$

The probability that the random variable will have a value within $\pm 3\sigma$ is very high (99.72%). In other words, the probability that the random variable will lie

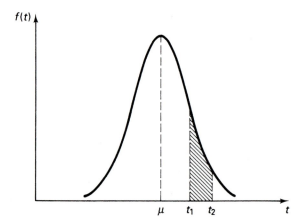

Figure 4.5 $P(t_1 < t < t_2)$ for the normal distribution.

outside $\pm 3\sigma$ is only 0.28%. Typically, it is assumed that $\pm 3\sigma$ will enclose all likely values of the random variable.

4.4.1 Moments

Let us differentiate both sides of

$$\int_{-\infty}^{\infty} \exp[-\alpha t^2]\, dt = \sqrt{\pi/\alpha}$$

k times with respect to α. We obtain

$$\int_{-\infty}^{\infty} t^{2k} \exp[-\alpha t^2]\, dt = \frac{1 \times 3 \times 5 \times \ldots \times (2k-1)}{2^k} \sqrt{\frac{\pi}{\alpha^{2k+1}}}$$

By letting $\alpha = 1/(2\sigma^2)$, we get

$$\int_{-\infty}^{\infty} t^{2k} \exp\left[-\frac{t^2}{2\sigma^2}\right] dt = 1 \times 3 \times 5 \times \ldots \times (2k-1)\sqrt{2\pi}\,\sigma^{(2k+1)} \quad (4\text{--}35)$$

Now let us consider a normal random variable t with zero mean and the density function

$$f(t) = \frac{1}{\sigma\sqrt{2\pi}} \exp\left[-\frac{t^2}{2\sigma^2}\right]$$

The nth moment of t is

$$E[t^n] = \frac{1}{\sigma\sqrt{2\pi}} \int_{-\infty}^{\infty} t^n \exp\left[-\frac{t^2}{2\sigma^2}\right] dt$$

If n is odd, the integrand is odd and the nth moment is zero; if n is even, we set $n = 2k$ in Equation (4–35). We obtain

$$E[t^n] = \begin{cases} 1 \times 3 \times 5 \times \ldots \times (n-1)\sigma^n & \text{for } n \text{ even} \\ 0 & \text{for } n \text{ odd} \end{cases} \quad (4\text{--}36)$$

■■   **Example 4–11**

An induction motor assembly plant purchases preformed stator coils from a local vendor. Past experience indicates that an average of 10% of the coils supplied by this vendor are defective. In a recent purchase of a lot of 200 coils, what is the probability of finding at least 30 defective coils?

$$P[\text{a coil is good}] = 0.9$$

$$P[\text{a coil is defective}] = 0.1$$

With $n = 200$, the expected number of defective coils is (200 × 0.1), or 20, and the standard deviation $\sigma = \sqrt{200 \times 0.9 \times 0.1} = 4.243$. Because of the large number of items involved, we are justified in using the normal approximation to the binomial distribution, with $\mu = 20$ and $\sigma = 4.243$. We have

$$z = \frac{30 - 20}{4.243} = 2.357$$

From the table for areas under normal curves,

$$\int_{2.357}^{\infty} f(z) \, dz = 0.5 - 0.4907 = 0.0093$$

Therefore,

$$P[\text{finding at least 30 defective coils}] = P[30 \text{ or more defective}]$$

$$= 0.0093, \text{ or } 0.93\%$$

The probability of finding 10 or more coils defective is calculated as follows:

$$z_1 = \frac{10 - 20}{4.243} = -2.357$$

$$\int_{-2.357}^{0} f(z) \, dz = \int_{0}^{2.357} f(z) \, dz = 0.4907$$

$$P[\text{finding 10 or more defective coils}] = 0.5 + 0.4907$$

$$= 0.9907, \text{ or } 99.07\%$$

Also,

$$P[20 \text{ or fewer coils defective}] = P[20 \text{ or more coils defective}]$$

$$= 0.5 \qquad \blacksquare\blacksquare$$

■■ **Example 4–12**
A certain electronic device has a normally distributed lifetime with a mean value of 2,000 hours. The design requirements of a system using such devices stipulate that 95% of them should last for at least 1,800 hours. Find the largest variance allowable for the lifetime of the devices.

Let T be the random variable called lifetime, in hours, for the device. We require that $P[T > 1,800] = 0.95$. Referring to the table of areas under normal curves, we see that $z = -1.65$ for

$$\int_{z}^{\infty} f(z) \, dz = 0.95$$

Therefore, $z = -1.65$ should correspond to $T = 1,800$ hours, and

$$\frac{1,800 - 2,000}{\sigma} = -1.65$$

or

$$\sigma = 121.2 \text{ hours}$$

So

$$\text{Maximum allowable variance} = \sigma^2 = 14,692.4 \text{ hours}^2 \qquad \blacksquare\blacksquare$$

■■   **Example 4–13**

The lifetimes of automobile batteries manufactured by a company are found to be normally distributed with a mean of 1,250 days and a standard deviation of 175 days. Taking one month as 30 days, for how many months can the manufacturer guarantee these batteries so that no more than 6% of them need replacement?

Referring to the table of areas under normal curves, we have

$$\int_0^{1.555} f(z)\,dz = 0.44$$

If d_1 is the number of days by which 6% of the batteries fail (and therefore need replacement), then

$$\frac{d_1 - 1,250}{175} = -1.555$$

and

$$d_1 = 977.9 \text{ days, or } 32.6 \text{ months}$$

So the manufacturer's guarantee should not exceed 32 months.   ■■

4.5 THE UNIFORM OR RECTANGULAR DISTRIBUTION

The *uniform* or *rectangular* distribution is defined by two parameters, t_1 and t_2. The density function has a constant nonzero value between t_1 and t_2 and is zero for all other values of t. The associated distribution function is

$$F(t) = Q(t) = \int_{t_1}^{t} f(\xi)\,d\xi = \frac{t - t_1}{t_2 - t_1} \tag{4-37}$$

The survival (or reliability) function is

$$1 - Q(t) = \frac{t_2 - t}{t_2 - t_1} = R(t) \tag{4-38}$$

and the hazard function is

$$\lambda(t) = \frac{f(t)}{R(t)} = \frac{1}{t_2 - t_1} \cdot \frac{t_2 - t_1}{t_2 - t} = \frac{1}{t_2 - t}, \quad t_1 < t < t_2 \tag{4-39}$$

The expected value is

$$E[t] = \int_{t_1}^{t_2} \frac{t\,dt}{t_2 - t_1}$$
$$= \frac{t_2^2 - t_1^2}{2(t_2 - t_1)} = \frac{t_2 + t_1}{2} \tag{4-40a}$$

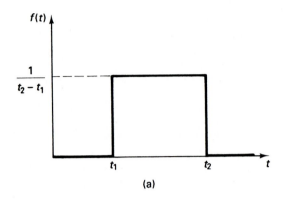

(a)

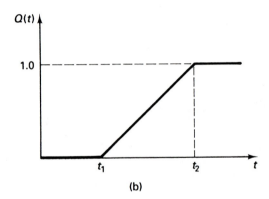

(b)

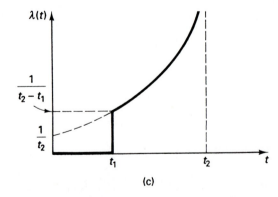

(c)

Figure 4.6 The uniform or rectangular distribution. (a) Failure density function. (b) Failure distribution function. (c) Hazard function.

and the second moment is

$$E[t^2] = \int_{t_1}^{t_2} \frac{t^2 \, dt}{t_2 - t_1}$$

$$= \frac{t_2^3 - t_1^3}{3(t_2 - t_1)} \tag{4-40b}$$

Therefore,

$$\text{Var } t = \sigma^2 = \frac{t_2^3 - t_1^3}{3(t_2 - t_1)} - \left(\frac{t_2 + t_1}{2}\right)^2$$

$$= \frac{4(t_2^3 - t_1^3) - 3(t_2 + t_1)^2(t_2 - t_1)}{12(t_2 - t_1)} \tag{4-41}$$

$$= \frac{(t_2 - t_1)^2}{12}$$

Figure 4.6 illustrates the nature of the functions associated with the rectangular distribution.

4.6 THE RAYLEIGH DISTRIBUTION

The Rayleigh distribution is defined in terms of a single parameter k and is useful in modeling a wearout characteristic. The corresponding hazard function increases linearly with time; that is,

$$\lambda(t) = kt \tag{4-42}$$

The remaining three reliability functions can easily be obtained by using the relationships given in Table 3.1. These functions are:

$$f(t) = kt \exp\left[-\frac{kt^2}{2}\right] \tag{4-43}$$

$$R(t) = \exp\left[-\frac{kt^2}{2}\right] \tag{4-44}$$

$$Q(t) = 1 - \exp\left[-\frac{kt^2}{2}\right] \tag{4-45}$$

Figure 4.7 illustrates the nature of the functions associated with the Rayleigh distribution.

We will show in the next section that the Rayleigh distribution can be considered a special case of the Weibull distribution. Nonetheless, it is a distribution in its own right, with applications in several areas of engineering, including re-

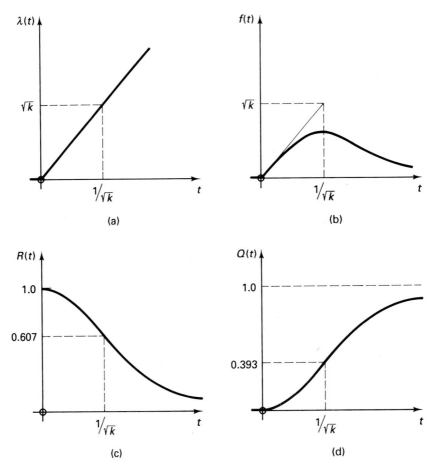

Figure 4.7 The Rayleigh reliability functions. (a) Hazard function. (b) Failure density function. (c) Reliability function. (d) Failure distribution function.

liability evaluation. Like the exponential distribution, the Rayleigh distribution is a single-parameter distribution and is therefore fairly easy to use.

The expected value and the variance of the Rayleigh distribution are easily calculated to be $\sqrt{\pi/(2k)}$ and $(2/k)[1 - (\pi/4)]$, respectively.

4.6.1 Linearly Decreasing Hazard Model

Initial failures of a batch of items can be modeled using a linearly decreasing hazard function. During the initial debugging period (also known as the infant mortality region of the associated hazard function), the hazard rate decreases as shown in Figure 4.8. The decreasing part, all by itself, cannot constitute a legitimate hazard model, since the area under the hazard function must go to infinity

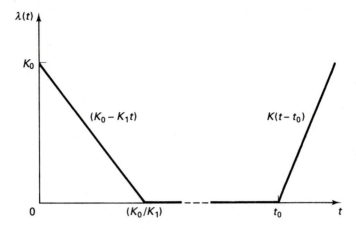

Figure 4.8 Linearly decreasing hazard model with linearly increasing portion appended.

as time goes to infinity. In consideration of this requirement, we include a linearly increasing hazard portion starting at time $t = t_0$, some large value well beyond the range of interest. Expressions for the hazard, density, and reliability functions for this model are as follows:

$$\lambda(t) = \begin{cases} K_0 - K_1 t & \text{for } 0 < t \leq (K_0/K_1) \\ 0 & \text{for } (K_0/K_1) < t \leq t_0 \\ K(t - t_0) & \text{for } t > t_0 \end{cases}$$

(4–46a)

$$f(t) = \begin{cases} (K_0 - K_1 t) \exp[-K_0 t + K_1(t^2/2)] & \text{for } 0 < t \leq (K_0/K_1) \\ 0 & \text{for } (K_0/K_1) < t \leq t_0 \\ K(t - t_0) \exp\left[-\dfrac{K_0^2}{2K_1} - \dfrac{K}{2}(t - t_0)^2\right] & \text{for } t > t_0 \end{cases}$$

(4–46b)

$$R(t) = \begin{cases} \exp[-K_0 t + K_1(t^2/2)] & \text{for } 0 < t \leq (K_0/K_1) \\ \exp[-K_0^2/(2K_1)] & \text{for } (K_0/K_1) < t \leq t_0 \\ \exp\left[-\dfrac{K_0^2}{2K_1} - \dfrac{K}{2}(t - t_0)^2\right] & \text{for } t > t_0 \end{cases}$$

(4–46c)

Next, we will make an attempt to compare the constant, linearly increasing, and linearly decreasing hazard models. In order to have a common basis for comparison, we will normalize these models in such a way that the areas under the hazard curves between $t = 0$ and $t = (1/\lambda)$ are equal. This is equivalent to making the average number of failures during the time interval $(0, 1/\lambda)$ the same for all three models. The following parameter assignments are necessary for the said normalization:

$$k = 2\lambda^2 \quad \text{in Equation (4–42)}$$

$$K_0 = 2\lambda \quad \text{in Equation (4–46a)}$$

$$K_1 = 2\lambda^2 \quad \text{in Equation (4–46a)}$$

The reliability functions valid during the time interval $(0, 1/\lambda)$ now become

$$\text{constant hazard:} \quad R(t) = \exp[-\lambda t]$$

$$\text{linearly increasing hazard:} \quad R(t) = \exp[-\lambda^2 t^2]$$

$$\text{linearly decreasing hazard:} \quad R(t) = \exp[-2\lambda t + \lambda^2 t^2]$$

The three reliability functions are sketched in Figure 4.9 against normalized time $\tau = \lambda t$. All three functions have the same value of 0.368 at $t = (1/\lambda)$, or $\tau = 1$, because of the basis selected for comparison.

Assumption of a constant hazard results in a reliability function that is in between the function for a linearly decreasing hazard and a linearly increasing hazard. With a linearly increasing hazard, the reliability function decreases slowly in the beginning, and the rate of decrease increases later. With a linearly decreasing hazard, the reliability function decreases rapidly in the beginning, and then it levels off.

■■ **Example 4–14**

The reliability of mechanical components can be expressed as the probability that the strength of the components is greater than their stress. That is,

$$\text{Reliability} = R = P[S > s] = P[s < S]$$

where S = strength and s = stress.

Both stress and strength can be modeled as random variables with Rayleigh distributions. We have, for stress,

$$f_1(s) = k_1 s \exp\left[-\frac{k_1 s^2}{2}\right] \quad 0 \le s < \infty$$

and for strength,

$$f_2(S) = k_2 S \exp\left[-\frac{k_2 S^2}{2}\right] \quad 0 \le S < \infty$$

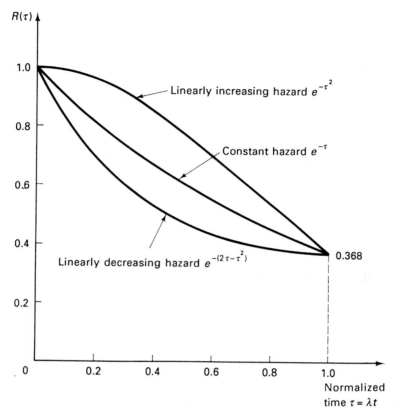

Figure 4.9 Comparison of the reliability functions for constant hazard, linearly increasing hazard, and linearly decreasing hazard models.

Let us consider two events, A, with strength $> \xi$, and B, with $\xi <$ stress $< (\xi + d\xi)$. Then

$$P[S > \xi] = \int_{\xi}^{\infty} f_2(S) \, dS$$

and

$$P[\xi < s < (\xi + d\xi)] = f_1(\xi) \, d\xi$$

Hence,

$$P[AB] = f_1(\xi) \left[\int_{\xi}^{\infty} f_2(S) \, dS \right] d\xi$$

The variance of the Weibull distribution is

$$\text{Var}[t] = \sigma^2 = \int_0^\infty t^2 f(t)\, dt - \mu^2 \tag{4-63}$$

where $f(t)$ and μ are given by Equations (4–48) and (4–57), respectively. Thus,

$$\sigma^2 = \alpha^2 \left[\Gamma\left(1 + \frac{2}{\beta}\right) - \Gamma^2\left(1 + \frac{1}{\beta}\right) \right] \tag{4-64}$$

The conditional (*a posteriori*) probability of failure, $Q_c(t)$, can be calculated using Equation (4–27). The result is

$$Q_c(t) = 1 - \exp\left[\frac{T^\beta - (T + t)^\beta}{\alpha^\beta} \right] \tag{4-65}$$

Figure 4.10 shows the nature of the functions associated with the Weibull distribution for $\beta = 0.5$, 2, and 3 and for $\alpha = 1$.

■■ **Example 4–15**
The time to failure of a component is Weibull distributed with a scale parameter α of 10,000 hours and a shape parameter β of 0.5. Find (1) the probability that the component will still be operational after one year and (2) the conditional probability that the component will fail during the second year of operation, given that it has survived the first year.

$$P[\text{successful operation for at least 1 year}] = R(8,760)$$
$$= \exp[-(8,760/10,000)^{0.5}]$$
$$= \exp[-0.9359]$$
$$= 0.3922$$

The conditional probability of failure in the second year of operation, given that the component has survived the first year, is the *a posteriori* probability $Q_c(T)$ with $T = 8,760$. We have

$$Q_c(8,760) = 1 - \exp\left[\frac{\sqrt{8,760} - \sqrt{17,520}}{\sqrt{10,000}} \right]$$
$$= 0.3214 \qquad\qquad ■■$$

■■ **Example 4–16**
The hourly wind speeds at a site in Hawaii during August have a mean value of 30 km/h and a variance of 45 (km/h)^2. Assuming that a Weibull model is to be used to represent the wind speeds, find α and β.

it follows that

$$f(t) = \left(\frac{1}{\alpha}\right) \exp\left(-\frac{t}{\alpha}\right) \tag{4–52}$$

and

$$\text{MTTF} = \alpha \tag{4–53}$$

Special case of $\beta = 2$. If $\beta = 2$, the Weibull distribution reduces to the Rayleigh distribution with $k = 2/\alpha^2$. And when

$$\lambda(t) = \left(\frac{2}{\alpha^2}\right) t \tag{4–54}$$

it follows that

$$f(t) = \left(\frac{2}{\alpha^2}\right) t \exp\left[-\left(\frac{t}{\alpha}\right)^2\right] \tag{4–55}$$

In general, a value of β less than 1 represents a decreasing hazard rate, β greater than 1 represents an increasing hazard rate, and β equal to 1 represents a constant hazard rate (exponential distribution).

The expected value of the Weibull distribution is obtained by taking the first moment of the random variable. We obtain

$$\mu = E[t] = \int_0^\infty tf(t)\,dt \tag{4–56}$$

where $f(t)$ is given by Equation (4–48). The integration can be performed by defining a new variable $y = (t/\alpha)^\beta$, and the result is

$$\mu = \alpha\Gamma\left(1 + \frac{1}{\beta}\right) \tag{4–57}$$

where Γ is the gamma function

$$\Gamma(x) = \int_0^\infty t^{x-1}e^{-t}\,dt \tag{4–58}$$

with the stipulation that $x > 0$.

For integer values of x and n, the following relationships apply:

$$\Gamma(x) = (x - 1)! \tag{4–59}$$

$$\Gamma(x + 1) = x\Gamma(x) \tag{4–60}$$

$$\Gamma(n + 1) = n!, \quad n = 0, 1, 2, \dots \tag{4–61}$$

$$\Gamma(n) = \frac{\Gamma(n + 1)}{n} \quad \text{for } n > 0 \tag{4–62}$$

The variance of the Weibull distribution is

$$\text{Var}[t] = \sigma^2 = \int_0^\infty t^2 f(t)\, dt - \mu^2 \tag{4-63}$$

where $f(t)$ and μ are given by Equations (4–48) and (4–57), respectively. Thus,

$$\sigma^2 = \alpha^2 \left[\Gamma\left(1 + \frac{2}{\beta}\right) - \Gamma^2\left(1 + \frac{1}{\beta}\right) \right] \tag{4-64}$$

The conditional (*a posteriori*) probability of failure, $Q_c(t)$, can be calculated using Equation (4–27). The result is

$$Q_c(t) = 1 - \exp\left[\frac{T^\beta - (T + t)^\beta}{\alpha^\beta} \right] \tag{4-65}$$

Figure 4.10 shows the nature of the functions associated with the Weibull distribution for $\beta = 0.5$, 2, and 3 and for $\alpha = 1$.

■■ **Example 4–15**
The time to failure of a component is Weibull distributed with a scale parameter α of 10,000 hours and a shape parameter β of 0.5. Find (1) the probability that the component will still be operational after one year and (2) the conditional probability that the component will fail during the second year of operation, given that it has survived the first year.

$$P[\text{successful operation for at least 1 year}] = R(8,760)$$
$$= \exp[-(8,760/10,000)^{0.5}]$$
$$= \exp[-0.9359]$$
$$= 0.3922$$

The conditional probability of failure in the second year of operation, given that the component has survived the first year, is the *a posteriori* probability $Q_c(T)$ with $T = 8,760$. We have

$$Q_c(8,760) = 1 - \exp\left[\frac{\sqrt{8,760} - \sqrt{17,520}}{\sqrt{10,000}} \right]$$
$$= 0.3214 \qquad\qquad ■■$$

■■ **Example 4–16**
The hourly wind speeds at a site in Hawaii during August have a mean value of 30 km/h and a variance of 45 (km/h)^2. Assuming that a Weibull model is to be used to represent the wind speeds, find α and β.

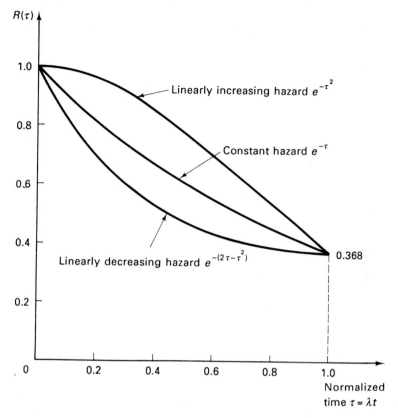

Figure 4.9   Comparison of the reliability functions for constant hazard, linearly increasing hazard, and linearly decreasing hazard models.

Let us consider two events, A, with strength $> \xi$, and B, with $\xi <$ stress $< (\xi + d\xi)$. Then

$$P[S > \xi] = \int_{\xi}^{\infty} f_2(S)\, dS$$

and

$$P[\xi < s < (\xi + d\xi)] = f_1(\xi)\, d\xi$$

Hence,

$$P[AB] = f_1(\xi) \left[\int_{\xi}^{\infty} f_2(S)\, dS \right] d\xi$$

Since ξ can have any value between 0 and ∞,

$$R = \int_0^\infty f_1(\xi) \left[\int_\xi^\infty f_2(S) \, dS \right] d\xi$$

$$= \int_0^\infty \left(k_1 \xi \exp\left[-\frac{k_1 \xi^2}{2} \right] \right) \left[\int_\xi^\infty k_2 S \exp\left[-\frac{k_2 S^2}{2} \right] dS \right] d\xi$$

$$= \int_0^\infty k_1 \xi \exp\left[-\frac{k_1 \xi^2}{2} \right] \exp\left[-\frac{k_2 \xi^2}{2} \right] d\xi$$

$$= k_1 \int_0^\infty \xi \exp\left[-\left(\frac{k_1 + k_2}{2} \right) \xi^2 \right] d\xi$$

$$= \frac{k_1}{k_1 + k_2}$$

■■

4.7 THE WEIBULL DISTRIBUTION

The Weibull distribution is a general two-parameter distribution. By adjusting a scale parameter α and a shape parameter β, a variety of shapes can be obtained to fit experimental data. Thus, this distribution is highly adaptable and is used widely in reliability engineering.

The hazard function for the Weibull distribution is

$$\lambda(t) = \frac{\beta t^{\beta-1}}{\alpha^\beta} \tag{4-47}$$

in which $\alpha > 0$, $\beta > 0$, and $t \geq 0$. The corresponding failure density function is

$$f(t) = \frac{\beta t^{\beta-1}}{\alpha^\beta} \exp\left[-\left(\frac{t}{\alpha} \right)^\beta \right] \tag{4-48}$$

The reliability function is

$$R(t) = \exp\left[-\left(\frac{t}{\alpha} \right)^\beta \right] \tag{4-49}$$

and the failure distribution function is

$$Q(t) = 1 - \exp\left[-\left(\frac{t}{\alpha} \right)^\beta \right] \tag{4-50}$$

Special case of $\beta = 1$.   If $\beta = 1$, the Weibull distribution reduces to the exponential distribution with a constant hazard rate of $1/\alpha$. And when

$$\lambda(t) = \frac{1}{\alpha} \tag{4-51}$$

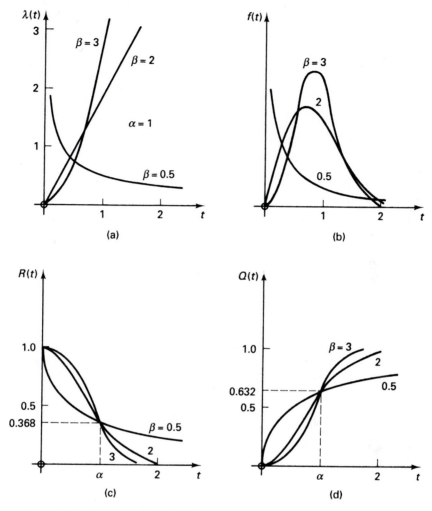

Figure 4.10   The Weibull model and the associated reliability functions. (a) Hazard function. (b) Failure density function. (c) Reliability function. (d) Failure distribution function.

We have the two equations

$$\mu = 30 = \alpha\Gamma\left(1 + \frac{1}{\beta}\right)$$

and

$$\sigma^2 = 45 = \alpha^2\left[\Gamma\left(1 + \frac{2}{\beta}\right) - \Gamma^2\left(1 + \frac{1}{\beta}\right)\right]$$

Solving these equations by trial and error (using tabulated values of gamma functions), we obtain

$$\beta = 5.2$$

and

$$\alpha = 32.6$$ ■■

■■ **Example 4–17**

The Weibull reliability function is

$$R(t) = \exp\left[-\left(\frac{t}{\alpha}\right)^{\beta}\right]$$

where α is the scale parameter and β is the shape parameter. Therefore,

$$\ln R(t) = -\left(\frac{t}{\alpha}\right)^{\beta}$$

and

$$\ln[-\ln R(t)] = \beta \ln\left(\frac{t}{\alpha}\right)$$

$$= \beta \ln t - \beta \ln \alpha$$

A plot of $\ln[-\ln R(t)]$ versus $\ln t$ is a straight line with slope equal to β and y-intercept equal to $(-\beta \ln \alpha)$. This suggests a graphical approach to finding α and β.

Suppose N items are tested and n of them fail in time, t or less. Then an estimate for the reliability function is

$$\hat{R}(t) = \frac{N - n}{N}$$

The available failure data can be used to draw a Weibull plot—that is, a plot of $\ln[-\ln \hat{R}(t)]$ versus $\ln t$—and the parameters α and β can be estimated from this plot.

Suppose that failure data obtained from testing 100 items are as given in the following tabulation:

Cumulative no. of failures (n):	4	17	35	54	73	85
Time in hrs (t):	6	12	18	24	30	36

Values of $R(t)$, $\ln[-\ln \hat{R}(t)]$, and $\ln t$ are as follows:

$\hat{R}(t)$:	0.96	0.83	0.65	0.46	0.27	0.15
$\ln[-\ln \hat{R}(t)]$:	-3.20	-1.68	-0.84	-0.25	0.27	0.64
$\ln t$:	1.79	2.48	2.89	3.18	3.40	3.58

A plot of $\ln[-\ln \hat{R}(t)]$ versus $\ln t$ (Figure 4.11) is a straight line, as expected. Its slope is equal to 2.15, and its y-intercept is equal to (-3.25×2.15), or -6.99. Therefore,

$$\beta = 2.15$$

and

$$\ln \alpha = \frac{6.99}{2.15} = 3.25$$

or

$$\alpha = \exp[3.25] = 25.79$$

We conclude that the Weibull density function for the failure time is

$$f(t) = \frac{2.15 t^{1.15}}{25.79^{2.15}} \exp\left[-\left(\frac{t}{25.79}\right)^{2.15}\right]$$

$$= (1.985 \times 10^{-3}) t^{1.15} \exp[-9.234 \times 10^{-4} t^{2.15}]$$

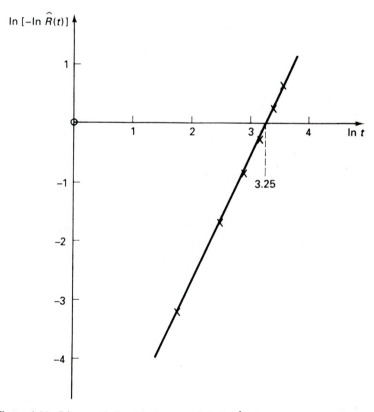

Figure 4.11   Linear relationship between $\ln[-\ln \hat{R}(t)]$ and $\ln t$ for the Weibull model.

The failure distribution function is

$$F(t) = 1 - \exp[-9.234 \times 10^{-4} t^{2.15}]$$

and the reliability function is

$$R(t) = \exp[-9.234 \times 10^{-4} t^{2.15}]$$

The probability that the item fails within one day is

$$F(24) = 1 - \exp[-0.8567]$$
$$= 0.575$$

The probability that the item operates successfully for 50 hours is

$$R(50) = \exp[-9.234 \times 10^{-4}(50)^{2.15}]$$
$$= \exp[-4.15]$$
$$= 0.0157, \text{ or } 1.57\%$$ ■■

4.8 THE GAMMA DISTRIBUTION

The gamma distribution is a two-parameter distribution that has properties similar to those of the Weibull distribution. The two parameters, a scale parameter α and a shape parameter β, can be adjusted to fit observed data. The failure density function is

$$f(t) = \frac{t^{\beta-1}}{\alpha^\beta \Gamma(\beta)} \exp\left(-\frac{t}{\alpha}\right) \tag{4-66}$$

where $\alpha > 0$, $\beta > 0$, and $t \geq 0$.

The corresponding failure distribution function is

$$Q(t) = \frac{1}{\Gamma(\beta)} \int_0^{t/\alpha} z^{\beta-1} e^{-z} \, dz \tag{4-67}$$

The integral in Equation (4–67) is known as the *incomplete gamma function*, and numerical values for it are available in mathematical tables.

The corresponding hazard function $\lambda(t)$ does not have a compact form. It can be obtained using

$$\lambda(t) = \frac{f(t)}{1 - Q(t)} \tag{4-68}$$

The expected value of the gamma distribution is

$$\mu = E[t] = \int_0^\infty t f(t) \, dt \tag{4-69}$$

where $f(t)$ is given by Equation (4–66). The integration can be performed by setting $x = t/\alpha$, and the result is

$$\mu = \alpha\beta \tag{4-70}$$

The second moment of the random variable is

$$E[t^2] = \int_0^\infty t^2 f(t)\, dt = \alpha^2\beta^2 + \alpha^2\beta \tag{4-71}$$

and the variance is

$$\text{Var}[t] = \sigma^2 = E(t^2) - \mu^2 = \alpha^2\beta \tag{4-72}$$

Special case of $\beta = 1$.   When $\beta = 1$, we have

$$f(t) = (1/\alpha) \exp\left(-\frac{t}{\alpha}\right) \tag{4-73}$$

Thus, for this case, the gamma distribution reverts back to the exponential distribution with $\lambda = 1/\alpha$ and MTTF $= \alpha$.

Special case of an integer β.   If β is an integer,

$$\Gamma(\beta) = (\beta - 1)! \tag{4-74}$$

and the failure density function becomes

$$f(t) = \frac{t^{\beta-1}}{\alpha^\beta(\beta - 1)!} \exp\left(-\frac{t}{\alpha}\right) \tag{4-75}$$

which is known as the *special Erlangian distribution*, used to represent systems with nonexponential distributions in terms of a number of identical stages in series (see Chapter 8). Another application for this distribution is in modeling ideal repair with exponentially distributed interfailure times, which is discussed later in this chapter.

The associated reliability function is

$$R(t) = \int_t^\infty f(\xi)\, d\xi \tag{4-76}$$

where $f(\xi)$ is given by Equation (4–75) with t replaced by ξ. The result of performing this integration with a change of variable $z = \xi/\alpha$ is

$$R(t) = [e^{-t/\alpha}] \sum_{j=0}^{\beta-1} \left(\frac{t}{\alpha}\right)^j \frac{1}{j!} \tag{4-77}$$

Special case of $\alpha = 2$ and $\beta = n/2$, n an integer.   Now,

$$f(t) = \frac{t^{n/2-1}}{2^{n/2}\Gamma\left(\dfrac{n}{2}\right)} \exp\left(-\frac{t}{2}\right) \tag{4-78}$$

This is known as the *chi-squared distribution*, and it is used in calculating confidence limits of random variables.

■■ **Example 4–18**

Failure times of commercial audiotape decks have a gamma distribution with $\beta = 2$ and $\alpha = 1,000$ hours. The failure density function is

$$f(t) = \frac{t^{2-1}}{1,000^2 \Gamma(2)} \exp(-t/1,000)$$

$$= 10^{-6} t \exp(-10^{-3} t)$$

The mean time to failure is

$$\text{MTTF} = \mu = \alpha\beta = 2,000 \text{ hours}$$

The probability of no failure in 1,000 hours of operation is

$$R(1,000) = \exp(-10^{-3} 10^3) \sum_{j=0}^{1} (1,000/1,000)^j (1/j!)$$

$$= [\exp(-1)]2 = 0.73576$$

If a new line of decks with an exponential failure density function and the same MTTF is available, then for this new line of decks, we have

$$\lambda = \frac{1}{\text{MTTF}} = 5 \times 10^{-4} \text{ hr}^{-1}$$

and

$$f(t) = (5 \times 10^{-4}) \exp(-5 \times 10^{-4} t)$$

The probability of no failure in 1,000 hours of operation for this new line of decks is then

$$R(1,000) = \exp(-5 \times 10^{-4} \times 10^3)$$

$$= 0.60653$$

This simple calculation leads us to conclude that the decks with gamma-distributed failure times are better than the new line of decks with exponentially distributed failure times, even though they both have the same MTTF.

If we desire a reliability of 0.95, then the operating times of the components should be limited to T_0, where, for decks with a gamma failure distribution,

$$0.95 = \int_{T_0}^{\infty} 10^{-6} t \exp(-10^{-3} t) \, dt$$

or

$$95 \times 10^4 = \left[\frac{\exp(-10^{-3} t)}{-10^{-3}} \left[t - \left(\frac{1}{-10^{-3}} \right) \right] \right]_{T_0}^{\infty}$$

or

$$950 = \left[\exp\left(\frac{-T_0}{1,000} \right) \right] [T_0 + 1,000]$$

Solving this by trial and error, we obtain

$$T_0 = 355 \text{ hours}$$

A similar calculation for the new line of decks with exponentially distributed failure times yields

$$0.95 = \exp(-5 \times 10^{-4}T_0)$$

or

$$T_0 = 102.6 \text{ hours}$$

Once again, the gamma-distributed decks come out ahead. However, for 3,000 hours of operation,

$$R(3,000) = [\exp(-3)][1 + 3]$$

$$= 0.199$$

for gamma-distributed decks, and

$$R(3,000) = \exp(-5 \times 10^{-4} \times 3,000)$$

$$= 0.223$$

for exponentially distributed decks. Thus, in the long run, the exponentially distributed decks show better performance than the gamma-distributed decks. These differences are obviously related to the shapes of the two failure density functions. ■■

4.9 THE LOGNORMAL DISTRIBUTION

The lognormal distribution assumes that the natural logarithm of the random variable is normally distributed with a mean value of μ and a standard deviation of σ. Here, it should be noted that μ and σ are the mean and standard deviation, not of the random variable t, but rather, of its natural logarithm. Repair times of repairable components can be modeled using this distribution. Thus, the repair density function is

$$f(t) = \frac{1}{t\sigma\sqrt{2\pi}} \exp\left[-\frac{(\ln t - \mu)^2}{2\sigma^2}\right] \tag{4-79}$$

where $t \geq 0$ is the repair time. The corresponding cumulative failure distribution function $Q(t)$ is

$$Q(t) = F(t) = \int_0^t \frac{1}{\xi\sigma\sqrt{2\pi}} \exp\left[-\frac{(\ln \xi - \mu)^2}{2\sigma^2}\right] d\xi \tag{4-80}$$

where the failure is construed to be failure to repair the component. Substitution of $z = (\ln \xi - \mu)/\sigma$ into Equation (4–80) yields

$$Q(t) = \frac{1}{\sqrt{2\pi}} \int_{-\infty}^{\frac{\ln t - \mu}{\sigma}} \exp\left(-\frac{z^2}{2}\right) dz \qquad (4–81)$$

clearly indicating that $(\ln t)$ is normally distributed.

The expected value of t is calculated as follows:

$$E[t] = \int_0^\infty tf(t)\, dt = \frac{1}{\sigma\sqrt{2\pi}} \int_0^\infty \exp\left[-\frac{(\ln t - \mu)^2}{2\sigma^2}\right] dt \qquad (4–82)$$

With the change of variable $x = \ln t$, we obtain   (4–83)

$$E[t] = \int_{-\infty}^\infty \frac{e^x}{\sigma\sqrt{2\pi}} \exp\left[-\frac{(x - \mu)^2}{2\sigma^2}\right] dx \qquad (4–84)$$

Now let

$$(x - \mu) = y \qquad (4–85)$$

Then

$$E[t] = e^\mu \int_{-\infty}^\infty \frac{e^y}{\sigma\sqrt{2\pi}} \exp\left[-\frac{y^2}{2\sigma^2}\right] dy \qquad (4–86)$$

Recognizing the expansion

$$e^y = 1 + y + \frac{y^2}{2!} + \frac{y^3}{3!} + \cdots \qquad (4–87)$$

we obtain

$$E[t] = e^\mu \left[E(y^0) + E(y^1) + \frac{E(y^2)}{2!} + \frac{E(y^3)}{3!} + \cdots\right] \qquad (4–88)$$

or

$$E[t] = e^\mu \left[1 + \frac{\sigma^2}{2} + \frac{\sigma^4}{8} + \frac{\sigma^6}{48} + \cdots\right]$$

$$= e^\mu [e^{\sigma^2/2}]$$

Alternatively,

$$E[t] = \exp\left(\mu + \frac{\sigma^2}{2}\right) \qquad (4–89)$$

Similar calculations yield the variance:

$$\text{Var}[t] = \sigma^2 = \exp(2\mu + 2\sigma^2) - \exp(2\mu + \sigma^2) \qquad (4–90)$$

Figure 4.12 illustrates the nature of the functions involved.

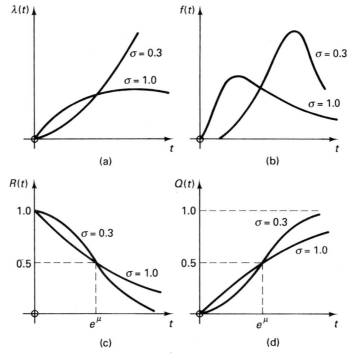

Figure 4.12   The lognormal reliability model. (a) Hazard function. (b) Failure density function. (c) Reliability function. (d) Failure distribution function.

■■   **Example 4–19**

The current gain of a transistor is proportional to $\ln \beta$, where β is the ratio of the output current to the input current. It has been found that current gains of a certain type of transistor are normally distributed with a mean value of 2 and a variance of 0.01. Find the probability that β is between 6 and 8.

Clearly, β has a lognormal distribution. Moreover,

$$z_1 = \frac{\ln 6 - 2}{0.1} = -2.08$$

$$z_2 = \frac{\ln 8 - 2}{0.1} = 0.794$$

Referring to the table of normal curve areas, we find that

$$\int_{-2.08}^{0.794} f(z)\, dz = 0.2863 + 0.4812 = 0.7675$$

Therefore, $P[6 \leq \beta \leq 8] = 76.75\%$.   ■■

■■ **Example 4–20**

The daily electric energy consumption in GWh of a city has a lognormal distribution with $\mu = 1.2$ and $\sigma = 0.5$. Then

$$\text{Average daily energy consumption} = \exp\left[1.2 + \frac{(0.5)^2}{2}\right] = 3.7622 \text{ GWh}$$

If the city's power plant has the capacity to supply a maximum of 8 GWh per day, compute the probability of not being able to supply the load on any given day.

We have

$$P[\text{not able to supply daily load}] = P[\text{daily consumption} > 8 \text{ GWh}]$$

Since, for this case, $z = [\ln 8 - 1.2]/0.5 = 1.759$, and, from the table of normal curve areas,

$$\int_{1.759}^{\infty} f(z)\, dz = 0.5 - 0.4607 = 0.0393$$

we see that

$$P[\text{not being able to supply the load}] = 0.0393, \text{ or } 3.93\% \quad \blacksquare\blacksquare$$

4.10 THE BETA DISTRIBUTION

The beta distribution is a two-parameter distribution that is rich enough to yield models for many random variables of practical importance that are bounded below and above by known values. Examples of such variables are the cloud cover factor affecting the total hourly insolation (incident solar radiation) on a horizontal surface for a particular hour of the day during a season, the distance from one end of a rod to the point where fracture occurs under stress, the fraction of land area spoiled by pests, and the fraction of an age group with a certain score in a test. No matter what the upper and lower bounds are, as long as they are finite, we can always derive a corresponding random variable that has values between 0 and 1 by a suitable normalization procedure. For example, if X is a random variable with $a \le x \le b$, then $(X - a)/(b - a)$ is the corresponding variable of interest with values between 0 and 1.

The density function for the beta distribution is

$$f(t) = \begin{cases} \dfrac{t^{\gamma}(1 - t)^{\beta}}{B(\gamma, \beta)} & \text{for } 0 \le t \le 1 \\ 0 & \text{otherwise} \end{cases} \tag{4–91a}$$

in which the beta function is defined as

$$B(\gamma, \beta) = \frac{\Gamma(\gamma + 1)\Gamma(\beta + 1)}{\Gamma(\gamma + \beta + 2)} \tag{4–91b}$$

with $\gamma, \beta > -1$. (See Appendix F for details.)

By choosing different values of γ and β, the following large variety of density functions can be generated:

$\gamma = 0$ and $\beta = 0$	uniform distribution
$\gamma = \beta$ and $\beta = \gamma$	symmetric about $t = 0.5$
$\gamma < 0$ or $\beta < 0$	infinitely large at $t = 0$ and $t = 1$, respectively
$\gamma < 0$ and $\beta \geq 0$	decreasing in t and concave
$\gamma \geq 0$ and $\beta < 0$	increasing in t and concave
$\gamma < 0$ and $\beta < 0$	U-shaped; infinitely large at $t = 0$ and $t = 1$
$\gamma > 0$ and $\beta > 0$	has a single hump

The corresponding cumulative distribution function is

$$Q(t) = \begin{cases} 1 & \text{for } t > 1 \\ \int_0^t \dfrac{\xi^{\gamma}(1 - \xi)^{\beta}}{B(\gamma, \beta)} \, d\xi & \text{for } 0 \leq t \leq 1 \\ 0 & \text{for } t < 0 \end{cases} \qquad (4\text{--}91c)$$

The integral in Equation (4–91c) is known as the *incomplete beta function*. Tables of values for this function are available in mathematical handbooks.

The mean and the variance of the beta distribution are, respectively,

$$\mu = \frac{\gamma + 1}{\gamma + \beta + 2} \qquad (4\text{--}92a)$$

and

$$\sigma^2 = \frac{(\gamma + 1)(\beta + 1)}{(\gamma + \beta + 2)^2(\gamma + \beta + 3)} \qquad (4\text{--}92b)$$

Expressions for γ and β can be derived in terms of μ and σ^2 from these relationships. They are:

$$\gamma = \mu \left[\frac{\mu(1 - \mu)}{\sigma^2} - 1 \right] - 1 \qquad (4\text{--}92c)$$

and

$$\beta = (1 - \mu) \left[\frac{\mu(1 - \mu)}{\sigma^2} - 1 \right] - 1 \qquad (4\text{--}92d)$$

■■ **Example 4–21**

A special beta distribution has the density function

$$f(t) = Kt^{11}(1 - t)$$

for $0 < t < 1$ and zero elsewhere. Find (1) the value of K, (2) the mean value, (3) the second moment, and (4) the variance for the distribution.

To find the value of K, we use the fact that the area under the density function should be unity. This yields $K = 156$.

The mean value is

$$\mu = \int_0^1 tf(t)\, dt$$

$$= 156 \int_0^1 t^{12}(1 - t)\, dt$$

$$= \frac{6}{7}$$

The second moment is

$$E[t^2] = \int_0^1 t^2 f(t)\, dt$$

$$= 156 \int_0^1 t^{13}(1 - t)\, dt$$

$$= \frac{78}{105}$$

The variance is

$$\sigma^2 = E[t^2] - (E[t])^2$$

$$= \frac{78}{105} - \left(\frac{6}{7}\right)^2$$

$$= \frac{2}{145}$$ ∎∎

4.11 EXTREME-VALUE DISTRIBUTIONS

In Section 2.10, we discussed the concepts of maximum and minimum extreme-value distributions for a set of n random variables, typically loads and capacities. If the number of random variables is large, as is usually the case, it is not convenient to use the expressions developed in that section. Moreover, the assumptions of identical distributions and independence of variables may not be valid. Real-world data on extreme values are very scarce. However, for the case of large n, certain asymptotic extreme-value distributions can be used to model maximum and minimum values.

Three types of asymptotic extreme-value distributions are commonly used. The type I or *Gumbel* distribution is used when the random variable can take on any value between $-\infty$ and $+\infty$. It can be shown that exponential and normal parent distributions result in type I maximum extreme-value distributions in the limiting case when n becomes large. Also, parent distributions with exponential-like tails give rise to minimum extreme-value distributions of type I. If the random variable is limited on the left by zero, then, as n becomes large, the maximum extreme-value distribution tends towards type II. Similarly, if the random variable is limited by zero on the right, then the minimum extreme-value distribution tends towards type II. Finally, if the random variable is limited on the right (left) by some finite value y_1, then the maximum (minimum) extreme-value distribution tends towards type III as n becomes large.

Type I.   Let y_m be the value of y for which the density function $f_Y(y)$ peaks, and let $\beta > 0$ be a constant that describes the dispersion of the random variable. Then

$$F_Y(y) = \exp\left[-\exp\left\{-\left(\frac{y - y_m}{\beta}\right)\right\}\right] \tag{4–93a}$$

$$F_Z(z) = 1 - \exp\left[-\exp\left(\frac{z - y_m}{\beta}\right)\right] \tag{4–93b}$$

for $-\infty \leq y$ or $z \leq \infty$ and for $\beta > 0$.

Type II

$$F_Y(y) = \exp\left[-\left(\frac{y}{\beta}\right)^{-m}\right] \tag{4–94a}$$

$$F_Z(z) = 1 - \exp\left[-\left(\frac{-z}{\beta}\right)^{-m}\right] \tag{4–94b}$$

for $y \geq 0$, $z \leq 0$, $\beta > 0$, and $m > 0$.

Type III

$$F_Y(y) = \exp\left[-\left(\frac{y_1 - y}{\beta}\right)^{m}\right] \tag{4–95a}$$

$$F_Z(z) = 1 - \exp\left[-\left(\frac{z - y_1}{\beta}\right)^{m}\right] \tag{4–95b}$$

for $y \leq y_1$, $y_1 \leq z$, $\beta > 0$, and $m > 0$.

Finding the corresponding density functions, mean values, and variances are left as exercises at the end of the chapter.

One special case that belongs to this type of distribution is given by the failure density function

$$f(t) = e^t e^{-e^t}$$

for $-\infty < t < +\infty$. The corresponding reliability function and hazard function are easily found to be

$$R(t) = 1 - \int_0^t e^{\xi} e^{-e^{\xi}} \, d\xi$$

or

$$R(t) = e^{-e^t} \qquad\qquad (4\text{--}96\text{a})$$

and

$$\lambda(t) = e^t \qquad\qquad (4\text{--}96\text{b})$$

A more general form of the hazard function is

$$\lambda(t) = Ke^{\alpha t}, \quad \alpha > 0 \qquad\qquad (4\text{--}97)$$

called the *exponential hazard model*. The exponential hazard function varies slowly and is nearly constant initially but increases rapidly later on, as shown in Figure 4.13. If a population is growing exponentially, and if some hazard is proportional to the population, then the exponential hazard model becomes appli-

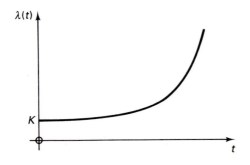

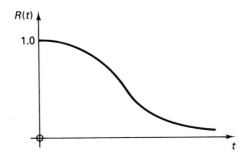

Figure 4.13 Exponential hazard model and associated reliability function.

cable. The associated density and reliability functions are

$$f(t) = Ke^{\alpha t} \exp\left[-\frac{K}{\alpha}(e^{\alpha t} - 1) \right] \qquad (4\text{--}98a)$$

and

$$R(t) = \exp\left[-\frac{K}{\alpha}(e^{\alpha t} - 1) \right] \qquad (4\text{--}98b)$$

4.12 THE HAZARD RATE MODEL DISTRIBUTION

The hazard rate model distribution has one shape parameter, b, and two scale parameters, β and λ_1. The hazard function is

$$\lambda(t) = k\lambda_1 \tanh \lambda_1 t + (1 - k)bt^{b-1}\beta e^{-\beta t^b} \qquad (4\text{--}99)$$

where $b, \beta, \lambda_1 > 0$; $0 \le k \le 1$; and $t \ge 0$. The corresponding reliability function is

$$R(t) = \exp[-k \ln \cosh \lambda_1 t + (1 - k)\{e^{-\beta t^b} - 1\}] \qquad (4\text{--}100)$$

A wide variety of shapes can be obtained by adjusting the three parameters.

4.13 THE GENERAL DISTRIBUTION

The general distribution has four parameters—two scale parameters (β, λ) and two shape parameters (b, c). With the proper choices of these four parameters, this distribution can be used to model the failure behavior of many components. The associated hazard function is

$$\lambda(t) = k\lambda c t^{c-1} + (1 - k)bt^{b-1}\beta e^{\beta t^b} \qquad (4\text{--}101)$$

where $b, c, \beta, \lambda > 0$; $0 \le k \le 1$; and $t \ge 0$. The corresponding reliability function is

$$R(t) = \exp[-k\lambda t^c - (1 - k)\{e^{\beta t^b} - 1\}] \qquad (4\text{--}102)$$

Many of the models discussed thus far can be obtained as special cases of the general distribution. Some of the important ones are as follows:

Parameter Values	Model
$k = 1$	Weibull
$k = 0, \quad b = 1$	Extreme value
$c = 0.5, b = 1$	Bathtub curve

4.14 OTHER POSSIBLE MODELS

Any of the hazard functions discussed thus far can be shifted in time by replacing t by $(t - t_0)$ to fit observed data. The resulting model is called a *shifted hazard model*, with t_0 being adjustable.

An actual hazard function can be approximated by piecewise-linear models. The accuracy of such a model can be improved by increasing the number of segments.

Power series models offer another approach. Let

$$\lambda(t) = K_0 + K_1 t + K_2 t^2 + \ldots + K_n t^n \qquad (4-103)$$

Then the corresponding reliability function becomes

$$R(t) = \exp\left[- \left(K_0 t + K_1 \frac{t^2}{2} + K_2 \frac{t^3}{3} + \ldots + K_n \frac{t^{n+1}}{n+1} \right) \right] \qquad (4-104)$$

In most cases, complex models are not needed. The real challenge is in keeping models of individual component simple enough that the task of evaluating the reliability of complex systems using these components does not become unwieldy. At the same time, the models must represent the failure mechanisms as faithfully as possible.

4.15 MODELING THE WEAROUT REGION

The wearout region corresponds to the rapidly increasing portion towards the end of the useful life of the bathtub hazard function. It was mentioned earlier that the Rayleigh distribution can be used to model this region. Another possibility is to use a normal distribution around the mean wearout life W of the component with suitably chosen parameters.

For the normal distribution, the hazard function has the general shape shown in Figure 4.14. If the mean wearout life of a component is W, with a standard

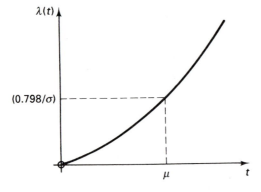

Figure 4.14 Typical hazard function associated with the normal distribution.

deviation of σ, then the associated density function can be expressed as

$$f(t) = \frac{1}{\sigma\sqrt{2\pi}} \exp\left[-\frac{(t - W)^2}{2\sigma^2}\right] \qquad (4\text{--}105)$$

where t is the age of the component. In practice, W could be much less than the MTTF value m computed using the constant failure rate during the component's useful life.

Consider a duration t in the lifetime of a component, starting at time T. During the time interval $(T, T + t)$, if failures are due only to chance (random) events, then the failure rate is constant and equal to the λ corresponding to the useful lifetime of the component, and the component's reliability is

$$R_c(t) = e^{-\lambda t} \qquad (4\text{--}106)$$

associated with the constant hazard (exponential) model. It should be noted that $R_c(t)$ is independent of the starting time T and depends only on λ and the duration t considered.

The probability of failure due to wearout during $(T, T + t)$, given survival up to time T, is the *a posteriori* probability

$$Q_w(t) = \frac{\displaystyle\int_T^{T+t} f(\xi)\,d\xi}{\displaystyle\int_T^{\infty} f(\xi)\,d\xi} \qquad (4\text{--}107)$$

Therefore, the probability of there being no wearout failure during $(T, T + t)$ is

$$R_w(t) = 1 - Q_w(t) = \frac{\displaystyle\int_{T+t}^{\infty} f(\xi)\,d\xi}{\displaystyle\int_T^{\infty} f(\xi)\,d\xi} \qquad (4\text{--}108)$$

or

$$R_w(t) \equiv \frac{R_{we}(T + t)}{R_{we}(T)} \qquad (4\text{--}109)$$

where $R_{we}(\tau)$ denotes the probability of no wearout failure at τ.

Next, consider both chance failures and wearout failures. The probability of there being no chance failures *and* no wearout failures during $(T, T + t)$ is

$$R(t) = R_c(t)R_w(t) = e^{-\lambda t}\frac{R_{we}(T + t)}{R_{we}(T)} \qquad (4\text{--}110)$$

Assuming that the component starts to operate at $T = 0$, since $R_{we}(0) = 1$, we have

$$R(t) = e^{-\lambda t}R_{we}(t) \qquad (4\text{--}111)$$

Figure 4.15 illustrates the variation of $R_{we}(t)$ with time. The variation of $R(t)$ with time depends significantly on the relative values of m and W. Two cases are of interest: (*i*) $m > W$ and (*ii*) $W > m$. These cases are shown in Figure 4.16. A careful observation of the figure reveals the following:

- If $m > W$, $R(t)$ follows the exponential curve $\exp(-\lambda t)$ up to a certain time (which can be much less than m), but subsequently, wearout failures dominate and reduce $R(t)$ to zero rapidly.
- If $W > m$, failures occur mainly by chance up to the MTTF m, beyond which wearout failures dominate.

Given the foregoing, it is clear what general guidelines should be followed to achieve high values of system reliability:

1. Subject the system to initial burn-in and the associated debugging procedures.
2. Follow strict preventive maintenance and replacement schedules to ensure that components do not enter their wearout region.

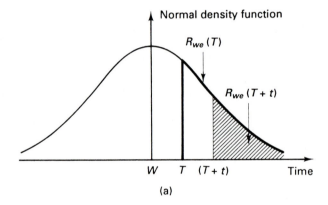

(a)

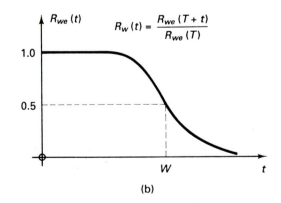

(b)

Figure 4.15 (a) Nature and (b) variation of $R_{we}(t)$ with time.

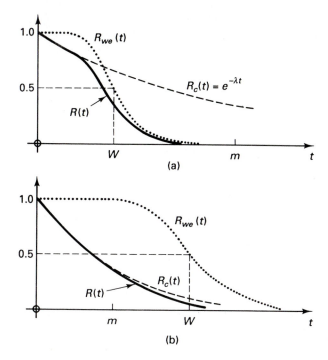

Figure 4.16 Reliability function with wearout considered for (a) $m > W$ and (b) $m < W$.

By replacing components periodically, the system is restored to the normal operating region of low failure probability. Periodic replacement allows system operation for long periods of time, much greater than the normal wearout life, with a high degree of reliability. However, if not done properly, this kind of preventive maintenance may reduce system reliability.

4.16 RELIABILITY AND MAINTENANCE

There are two basic categories of maintenance:

1. Scheduled (or preventive) maintenance
2. Forced (or corrective) maintenance.

Scheduled maintenance is performed at constant intervals of time, even if the system is still working satisfactorily. Such a process prolongs the life of components, decreases the number of failures, and increases the MTTF of the system.

Corrective maintenance follows in-service failures. In other words, nothing is done until the system fails. As soon as this occurs, needed replacement, adjustment, or repair of components is done to restore the system to normal operation. In a way, we can think of corrective maintenance as repair.

In the case of parallel redundant systems, all the components must fail for the system to fail. Failure of one or more redundant components will not be detected and rectified, unless there is periodic inspection and preventive maintenance. It is quite obvious that the MTTF will increase with increasing frequency of (i.e., decreasing time between) inspections. Without inspection (i.e., the time between inspections is infinity), the MTTF is constant and is equal to the value computed for a parallel redundant system. With proper inspection and preventive maintenance, the MTTF can be increased dramatically. By contrast, if maintenance is not done properly, it is possible for the MTTF to be less than the value computed for the parallel redundant system.

4.16.1 Modeling Ideal Scheduled Maintenance

Consider a component that is not repairable, but is amenable to periodic preventive (scheduled) maintenance. Such maintenance is said to be ideal if it takes zero time (practically speaking, very little time as compared to the time between instances of maintenance) to complete and if the component is restored to an "as new" condition after the maintenance is completed. Although the component is not repairable and is thrown out once it fails, the reasoning for scheduled maintenance is still to prolong the lifetime of the component and postpone its failure.

If the component has a constant hazard, its time to failure has an exponential distribution. In other words, the probability of failure during the next time increment Δt remains unchanged throughout the lifetime of the component, indicating that it is as good as new no matter how long it has operated. In this case, preventive maintenance is irrelevant.

If the component has a decreasing hazard, it is improving as time progresses, and therefore, maintenance aimed at restoring it to an "as new" condition is actually disadvantageous and is not advisable.

Scheduled maintenance is worthwhile only if the component has an increasing hazard. As most mechanical systems come under this category, the rest of this discussion assumes that the component has an increasing hazard. Also, maintenance is performed only on working components.

Let

$$f_T(t) = \text{failure density function}$$

$$T_M = \text{fixed time interval between maintenances}$$

$$f_1(t) = \begin{cases} f_T(t) & \text{for } 0 < t \le T_M \\ 0 & \text{otherwise} \end{cases} \tag{4-112}$$

and

$$R(t) = \text{reliability function for the component}$$

Then the density function $f_T^*(t)$ for the component, after incorporating maintenance, can be written as

$$f_T^*(t) = \sum_{k=0}^{\infty} f_1(t - kT_M)R^k(T_M) \qquad (4\text{--}113)$$

A typical $f_T^*(t)$ is shown in Figure 4.17. The time scale is divided into equal segments of duration T_M. The function $f_T^*(t)$ in each segment is a scaled-down version of the function in the previous segment, with the scaling factor equal to $R(T_M)$. The scaling factor is also equal to the fraction of the components entering a segment that will survive into the next segment.

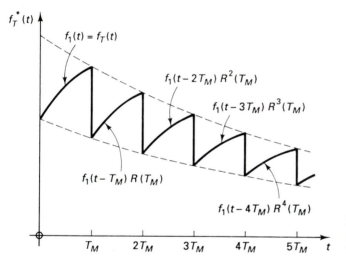

Figure 4.17 Density function with ideal scheduled maintenance incorporated.

A close examination of the figure shows that density functions of lifetimes of components with preventive maintenance exhibit an exponential tendency. The important effect of periodic preventive maintenance is to alter the failure density function from its original shape to one with an exponential character. This alteration is one of the justifications for the widespread use of exponential distributions to model lifetimes of components. Without proper periodic scheduled maintenance, such an assumption is suspect.

It should be emphasized that in Equation (4–113), $k = 0$ is to be used only between $t = 0$ and $t = T_M$, $k = 1$ is to be used only between $t = T_M$ and $t = 2T_M$, and so on.

■■ **Example 4–22**

Let us consider a component with a uniformly distributed lifetime as given by

$$f(t) = 0.25 \text{ year}^{-1}, \quad 0 < t \le 4 \text{ years}$$

The component undergoes regular maintenance (assumed to be ideal) once per year. Find the modified density function when maintenance is incorporated.

Figure 4.18 shows the (a) density, (b) distribution, (c) reliability, and (d) hazard functions for the component without considering maintenance. The expressions associated with the last three are, respectively,

$$F(t) = 0.25t, \quad 0 < t \le 4$$

$$R(t) = 1 - 0.25t, \quad 0 < t \le 4$$

and

$$\lambda(t) = \frac{0.25}{1 - 0.25t}$$

$$= \frac{1}{4 - t}, \quad 0 < t < 4$$

The MTTF without maintenance is

$$\text{MTTF} = \int_0^4 R(t)\,dt = 2 \text{ years}$$

We are given that $T_M = 1$ year, and therefore,

$$R(T_M) = 0.75$$

Now, using Equation (4–113), we obtain

$$f_T^*(t) = \sum_{k=0}^{\infty} (0.25)(0.75)^k$$

This function, along with the corresponding hazard function $\lambda^*(t)$, is shown in Figure 4.19. The average value of $\lambda^*(t)$ is computed as follows:

$$\overline{\lambda^*(t)} = \frac{1}{1} \int_0^1 \frac{dt}{4 - t}$$

$$= \int_3^4 \frac{d\xi}{\xi}$$

where

$$\xi \equiv 4 - t$$

Therefore,

$$\overline{\lambda^*(t)} = \ln 4 - \ln 3 = 0.2877$$

Periodic maintenance has replaced the original density function $f(t)$ by one with an exponential tendency, as shown by the exponential approximation to $f_T^*(t)$ in Figure 4.19. Based on this exponential approximation, the MTTF is, with the effect of maintenance included, equal to 1/0.2877, or 3.476, years, a significant increase from the no-maintenance value of 2 years.

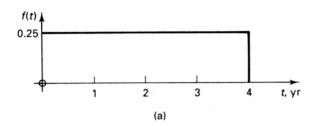

(a)

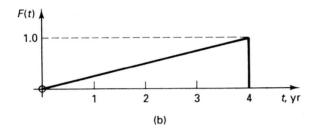

(b)

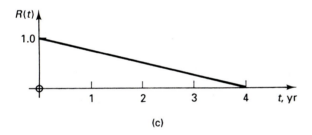

(c)

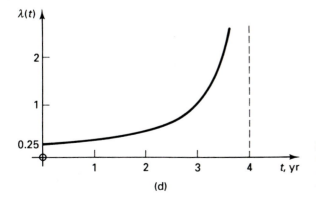

(d)

Figure 4.18 Reliability functions for Example 4–22 without considering ideal maintenance. (a) Density, (b) distribution, (c) reliability, (d) hazard.

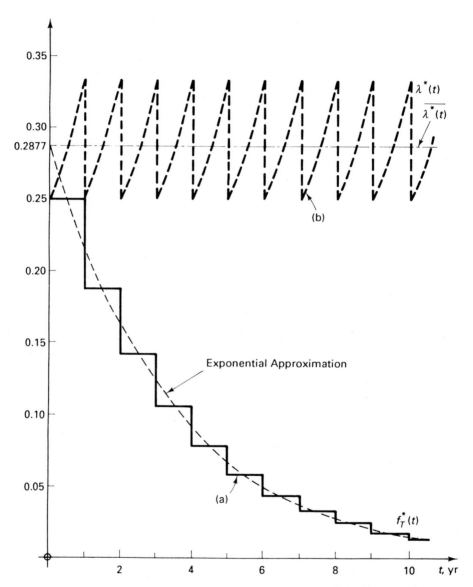

Figure 4.19 (a) Density and (b) hazard functions for Example 4–22 with maintenance included.

The exact value of the MTTF with maintenance included is

$$\text{MTTF}^* = \int_0^\infty t f_T^*(t)\, dt$$

$$= \frac{1}{8} \sum_{k=0}^{\infty} (0.75)^k (2k + 1)$$

$$= 3.5 \text{ years}$$

∎∎

4.17 IDEAL REPAIR

Ideal repair presupposes two conditions:

1. The duration of repair after each failure is so small compared to the time between failures that it can be assumed to be zero.
2. After repair, the component is restored to an "as new" condition.

A good example of ideal repair is the replacement of the failed component by a new one, with the process taking negligible time. There is a fundamental difference between ideal repair and ideal scheduled maintenance: While ideal preventive maintenance takes place at predetermined intervals during which the component is still in working condition, ideal repair always follows a failure, and failure times are not predetermined (they are random).

We will assume the lifetime T of the component to be a continuous random variable with the density function

$$f_T(t) = \lim_{\Delta t \to 0} \frac{1}{\Delta t} P[t < T \leq (t + \Delta t)] \qquad (4\text{--}114)$$

Clearly, the density function $f_1(t)$ for the continuous random variable called time to the first failure is the same as $f_T(t)$. The question to be answered now is, What is the density function $f_2(t)$ for the continuous random variable called total time to the second failure?

Let us assume that the first failure occurs in the neighborhood of τ. Then the probability of a second failure in $(t,\ t + \Delta t)$, $t > \tau$, for a given τ, is

$$f_2(t)\, \Delta t \simeq [f_1(\tau)\, \Delta \tau][f_1(t - \tau)\, \Delta t] \qquad (4\text{--}115)$$

since the duration of the second lifetime is $t - \tau$.

In the limit, considering all possible values of τ less than t, we obtain

$$f_2(t) = \int_0^t f_1(\tau) f_1(t - \tau)\, d\tau \qquad (4\text{--}116)$$

Similar arguments lead us to the density function for the continuous random variable called total time to the kth failure, namely,

$$f_k(t) = \int_0^t f_{k-1}(\tau)f_1(t - \tau) \, d\tau; \quad k \geq 2 \tag{4-117}$$

Notice that $f_k(t)$ is the k-fold convolution of $f_1(t)$.

When we consider all failures—first, second, third, etc.—the probability of some failure occurring in $(t, t + \Delta t)$ is the sum of the probabilities of a first, second, third, etc., failure occurring in the time interval considered. Let $L(t)$ be the density function of some failure occurring with ideal repair. Then

$$L(t) \, \Delta t = \text{probablity of some failure occurring in } (t, t + \Delta t)$$

$$L(t) = \lim_{\Delta t \to 0} \frac{1}{\Delta t} P[\text{some failure in } (t, t + \Delta T)] \tag{4-118}$$

$$= \sum_{k=1}^{\infty} f_k(t)$$

or

$$L(t) = f_1(t) + \sum_{k=2}^{\infty} \int_0^t f_{k-1}(\tau)f_1(t - \tau) \, d\tau \tag{4-119}$$

Figure 4.20 illustrates the general nature of $f_k(t)$ for $k = 1, 2, 3, \ldots$ and $L(t)$. The higher the value of k, the broader and flatter is the shape of $f_k(t)$, indicating an accumulation of all the uncertainties in the times to previous failures.

Special case of exponentially distributed interfailure times. If the interfailure times are exponentially distributed, then $f_k(t)$ becomes the special Erlangian distribution, which is the gamma distribution with an integer value of

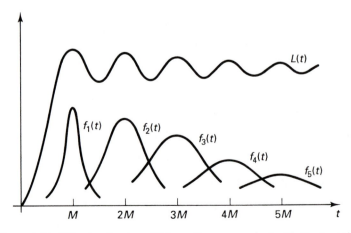

Figure 4.20   Effective density of failures for a component with ideal repair.

β. We derive this distribution as follows:

$$f_1(t) = f_T(t) = \lambda e^{-\lambda t} \tag{4-120}$$

$$f_2(t) = \int_0^t (\lambda e^{-\lambda \tau})\lambda e^{-\lambda(t-\tau)}\, d\tau$$

$$= \lambda^2 t e^{-\lambda t} \tag{4-121}$$

$$f_3(t) = \int_0^t (\lambda^2 \tau e^{-\lambda \tau})\lambda e^{-\lambda(t-\tau)}\, d\tau$$

$$= \lambda^3 \frac{t^2}{2} e^{-\lambda t} \tag{4-122}$$

The pattern can now be recognized, and we have

$$f_k(t) = \lambda^k \frac{t^{k-1}}{(k-1)!} e^{-\lambda t} \tag{4-123}$$

Equation (4–123) is the same as Equation (4–75) with β replaced by k and α replaced by $1/\lambda$. The probability density function $L(t)$ of some failure occurring with ideal repair for this special case is

$$L(t) = \sum_{k=1}^{\infty} f_k(t) = (\lambda e^{-\lambda t}) \sum_{k=1}^{\infty} \frac{(\lambda t)^{k-1}}{(k-1)!}$$

$$= \lambda e^{-\lambda t} e^{\lambda t}$$

or

$$L(t) = \lambda \tag{4-124}$$

As expected, the failure density for a component with an exponentially distributed lifetime and ideal repair is constant and equal to λ, the reciprocal of its mean lifetime. This λ is the constant failure rate (or hazard function) that led us to exponentially distributed lifetimes.

■■ **Example 4–23**
Referring back to Example 4–22, if ideal repair is carried out after each failure, the various density functions can be calculated as follows:

$$f_1(t) = f_T(t) = 0.25$$

$$f_2(t) = \int_0^t (0.25)(0.25)\, d\tau = \left(\frac{1}{4}\right)^2 t$$

$$f_3(t) = \int_0^t \left(\frac{1}{4}\right)^2 \tau \left(\frac{1}{4}\right) d\tau = \left(\frac{1}{4}\right)^3 [t^2/2]$$

Proceeding in a similar fashion, we conclude that

$$f_k(t) = \left(\frac{1}{4}\right)^k \left[\frac{t^{k-1}}{(k-1)!}\right]$$

Therefore,

$$L(t) = \sum_{k=1}^{\infty} f_k(t)$$

$$= \left(\frac{1}{4}\right) \sum_{k=1}^{\infty} \frac{(t/4)^{k-1}}{(k-1)!}$$

or

$$L(t) = 0.25 \exp\left[\frac{t}{4}\right]$$

■■

4.18 IDEAL REPAIR AND PREVENTIVE MAINTENANCE

We saw earlier that, for components with increasing hazard functions, periodic preventive maintenance increased the MTTF and resulted in failure density functions that had an exponential tendency. If, in addition, the possibility of ideal repair is included, the net result of maintenance would be to reduce the frequency of repairs. Assuming ideal maintenance at periodic intervals T_M, the frequency of repair f_R will be equal to the average density of failures over a time duration T_M. That is,

$$f_R = \frac{1}{T_M} \int_0^{T_M} L(t)\, dt \qquad (4\text{--}125)$$

where $L(t)$ is given by Equation (4–119). As the frequency of maintenance increases, T_M decreases and f_R decreases. In other words, the effective MTTF, which is equal to the reciprocal of f_R, increases. If this is not the case, maintenance is of no use.

■■ **Example 4–24**

If the component in Examples 4–22 and 4–23 undergoes preventive maintenance at intervals T_M and is subjected to ideal repair, then, using Equation (4–125), we obtain

$$\text{frequency of repair} = f_R = \frac{1}{T_M} \int_0^{T_M} 0.25 e^{t/4}\, dt$$

or

$$f_R = \left(\frac{1}{T_M}\right)\left[\exp\left(\frac{T_M}{4}\right) - 1\right]$$

For $T_M = 1$ year,

$$f_R = e^{1/4} - 1$$

$$= 0.284 \text{ per year}$$

If we did not subject the component to scheduled maintenance, with only ideal repair, the original density function would be valid, and therefore, the MTTF would be two years. From this, we conclude that the frequency of repair is 0.5 per year. Clearly, scheduled maintenance performed once a year reduces the frequency of failure (repair) from 0.5 to 0.284 per year.   ■■

4.19 SUMMARY

In reliability work, we often are dealing with random variables, both discrete and continuous. Some of these variables are time to failure, time to repair, duration of a storm, maintenance time, peak load, wind speed, and number of failures. This chapter has discussed the important distributions that have found applications in modeling these random variables. Although no one distribution is discussed in any great depth, the information included is sufficient for later use and constitutes a good starting point for detailed studies. Over 15 distributions and techniques to alter them to fit available data have been presented. Applications in modeling the wearout region, scheduled maintenance, and ideal repair have also been discussed.

PROBLEMS

4.1 It has been determined from past experience that 2% of all of a certain type of diodes are bad when manufactured. In a bag of 100 such diodes, what is the probability that (*i*) exactly two diodes are defective and (*ii*) two or more diodes are defective?

4.2 In a 5,000-hour stress test on a certain type of electronic component, the probability of a component failing has been found from past experience to be 1%. If 50 such components randomly selected are stress tested, calculate (a) the probability of no failures, (b) the probability of at least one failure, and (c) the probability of exactly two failures.

4.3 The probability of a jet fighter surviving an intense combat mission is 0.8, and three fighters are needed for the success of the mission. How many fighters must be sent on the mission in order to achieve a 95% probability of success?

4.4 What is the requirement for the binomial distribution to be symmetric? Show that the kurtosis of a symmetric binomial distribution is

$$\frac{3}{16} n^2 - \frac{n}{8}$$

4.5 Two options are being evaluated to supply a load of 150 MW. Option no. 1 is to have two 100-MW units with a 1% probability of failure for each. Option no. 2 is to have one 100-MW unit identical to the one considered in option no. 1 and two 50-MW units with 1.5% failure probability for each. Which option is preferable, based on the loss of load expectancy?

4.6 Among a group of distribution transformers purchased by a public utility company, one unit is expected to be bad with a standard deviation of 0.9975. What is the probability of exactly two defects out of a lot of 50 such transformers?

4.7 In a game involving the toss of a pair of dice, if the same number of dots appear on both, then the toss is considered to be a success.
  (*i*) What is the probability of no success in 3 tosses?
  (*ii*) What is the expected number of successes in 30 tosses?
  (*iii*) What is the standard deviation for the experiment of part (*ii*)?
  (*iv*) What is the probability of more than one success in 5 tosses?

4.8 A range of domestic washing machines have a mean failure rate of one fault in 5 years, and this rate remains sensibly constant and time independent over a 10-year period of normal usage. If a customer purchases a new machine of this type, what is the probability that
  (*i*) at least one failure will have occurred by the end of the first year,
  (*ii*) one failure occurs in the first year of use,
  (*iii*) more than one failure occurs during the first year, and
  (*iv*) one failure occurs by the end of the 10th year?

4.9 Derive the following recursive formula for the Poisson distribution:

$$P_{x+1} = \left(\frac{\mu}{x+1}\right) P_x$$

4.10 Faults in a power system trip a particular circuit breaker once in every 2,000 hours of operation, on average.
  (a) What is the probability of one trip in 4,000 hours of operation?
  (b) If at $t = 0$, the breaker opens and is reset, what is the probability that the waiting time till the next trip is 3,000 hours?

4.11 One percent of a certain type of capacitor are defective when manufactured. The number of defects can be assumed to have a Poisson distribution. In a capacitor bank consisting of 100 such capacitors, find the probability that (*i*) exactly 3 capacitors are defective, (*ii*) 3 or more capacitors are defective, and (*iii*) all of the capacitors are good.

4.12 A computer wholesaler gets one order per day, seven days a week, on average.
  (a) What is the probability of receiving no orders in two days?
  (b) What is the probability of receiving two orders in one day?
  (c) What is the probability of receiving more than two orders in one day?

4.13 Only 0.5% of the graduating high school seniors are expected to join the six-year integrated medical school program. In a graduating class of 472 students, what is the probability of 3 or more students entering this program?

4.14 On average, 19 tornadoes touch down in a certain region of the United States each decade. What is the probability that in a given year the number of tornadoes touching down in this region is (*i*) 0, (*ii*) exactly 3, and (*iii*) 2 or more?

4.15 The probability of a disk drive failing an inspection is 0.8%. In a batch of 250 disk drives, what is the probability that there will be (*i*) 1 failure, (*ii*) no failures, and (*iii*) more than 3 failures?

4.16 The average demand for a certain spare part is 20 per year. A new stock of these spare parts can be purchased once a month. How large should a store's stock be to satisfy customer demand with 90% or higher probability?

4.17 An automobile repair garage can service four cars on any one day. The number of customers arriving has a Poisson distribution with an expected value of three per day.
(a) What is the probability of turning away customers on any given day?
(b) Calculate the expected number of (*i*) cars serviced per day and (*ii*) customers turned away per day.

4.18 What should be the failure rate of a component during its useful life in order to achieve 99% reliability for a mission time of 5,000 hours? Express your result in %/K. If the failure rate is doubled, by how much does the mission reliability decrease?

4.19 N identical constant-hazard components are subjected to testing, and n of them fail at the end of the test time t_0. Show that the MTTF can be estimated using

$$\text{MTTF} = t_0/\ln\left(\frac{1}{1 - n/N}\right)$$

4.20 For components with exponentially distributed failure times, derive an expression for the time t_f taken to expect a percentile p_f to fail.

4.21 NAND gates manufactured by a company have an exponential failure rate of 210 PPM/K.
(a) What is the probability of a gate surviving two years of use?
(b) What is the probability that a gate will fail in the next two years?
(c) What are the mean and median times to failure?
(d) What is the time taken for 35% of the gates to fail?

4.22 A manufacturer gives a one-year guarantee on an item based on the assumption that no more than 6% of the items will be brought back by the customers. If the items can be assumed to have a constant hazard, what is the maximum tolerable failure rate in FIT?

4.23 The time to failure of a component of a machine can be assumed to follow an exponential distribution. The probability that the component would not survive for more than 50 days is 0.92. How often can one expect to replace this component?

4.24 In a high-stress environment, the lifetime of a particular microchip is estimated to be 40 hours. What is the probability that a microprocessor chip of this type would survive for one week when subjected to a similar environment?

4.25 Eighty percent of a particular group of components fail after 30 days of operation. Assuming a constant hazard, calculate how often the components are expected to be replaced.

4.26 The lifetimes of a certain class of electric motor are normally distributed with a mean value of μ and a standard deviation of σ.
(*i*) What percentage of the motors will survive longer than $(\mu + 2\sigma)$?
(*ii*) What percentage will fail within the range $(\mu \pm \sigma)$?
(*iii*) What fraction will fail in less than $(\mu - \sigma)$?

(*iv*) What percentage will fail outside the range ($\mu \pm 3\sigma$)?

(*v*) Find the time at which 90% of the motors would have failed.

4.27 A machine produces components that are 6% defective, on average, and the number of defects is found to be normally distributed. In a random sample of 300 components produced by this machine, what is the probability that (*i*) 25 or more are defective, (*ii*) at most 20 are defective, and (*iii*) between 10 and 20 are defective?

4.28 The lifetimes of automobile batteries are normally distributed with a mean of 1,458 days and a standard deviation of 135 days. The manufacturer guarantees the batteries for 36 months. Assuming that all months have 30 days, what fraction of the batteries will require replacement?

4.29 Failures of CD players have a normal distribution with a mean value of 50 months and a standard deviation of 5 months. What is the design life for a reliability of 96%?

4.30 A robot arm can reach a specified point in the *xy*-plane with normally distributed errors in the *x*- and *y*-directions. Both errors have zero mean and the same standard deviation σ. Under these conditions, it can be shown that the radial error R_e has the distribution

$$F(r_e) = 1 - \exp\left[-\frac{r_e^2}{2\sigma^2}\right]$$

If $\sigma = 1$ mm, what is the probability that the arm will come closer than 3 mm to the targeted point? If an error rate of less than 1 in 2,000 is required, how far should the arm be allowed to come from the intended location?

4.31 The failure times of a class of items are found to be uniformly distributed between 1,000 and 1,500 hours.

(a) Find the probability of an item (*i*) surviving for 1,000 hours, (*ii*) failing in 1,100 hours, (*iii*) surviving beyond 1,500 hours, and (*iv*) failing within 1,200 hours.

(b) Given that a particular item has survived for 1,200 hours, what is the probability that it will fail during the next 100 hours?

(c) If the component has survived for 1,498 hours, what is the probability of its failing during the next hour?

4.32 (a) A random variable is uniformly distributed between the values of t_1 and t_2. Derive expressions for t_1 and t_2 in terms of μ and σ.

(b) Resistors manufactured by a factory have a mean resistance of 1,000 ohms and a standard deviation of 28.87 ohms. Assuming uniform distribution, find the tolerance limits for the resistors.

4.33 If failure times are Rayleigh distributed as given in Equation (4–43), find the MTTF and T_{50}.

4.34 The lifetimes of a system are found to have Rayleigh density with $k = 0.005$ when time t is considered in weeks.

(*i*) What is the probability that the system will fail in 2 weeks?

(*ii*) What is the probability that the lifetime of the system will be greater than one year (52 weeks)?

(*iii*) Find the MTTF and T_{50}.

4.35 One group of components has a constant failure rate of λ_0, and the failures of another group of components follow a Rayleigh distribution with $\lambda(t) = kt$. If the two groups have the same MTTF, derive the relationship between k and λ_0.

4.36 Derive an expression for the *a posteriori* failure probability for the Rayleigh distribution. Is this distribution memoryless? Explain your answer clearly.

4.37 What is the probability of failure of a component with a Weibull-distributed hazard at a time equal to the scale parameter? Does this probability depend on the shape parameter? For a shape parameter of 1, calculate the probability of no failure at times equal to integer multiples of the scale parameter, up to four.

4.38 Derive the following expressions for the Weibull distribution with a scale parameter α and a shape parameter β:

$$\beta = \frac{\ln[-\ln(1 - Q(t))]}{\ln(t/\alpha)}$$

$$t = \alpha[-\ln(1 - Q(t))]^{1/\beta}$$

$$\alpha = \frac{t}{[-\ln(1 - Q(t))]^{1/\beta}}$$

4.39 Redefine the scale parameter in the Weibull distribution as follows:

$$\text{scale parameter} = \theta \equiv \alpha^{\beta}$$

Obtain expressions for all the reliability functions in terms of this new scale parameter.

4.40 An engineer is asked to make a choice between two designs for a critical component. Extensive testing of prototypes has revealed the following information:

- The time to failure is Weibull distributed in both cases.
- Design I costs \$4,500 to build and has $\beta = 3$ and $\alpha = 190$ hours.
- Design II costs \$4,000 to build and has $\beta = 4$ and $\alpha = 100$ hours.

Which design should be chosen if the component must have (*i*) a 12-hour guaranteed life and (*ii*) a 24-hour guaranteed life?

4.41 If a continuous random variable has a gamma distribution with a scale parameter of 3 and a shape parameter of 2, what is the probability that the value of the variable is between 2 and 12? Also, calculate the mean, variance, and standard deviation of the variable.

4.42 If the failure times for a certain device are gamma distributed with $\beta = 3$ and $\alpha = 600$, what is the reliability of the device for an operating time of (*i*) 600 hours and (*ii*) 1,200 hours?

4.43 Derive Equation (4–77) for the reliability associated with the special Erlangian distribution.

4.44 It has been found that maintenance times (in minutes) for a system follow a lognormal distribution with $\mu = 2.5$ and $\sigma = 1$.
 (*i*) What is the probability that a maintenance will take longer than 30 minutes?
 (*ii*) What is the probability that a maintenance takes less than the average time?

4.45 If failure times have a lognormal distribution, express the median time to failure, T_{50}, in terms of μ. Suppose the failures of a population of devices follow a lognormal distribution with $T_{50} = 4,500$ hours and $\sigma = 0.72$. What fraction of the devices are expected to fail in 1,500 hours? What fraction are expected to survive one year (8,760 hours) or longer?

4.46 The failure times (in cycles) of a mechanical component obtained from fatigue testing fit a lognormal distribution with a median value of 15 million cycles and $\sigma = 2.1$. What should be the design life in cycles to keep the failures within 2%? If the design life is set at one-half of this value, what is the probability of failure of the component?

4.47 For the lognormal distribution, express MTTF and the variance σ^2 in terms of T_{50}.

4.48 For the general beta density function

$$f(x) = \begin{cases} Ax^{\gamma}(1 - x)^{\beta} & \text{for } 0 \leq x \leq 1 \\ 0 & \text{otherwise} \end{cases}$$

find A and show that $f(x)$ is a maximum for $x = [\gamma/(\gamma + \beta)]$.

4.49 If the parameters γ and β in the beta distribution of Problem 4.48 are integers, show that

$$E[x] = \frac{\gamma + 1}{\gamma + \beta + 2}$$

and

$$f(x) = (\beta + \gamma + 1)_{(\beta + \gamma)}C_{\gamma}x^{\gamma}(1 - x)^{\beta}$$

4.50 One form of the extreme value distribution for $t \leq 0$ is

$$F(t) = 1 - \exp[-(-t)^a]$$

How would you modify this equation for $t \geq 0$? Show that the modified equation is a Weibull distribution, and find the corresponding scale and shape factors.

4.51 Given that a random variable has the extreme value distribution of Gumbel type 1 of the form

$$F(t) = 1 - \exp[-e^t] \quad \text{for } -\infty < t < +\infty$$

find the mean and median values.

4.52 Consider a Weibull-distributed random variable t with the distribution

$$F(t) = 1 - \exp\left[-\left(\frac{t}{k}\right)^n\right]$$

and define a new random variable X, related to t by

$$x = \ln t$$

Show that the new random variable X has a Gumbel type 1 extreme value distribution given by

$$F(x) = 1 - \exp\left[-\exp\left\{\frac{x - a}{b}\right\}\right]$$

where $a = \ln k$ and $b = 1/n$.

4.53 Derive Equation (4–100).

4.54 Derive Equation (4–102).

4.55 The mode of a density function is the point at which the function is a maximum. Show that the mode of the shifted Rayleigh density function

$$f(t) = k(t - a) \exp\left[-\left(\frac{k}{2}\right)(t - a)^2\right], \quad t \geq a$$

occurs at $t = [a + (1/\sqrt{k})]$ and that the maximum value is $0.6065 \sqrt{k}$.

4.56 Often, a certain threshold amount of wear may have to take place before any failures can occur. Such failures are modeled by shifting the hazard function by a suitable amount of time t_0 along the time axis. For the shifted (also known as the three-parameter) Weibull hazard

$$\lambda(t) = \frac{\beta}{\alpha}\left(\frac{t - t_0}{\alpha}\right)^{\beta - 1}, \quad t \geq t_0$$

and $\lambda(t) = 0$ for $t < t_0$, obtain the failure density and distribution functions. What is the MTTF? Does shifting affect the value of the variance?

4.57 For the piecewise-linear hazard model shown in Figure P 4.57, derive an expression for the *a posteriori* failure probability $Q_c(t)$ for $T > T_0$.

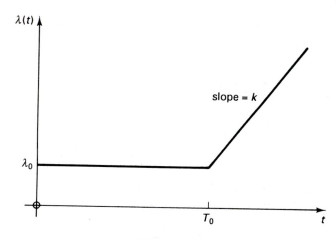

Figure P 4.57

4.58 Approximate the exponential hazard model

$$\lambda(t) = K \exp[\alpha t], \quad \alpha > 0$$

by the first three terms of its power series, and find the failure density, reliability, and unreliability functions.

4.59 Let the mean wearout life of a component be W, with a standard deviation of $0.15W$. If a mission is started at $0.9W$ in the component's life, how long can the mission last if the decrease in reliability due to wearout is not to exceed a factor of 0.95? Give your answer as a fraction of W. Repeat the calculation if the standard deviation is $0.05W$.

4.60 A small motor has a failure rate of 8%/K. Its mean wearout life is 8,500 hours, with a standard deviation of 1,500 hours. What is the reliability of the motor for a 200-hour run if it has already operated for (*i*) 8,000 hours and (*ii*) 9,000 hours?

4.61 Consider a device having a decreasing failure rate characterized by a two-parameter Weibull hazard with a scale parameter of 200 months and a shape parameter of 0.5. It is desired to have a reliability of at least 0.9 during the design life of the device.
   (a) With no initial wearin period, what is the design life of the device?
   (b) With an initial wearin period of 10 days, what is the improvement in the design life of the device?

4.62 Repeat Example 4–22 for

$$f(t) = 0.2 \text{ year}^{-1}, \quad 0 < t \le 5 \text{ years}$$

4.63 A component with a linearly increasing hazard

$$\lambda(t) = 0.1t$$

for *t* in years undergoes regular maintenance once every two years. Assuming the maintenance to be ideal, estimate the improvement realized in the MTTF. If the maintenance schedule is changed to once a year, what is the estimate of the new MTTF with maintenance?

4.64 The lifetime of a system has a uniform distribution

$$f_T(t) = 0.1 \text{ year}^{-1}, \quad 0 < t \le 10 \text{ years}$$

The system undergoes regular maintenance once in 2 years, and after each failure, ideal repair is carried out. Calculate the average density of failures for this system.

4.65 If the failure rate of a component can be assumed to be a constant and equal to one in every five years, what is the probability of two failures in three years? Assume ideal repair.

4.66 Components with a constant failure rate of 100 PPM/K are used in a system. If ideal repair is carried out after each failure, derive an expression for the probability of *k* failures in *t* hours.

4.67 Find the density function and the mean value for the maximum extreme-value distribution of type 1 given in Equation (4–93a).

4.68 Show that the mean value and variance corresponding to the maximum extreme-value distribution of type II [Equation (4–94a)] are:

$$\text{mean value} = \beta\Gamma\left(1 - \frac{1}{m}\right)$$

$$\text{variance} = \beta^2\left[\Gamma\left(1 - \frac{2}{m}\right) - \Gamma^2\left(1 - \frac{1}{m}\right)\right]$$

4.69 Derive approximate expressions for the maximum and minimum extreme-value distributions of type III by taking only the first two terms of the series expansion for the exponential terms. Under what conditions are these approximations valid?

4.70 The minimum extreme-value distribution of type III given in Equation (4–95b) is also known as the *three-parameter Weibull distribution* and is quite commonly used. Find the corresponding mean and variance.

4.71 Derive the mean and variance of the beta distribution given in Equations (4–92a) and (4–92b).

4.72 Confirm the validity of the expressions for γ and β given in Equations (4–92c) and (4–92d).

4.73 If the mean and variance of a beta-distributed sample are 0.1 and 0.008, respectively, estimate the γ and β parameters. Comment on the shape of the resulting density function.

4.74 Assuming that $\gamma = 4$ and $\beta = 9$, sketch the corresponding beta density function. Repeat for $\beta = 1$.

4.75 Repeat Problem 4.74 for
    (a) $\gamma = 4$ and $\beta = -0.5$.
    (b) $\gamma = -0.5$ and $\beta = 9$.
    (c) $\gamma = 3$ and $\beta = 3$.

4.76 For the special case of integer values of γ and β, consider the distribution function $Q(t)$ given in Equation (4–91c) as $Q(\gamma, \beta)$, and show that

$$Q(\gamma, \beta) = \frac{\Gamma(\gamma + \beta + 2)(1 - t)^{\beta} t^{\gamma + 1}}{\Gamma(\gamma + 2)\Gamma(\beta + 1)} + Q(\gamma + 1, \beta - 1)$$

4.77 By the repeated use of the result of Problem 4.76, show that

$$Q(\gamma, \beta) = \sum_{i=\gamma+1}^{\gamma+\beta+1} \frac{\Gamma(\gamma + \beta + 2) t^{i}(1 - t)^{\gamma + \beta + 1 - i}}{\Gamma(i + 1)\Gamma(\gamma + \beta + 2 - i)}$$

4.78 Suppose X is a binomially distributed discrete random variable with p replaced by t and n replaced by $(\gamma + \beta + 1)$ in Equation (4–2). Show that

$$Q(\gamma, \beta) = 1 - F_X(\gamma)$$

(*Hint:* Use the result of Problem 4.77.)

5

Combinatorial Aspects of System Reliability

5.1 INTRODUCTION

Complex systems are usually decomposed into functional entities composed of units, subsystems, or components for the purpose of reliability analysis. Network modeling techniques and combinatorial aspects of reliability analysis are employed to connect the components in series, parallel, series-parallel, or meshed structures, or in any combination of these. Probability concepts are then employed to compute the reliability of the system in terms of the reliabilities of its subunits.

5.2 SERIES OR CHAIN STRUCTURE

A set of components is said to be *in series* from a reliability point of view if the success of the system depends on the success of all the system components. The components themselves need not be physically or topologically in series; what is relevant is only the fact that all of them must succeed for the system to succeed. The block diagram of such a system is shown in Figure 5.1. Referring to this figure, let x_i denote the event that the ith unit is successful and $\bar{x}_i$ denote the event that the ith unit is not successful. Then the system reliability, which is the prob-

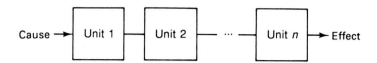

Figure 5.1 Series or chain structure.

ability of system success, can be expressed as

$$R_s = P(x_1 x_2 \ldots x_n) \tag{5-1a}$$

$$= P(x_1)P(x_2 \mid x_1)P(x_3 \mid x_1 x_2) \ldots P(x_n \mid x_1 x_2 \ldots x_{n-1}) \tag{5-1b}$$

If the units do not interact, then the events are all independent, and

$$R_s = P(x_1)P(x_2) \ldots P(x_n) \tag{5-2a}$$

$$= \prod_{i=1}^{n} P(x_i) \equiv \prod_{i=1}^{n} R_i \tag{5-2b}$$

$$\text{System unreliability} = Q_s = 1 - R_s = 1 - \prod_{i=1}^{n} R_i \tag{5-3a}$$

$$= 1 - \prod_{i=1}^{n} (1 - Q_i) \tag{5-3b}$$

If the components interact, then the conditional probabilities must be evaluated with care. In general, we can express the system unreliability as

$$Q_s = P(\bar{x}_1 + \bar{x}_2 + \ldots + \bar{x}_n) \tag{5-4}$$

where the plus $(+)$ symbol denotes the union operation. In practice, the reliability of a series system is only slightly affected by component dependence. Moreover, the reliability of a series system is always worse than that of the poorest component.

■■   **Example 5–1**

Consider a system consisting of 10 identical components, all of which must be good for the system to be good. If the reliability of each component is 0.97, then the system reliability is $(0.97)^{10}$, or 0.737424. If we want a system reliability of 0.95, what should be the minimum reliability of each component?

Since we want $0.95 = R^{10}$, we conclude that $R = (0.95)^{0.1} = 0.994884$

■■

5.3 PARALLEL STRUCTURE

A set of *n* components is said to be *in parallel* from a reliability point of view if the system can succeed when at least one component succeeds. The block diagram of such a system is shown in Figure 5.2. The reliability of a parallel system can be expressed as

$$R_p = P(x_1 + x_2 + \ldots + x_n) = 1 - P(\bar{x}_1\bar{x}_2 \ldots \bar{x}_n) \qquad (5\text{--}5a)$$

or

$$R_p = 1 - P(\bar{x}_1)P(\bar{x}_2 \mid \bar{x}_1)P(\bar{x}_3 \mid \bar{x}_1\bar{x}_2) \ldots P(\bar{x}_n \mid \bar{x}_1\bar{x}_2 \ldots \bar{x}_{n-1}) \qquad (5\text{--}5b)$$

If the unit failures are independent, then we have

$$R_p = 1 - P(\bar{x}_1)P(\bar{x}_2) \ldots P(\bar{x}_n) \qquad (5\text{--}6a)$$

$$= 1 - \prod_{i=1}^{n} Q_i \qquad (5\text{--}6b)$$

and

$$\text{system unreliability} = Q_p = \prod_{i=1}^{n} Q_i \qquad (5\text{--}7)$$

In general, parallel structure improves system reliability. If we have a system consisting of *n* elements with information on the reliability of each element but little or no information about its interconnection with other elements, then the lower bound for reliability is given by the series configuration and the upper bound is given by the parallel configuration.

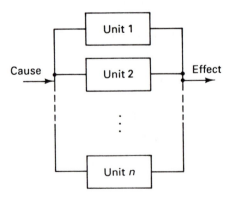

Figure 5.2   Parallel structure.

■■   **Example 5–2**

Consider components with individual reliabilities of 0.7. With two such components in parallel,

$$R_p = 1 - (0.3)(0.3) = 0.91$$

Thus, a 30% improvement in reliability is achieved over that of one component.
With three such components in parallel,

$$R_p = 1 - (0.3)(0.3)(0.3) = 0.973$$

a 39% improvement compared to the reliability of one component.
With four such components in parallel,

$$R_p = 1 - (0.3)^4 = 0.9919$$

a 41.7% improvement compared to the reliability of one component. Note that the biggest jump in reliability is obtained by going from one component to two components in parallel.

Next, we find the number *n* of components in parallel needed to achieve an overall system reliability of 0.9999. Since

$$0.9999 = 1 - (0.3)^n$$

it follows that

$$(0.3)^n = 0.0001$$

and

$$n = 7.6, \text{ or 8 components} \qquad\qquad ■■$$

■■   **Example 5–3**

Suppose some components have a time-dependent reliability given by the function $e^{-\lambda t}$. Then an expression for the reliability of a system consisting of two such components in parallel is

$$R(t) = 1 - (1 - e^{-\lambda t})^2$$
$$= 1 - (1 - 2e^{-\lambda t} + e^{-2\lambda t})$$
$$= 2e^{-\lambda t} - e^{-2\lambda t} \qquad\qquad ■■$$

5.4 AN *r*-OUT-OF-*n* STRUCTURE

We have considered this structure before as an application of the binomial distribution. If *p* is the probability of success of each component, then the system reliability *R* can be expressed as

$$R = \sum_{k=r}^{n} {}_nC_k p^k (1 - p)^{n-k} \qquad\qquad (5\text{–}8)$$

Note that for $r = 1$, the structure becomes a parallel system, and for $r = n$, it becomes a series system.

■■ **Example 5–4**
Let $n = 4$, $r = 2$, and $p = e^{-\lambda t}$. Then

$$R(t) = {}_4C_2(e^{-\lambda t})^2(1 - e^{-\lambda t})^2$$

$$+ {}_4C_3(e^{-\lambda t})^3(1 - e^{-\lambda t})^1$$

$$+ {}_4C_4(e^{-\lambda t})^4(1 - e^{-\lambda t})^0$$

$$= 6e^{-2\lambda t} - 8e^{-3\lambda t} + 3e^{-4\lambda t}$$ ■■

5.5 STAR AND DELTA STRUCTURES

A system consisting of three components may be connected in a star or a delta configuration, as illustrated in Figure 5.3. For these configurations to be equivalent, the reliabilities computed across the terminals AB, BC, and CA should be the same for both.

For the reliabilities computed across AB, we have

$$R_A R_B = 1 - (1 - R_{AB})(1 - R_{BC} R_{CA}) \qquad (5-9)$$

Similarly, the reliabilities across BC and CA are

$$R_B R_C = 1 - (1 - R_{BC})(1 - R_{CA} R_{AB}) \qquad (5-10)$$

and

$$R_C R_A = 1 - (1 - R_{CA})(1 - R_{AB} R_{BC}) \qquad (5-11)$$

These three equations can be solved simultaneously to express the star quantities R_A, R_B, and R_C in terms of the delta quantities R_{AB}, R_{BC}, and R_{CA}, or vice versa. Doing the former, we have

$$R_A = \sqrt{\frac{(R_A R_B)(R_C R_A)}{R_B R_C}}$$

$$= \sqrt{\frac{[1 - (1 - R_{AB})(1 - R_{BC} R_{CA})][1 - (1 - R_{CA})(1 - R_{AB} R_{BC})]}{1 - (1 - R_{BC})(1 - R_{CA} R_{AB})}} \qquad (5-12)$$

An expression similar to the rightmost side of Equation (5–12) can be obtained for R_B and R_C using the relationships

$$R_B = \sqrt{\frac{(R_A R_B)(R_B R_C)}{R_A R_C}} \qquad (5-13)$$

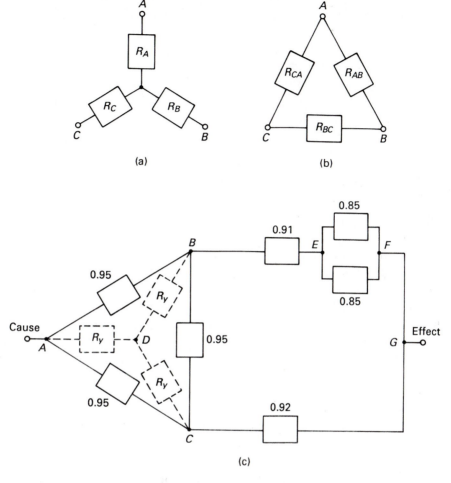

Figure 5.3 (a) Star (wye) configuration. (b) Delta (mesh) configuration. (c) System that includes a delta.

and

$$R_C = \sqrt{\frac{(R_B R_C)(R_C R_A)}{R_A R_B}} \tag{5-14}$$

For the special case of

$$R_{AB} = R_{BC} = R_{CA} \equiv R_\Delta \tag{5-15}$$

and

$$R_A = R_B = R_C = R_y \tag{5-16}$$

we get

$$R_y = \sqrt{R_\Delta^2 + R_\Delta - R_\Delta^3} \tag{5-17}$$

If $R_\Delta = 0.9$, Equation (5–17) yields $R_y = 0.99045$. Thus, we need better components (i.e., components with higher reliability) in the star configuration, as compared to the components in the delta configuration, to achieve the same level of system reliability.

5.6 SERIES-PARALLEL STRUCTURE

By combining the appropriate series and parallel branches of the network model systematically, we can reduce the entire system to one single equivalent element. The reliability of this equivalent element is the reliability of the original system. Such a procedure is known as the *network reduction* technique.

■■ **Example 5–5**

Let us consider the system shown in Figure 5.3(c). The individual component reliabilities are as indicated. To evaluate the system reliability, we will first replace the delta *ABC* by an equivalent star, shown in dashed lines. Then we will employ series-parallel network reduction techniques to obtain the desired result. We have

$$R_y = \sqrt{(0.95)^2 + (0.95) - (0.95)^3}$$

$$= 0.99756$$

The two components in parallel across *EF* can be replaced by an equivalent component with a reliability of $[1 - (1 - 0.85)^2]$, or 0.9775. Then,

$$\text{Reliability of path } DBEFG = (0.99756)(0.91)(0.9775)$$

$$= 0.887355$$

$$\text{Reliability of path } DCG = (0.99756)(0.92)$$

$$= 0.917755$$

For the subsystem consisting of paths *DBEFG* and *DCG* in parallel,

$$\text{Reliability} = 1 - (1 - 0.887355)(1 - 0.917755)$$

$$= 0.99074$$

The system is now reduced to a component with reliability R_y in *AD* in series with an equivalent component having a reliability of 0.99074. Therefore,

$$\text{System reliability} = (0.99756)(0.99074)$$

$$= 0.988323 \qquad\qquad ■■$$

5.7 TRIPLE MODULAR REDUNDANCY

Triple modular redundancy (TMR) is often used in the design of computer systems. The basic TMR configuration consists of three identical units feeding into a majority voter system, as shown in Figure 5.4. Typically, the three identical units represent a single logic (binary) variable, and the output logic variable is determined on the basis of majority voting. The corresponding voter output table is as follows:

Output of Units			
#1	#2	#3	Voter Output
0	0	0	0
0	0	1	0
0	1	0	0
0	1	1	1
1	0	0	0
1	0	1	1
1	1	0	1
1	1	1	1

The advantage of using TMR is that if the output of one of the three units is in error, it is masked and the output remains correct.

Assuming a perfect voter, we need two out of the three units to be good for the system to succeed. Therefore,

$$R_{TMR} = \sum_{k=2}^{3} {}_3C_k R^k (1 - R)^{3-k}$$

or

$$R_{TMR} = R^2(3 - 2R) \tag{5-18}$$

where R is the reliability of one unit. If the voter is imperfect, with a reliability of R_v, then,

$$R_{TMR} = R_v R^2(3 - 2R) \tag{5-19}$$

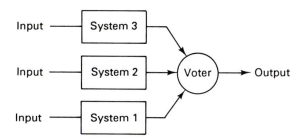

Figure 5.4 Triple modular redundancy configuration.

5.8 *N*-TUPLE MODULAR REDUNDANCY

In *N*-tuple modular redundancy (NMR), we have $N = (2n + 1)$ replicated units, feeding into an $(n + 1)$-out-of-N voter. For $n = 1$, NMR reduces to TMR. For successful operation, at least $(n + 1)$ units out of N units must survive. If we neglect the effect of compensating failures, then the reliability of an NMR configuration can be written as

$$R_{NMR} = \sum_{k=(n+1)}^{N} {}_NC_k R^k (1 - R)^{N-k} \qquad (5\text{--}20)$$

By letting $i = (N - k)$, we obtain $k = (N - i)$, and Equation (5–20) can be written as

$$R_{NMR} = \sum_{i=0}^{n} {}_NC_i (1 - R)^i R^{N-i} \qquad (5\text{--}21)$$

5.9 STANDBY REDUNDANCY

In the case of parallel redundancy, all the components are operating simultaneously. In the case of standby redundancy, the standby unit is brought into operation only when a normally operating unit fails. The differences between these two cases are illustrated in block diagram form in Figure 5.5. The standby redundancy mode is appropriate if the failure rate in the nonoperating standby mode is lower than the failure rate in the continuously operating mode.

CASE (*i*)

Perfect switching. In this case, the switch is assumed never to fail and will switch on the standby unit the instant the normally operating unit fails. Also, the failure rate in the standby mode is assumed to be zero. In other words, *B* can fail only if *A* has already failed (and therefore, *B* is operating).

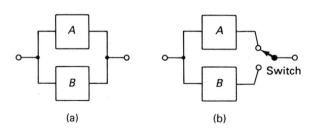

(a)   (b)

Figure 5.5 (a) Parallel and (b) standby redundant systems.

Hence,

$$\text{Probability of system failure} = Q = Q(A)Q(B \mid \overline{A}) \qquad (5\text{-}22)$$

If the units are independent, then

$$Q = Q(A)Q(B) \equiv Q_A Q_B \qquad (5\text{-}23)$$

Since B is used only when A fails, Q_B is much smaller than the value it would have if B were operating continuously.

CASE (ii)

Imperfect switching.   In this case, we include the possibility of the switch failing to switch on B when A has failed.

Let P_s be the probability of a successful changeover. Then,

$P(\text{system failure}) = P(\text{system failure } given \text{ successful changeover})$

$\qquad\qquad \times\ P(\text{successful changeover})$

$\qquad\qquad +\ P(\text{system failure } given \text{ unsuccessful changeover})$

$\qquad\qquad \times\ P(\text{unsuccessful changeover})$

$$Q = Q_A Q_B P_s + Q_A(1 - P_s) \qquad (5\text{-}24a)$$

$$= Q_A - Q_A P_s(1 - Q_B) \qquad (5\text{-}24b)$$

If the switch can fail in the initial operating position, in addition to failing to changeover when required, we have to modify the system block diagram used in reliability evaluation. The modified diagram is given in Figure 5.6. Referring to this figure, we obtain

$$R = \{1 - [Q_A - Q_A P_s(1 - Q_B)]\}R_s \qquad (5\text{-}25)$$

and

$$Q = 1 - R \qquad (5\text{-}26)$$

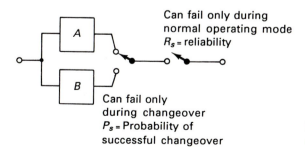

Can fail only during
normal operating mode
R_s = reliability

Can fail only
during changeover
P_s = Probability of
successful changeover

Figure 5.6 Standby redundancy with imperfect switching.

5.10 GENERAL TECHNIQUES FOR EVALUATING THE RELIABILITY OF COMPLEX SYSTEMS

Many practical systems have complex structures that are not purely series, parallel, or series-parallel. Evaluation of the reliability of such systems requires some general and powerful techniques.

5.10.1 Inspection

This approach is useful only when a small number of components are involved. Let us consider some examples.

   a. For the system illustrated in Figure 5.7, let $P(x_1) \equiv$ probability that component (path) x_1 is good and $P(\bar{x}_1) \equiv$ probability that component (path) x_1 is bad. Then,

$$\text{System reliability} = R_s = 1 - [P\{(\bar{x}_1 + \bar{x}_2)\bar{x}_3\}] \tag{5–27}$$

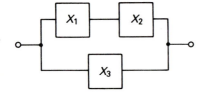
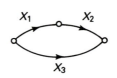

Figure 5.7   A simple system configuration.

If the units are identical and independent, and if p is the probability of success of any one unit, then

$$R_s = 1 - (1 - p^2)(1 - p) = (p + p^2 - p^3) \tag{5–28}$$

   b. For the system shown in Figure 5.8, we get

$$R_s = P\{(x_1 + x_2)x_3\} \tag{5–29}$$

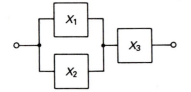
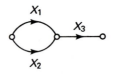

Figure 5.8   System with three components.

If the units are identical and independent, then

$$R_s = [1 - (1 - p)^2]p = (2p^2 - p^3) \tag{5–30}$$

   c. Now consider the system shown in Figure 5.9, for which

$$R_s = P(x_1 + x_2 + x_3x_4) \tag{5–31}$$

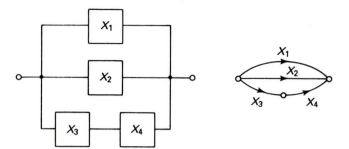

Figure 5.9   System with four units.

If the units are identical and independent, then

$$R_s = 1 - (1 - p)^2(1 - p^2) = 1 - (1 - 2p + p^2)(1 - p^2)$$

$$= 1 - (1 - 2p + p^2 - p^2 + 2p^3 - p^4) \qquad (5\text{–}32)$$

$$= (2p - 2p^3 + p^4)$$

d. The reliability of the system shown in Figure 5.10 is

$$R_s = P[x_1(x_2 + x_3x_4)] \qquad (5\text{–}33)$$

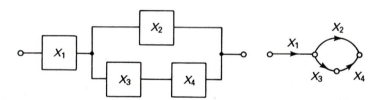

Figure 5.10   An alternate system with four units.

If the units are identical and independent, then

$$R_s = p[1 - (1 - p)(1 - p^2)] \qquad (5\text{–}34)$$

$$= (p^2 + p^3 - p^4)$$

5.10.2 Event-Space Method

In this method, all of the logically possible occurrences are listed systematically, and the list is separated into favorable and unfavorable events. Then the system reliability is obtained by summing the probabilities of occurrence of all the favorable (successful) events. If the list is prepared properly, then no event will be left out, and all the listed events will be mutually exclusive.

For a system consisting of n components, with each component possibly being good or bad, there will be a total of 2^n events in all, and the procedure becomes unwieldy if n is greater than 5 or 6.

If the number of unfavorable events is less than the number of favorable events, then the system reliability can be computed by subtracting from 1 the sum of the probabilities of occurrence of all the unsuccessful events.

■■ **Example 5–6**

For the system illustrated in Figure 5.11, let us use a and $\bar{a}$ to indicate that the component a is good and bad, respectively. The event space consists of 2^4 or 16 entries or events. They are listed systematically as follows:

Group 0	No failures		$abcd$		
Group 1	One failure	$\bar{a}bcd$	$a\bar{b}cd$	$ab\bar{c}d$	$abc\bar{d}$
Group 2	Two failures	$\overline{(\bar{a}\bar{b}cd)}$	$\bar{a}b\bar{c}d$	$\bar{a}bc\bar{d}$	
		$a\bar{b}\bar{c}d$	$a\bar{b}c\bar{d}$		
		$ab\bar{c}\bar{d}$			
Group 3	Three failures	$\overline{(\bar{a}\bar{b}\bar{c}d)}$	$\overline{(\bar{a}\bar{b}c\bar{d})}$		
		$\overline{(\bar{a}b\bar{c}\bar{d})}$			
		$a\bar{b}\bar{c}\bar{d}$			
Group 4	Four failures	$\overline{(\bar{a}\bar{b}\bar{c}\bar{d})}$			

All the unsuccessful events are circled. Since there are only five such events, we will use

$$R_s = 1 - \sum(\text{probabilities of occurrence of all unsuccessful events})$$

$$= 1 - [p^2(1 - p)^2 + 3p(1 - p)^3 + (1 - p)^4]$$

or

$$R_s = (p + 2p^2 - 3p^3 + p^4) \tag{5–35}$$

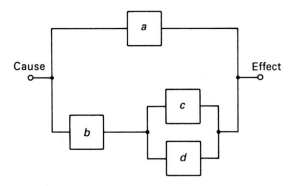

Figure 5.11 System with four components.

Also,

$$R_s = \sum(\text{probabilities of occurrence of all successful events})$$

$$= p^4 + 4p^3(1 - p) + 5p^2(1 - p)^2 + p(1 - p)^3$$

$$= p + 2p^2 - 3p^3 + p^4, \text{ as before} \qquad \blacksquare\blacksquare$$

5.10.3 Path-Tracing Method

In this method, we suppose all the blocks in the reliability block diagram to be missing. We then replace the units (blocks) singly, in pairs, in triplets, in quadruplets, etc., and note all the successful paths from cause (in) to effect (out). Since each successful path constitutes a favorable event, we conclude that

$$R_s = P(\text{union of all the favorable events})$$

If there are n favorable paths, then, in general, we will have $(2^n - 1)$ terms in the expansion of $P(\text{union})$. Note that not all of the favorable events will, in general, be mutually exclusive. After arriving at the probabilities of all the terms, we simplify the expression to get the final expression for R_s.

This procedure involves more algebra and becomes unwieldy if $n > 5$.

$\blacksquare\blacksquare$ **Example 5–7**

Referring again to Figure 5.11, we note that there are three successful paths:

$$\mathscr{P}_1 = a; \quad \mathscr{P}_2 = bc; \quad \mathscr{P}_3 = bd$$

So

$$R_s = P(\mathscr{P}_1 \cup \mathscr{P}_2 \cup \mathscr{P}_3) = P(\mathscr{P}_1) + P(\mathscr{P}_2) + P(\mathscr{P}_3)$$

$$- [P(\mathscr{P}_1\mathscr{P}_2) + P(\mathscr{P}_2\mathscr{P}_3) + P(\mathscr{P}_3\mathscr{P}_1)] + P(\mathscr{P}_1\mathscr{P}_2\mathscr{P}_3)$$

Assuming that all of the components are identical with a reliability of p, we get

$$R_s = p + p^2 + p^2 - [p{\cdot}p^2 + p^2{\cdot}p + p^2{\cdot}p] + [p{\cdot}p^2{\cdot}p]$$

$$= (p + 2p^2 - 3p^3 + p^4), \text{ as before} \qquad \blacksquare\blacksquare$$

5.10.4 Decomposition Method

In this method, also known as the conditional probability approach, we find R_s through the successive application of the conditional probability theorem. First, select a *keystone component* that appears to bind together the reliability structure of the problem, and let that component be A. Then the system reliability is

$$R_s = [P(\text{system is successful} \mid A \text{ is good})]P(A) \tag{5–36}$$
$$+ [P(\text{system is successful} \mid A \text{ is bad})]P(\overline{A})$$

Similarly, the system unreliability can be expressed as

$$Q_s = (1 - R_s) = [P(\text{system fails} \mid A \text{ is good})]P(A) \tag{5–37}$$
$$+ [P(\text{system fails} \mid A \text{ is bad})]P(\overline{A})$$

Thus, we decompose one difficult problem into two easier ones. In the case of complex structures, we may have to repeat the decomposition process on the substructures formed after the first decomposition.

If the keystone component is selected properly, the decomposition will be dramatic, and the resulting two simpler problems can be solved easily. However, the method itself is valid no matter which component is chosen as the keystone component.

■■ Example 5–8

All of the units in the system shown in Figure 5.12 are identical, are independent, and have a reliability of p. We derive an expression for the reliability of the system by selecting unit 3 as the keystone component. Then

$$R_s = [P(\text{system good} \mid 3 \text{ is good})]p + [P(\text{system good} \mid 3 \text{ is bad})](1 - p)$$

$$\equiv P_1 p + P_2 (1 - p)$$

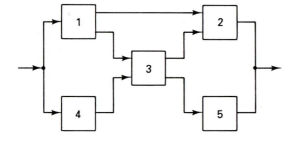

Figure 5.12 A five-component system.

To calculate P_2, we refer to Figure 5.13(a) and obtain $P_2 = p^2$.

To calculate P_1, we refer to Figure 5.13(b) and select 1 as the keystone component. Then

$$P_1 = [P(\text{system good} \mid 1 \text{ is good})]p + [P(\text{system good} \mid 1 \text{ is bad})](1 - p)$$

$$\equiv P_3 p + P_4 (1 - p)$$

To calculate P_3, we refer to Figure 5.13(c) and see that it makes no difference whether component 4 is good or bad. Therefore,

$$P_3 = 1 - (1 - p)^2 = 2p - p^2$$

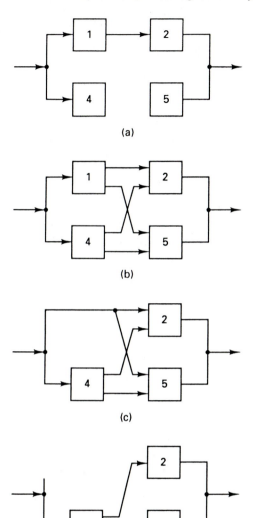

(a)

(b)

(c)

(d)

Figure 5.13 Different special configurations of the system shown in Figure 5.12. (a) For calculation of P_2. (b) For calculation of P_1. (c) For calculation of P_3. (d) For calculation of P_4.

To calculate P_4, we note that the system reduces to Figure 5.13(d), and

$$P_4 = p[1 - (1 - p)^2] = (2p^2 - p^3)$$

P_1 can now be calculated. We have

$$P_1 = p(2p - p^2) + (1 - p)(2p^2 - p^3)$$

$$= 4p^2 - 4p^3 + p^4$$

Finally,

$$R_s = (4p^2 - 4p^3 + p^4)p + p^2(1 - p)$$
$$= p^5 - 4p^4 + 3p^3 + p^2$$

If each component has a time-dependent reliability given by

$$p = e^{-\lambda t}$$

then the time-dependent system reliability can be expressed as

$$R_s(t) = e^{-5\lambda t} - 4e^{-4\lambda t} + 3e^{-3\lambda t} + e^{-2\lambda t}$$ ■■

5.10.5 Minimal Cut Set Method

The minimal cut set method is a powerful technique that forms the basis for many of the network evaluation methods of computing system reliability. It may easily be programmed on a digital computer.

Definition.   A *minimal cut set* is a set of system components such that if all the components fail, system failure results, but if any *one* component has not failed, no system failure results. Simply put, the definition says that *all* the components of a minimal cut set must fail for the system to fail. Let the cut sets of the system be denoted $C_1, C_2, \ldots, C_n$, and let $P(C_i)$ be the probability of all the components of C_i failing. Then the system unreliability is

$$Q_s = P(C_1 + C_2 + \ldots + C_n) \tag{5-38}$$

where "$+$" denotes the logical sum (or union) of the operands. The system reliability is

$$R_s = 1 - Q_s \tag{5-39}$$

The expansion of the probability of the union of n nondisjoint events contains $(2^n - 1)$ terms. However, an upper bound for Q_s can be found by using the approximation

$$P(C_1 + C_2 + \ldots + C_n) \cong P(C_1) + P(C_2) + \ldots + P(C_n) \tag{5-40}$$

Thus, a lower bound for the system reliability is

$$R_s \geq 1 - [P(C_1) + P(C_2) + \ldots + P(C_n)] \tag{5-41}$$

This lower bound is a good approximation in the high-reliability region of components, the region with component reliabilities close to unity.

■■ **Example 5–9**
For the four-component system of Figure 5.11, there are two minimal cut sets: $C_1 = ab$ and $C_2 = acd$.

Therefore,

$$Q_s = P(C_1 + C_2) = P(C_1) + P(C_2) - P(C_1C_2)$$

$$= P(\overline{a}\overline{b}) + P(\overline{a}\overline{c}\overline{d}) - P(\overline{a}\overline{b}\overline{c}\overline{d})$$

or

$$Q_s = q_aq_b + q_aq_cq_d - q_aq_bq_cq_d$$

If all the components are identical, are independent, and have a failure probability of q, then

$$Q_s = q^2 + q^3 - q^4$$

and

$$R_s = (1 - Q_s) = 1 - q^2(1 + q - q^2)$$

For $q = 0.05$,

$$R_s = 0.997381$$

The lower bound approximation gives

$$R_s \geq 1 - [P(C_1) + P(C_2)]$$

or

$$R_s \geq 1 - [(0.05)^2 + (0.05)^3]$$

which gives

$$R_s \geq 0.997375$$

Thus, the difference between the lower bound approximation and the correct value is only 6×10^{-6}, or 0.0006016%. ■■

5.10.6 Minimal Tie Set Method

The minimal tie set method is, for all practical purposes, a complement of the minimal cut set method. Since it does not directly identify the failure modes of the system, as does the minimal cut set method, it is not used as frequently.

   Definition.   A *minimal tie set* is a group of branches which forms a connection between the input and the output when traversed in the direction of the arrow, with no node encountered more than once.

   The components of a tie set are connected in series, and therefore, a tie set fails if any one of its components fails. However, we need only one of the tie sets to be intact for the system to succeed. In effect, this is nothing but a formal statement of the path-tracing method.

   Let $T_1, T_2, \ldots, T_n$ be the minimal tie sets of a system, and let $P(T_i)$ be the probability of all the components in T_i succeeding. Then the system reliability is

$$R_s = P(T_1 + T_2 + \ldots + T_n) \tag{5-42}$$

where "$+$" denotes the logical sum (or union) of the operands. If the tie sets are not disjoint, the expansion for R_s will contain $(2^n - 1)$ terms. However, an upper bound for R_s can be found by assuming the tie sets to be disjoint. In other words,

$$R_s \le P(T_1) + P(T_2) + \ldots + P(T_n) \tag{5-43}$$

This upper bound is a good approximation in the low-reliability region of components, the region with component reliabilities close to zero.

■■ **Example 5–10**
In the four-component system of Figure 5.11, the minimal tie sets are $T_1 = a$, $T_2 = bc$, and $T_3 = bd$.
Therefore,

$$\begin{aligned}
R_s = P(T_1 + T_2 + T_3) &= P(T_1) + P(T_2) + P(T_3) \\
&\quad - [P(T_1T_2) + P(T_2T_3) + P(T_3T_1)] \\
&\quad + P(T_1T_2T_3) \\
&= P(a) + P(bc) + P(bd) \\
&\quad - [P(abc) + P(bcd) + P(abd)] \\
&\quad + P(abcd)
\end{aligned}$$

$$R_s = p_a + p_bp_c + p_bp_d - p_ap_bp_c - p_bp_cp_d - p_ap_bp_d + p_ap_bp_cp_d$$

If all the components are identical, are independent, and have a success probability of p, then

$$R_s = (p + 2p^2 - 3p^3 + p^4)$$

which is equivalent to $[1 - q^2(1 + q - q^2)]$, where $(p + q) = 1$.
     If $p = 0.2$, then $R_s = [1 - (0.8)^2(1 + 0.8 - 0.64)] = 0.2576$, and the upper bound approximation gives

$$R_s \le P(T_1) + P(T_2) + P(T_3)$$

$$R_s \le 0.28$$

The difference between the upper bound approximation and the correct value is only 0.0224, or 8.696%, because of the low value assumed for p.   ■■

5.10.7 Connection Matrix Techniques

The connection matrix technique involves the construction of a connection matrix **M** for a system and then employing node removal or matrix multiplication to obtain the transmission from input to output. The following example illustrates the procedure.

■■   **Example 5–11**
Consider again the four-component system of Figure 5.11, which is redrawn in Figure 5.14 with nodes 1, 2, and 3 as shown. The connection matrix is constructed by entering the element connected between different nodes, as follows:

$$
\begin{array}{cc}
& \text{To Node} \\
\text{From Node} \begin{array}{c} 1 \\ 2 \\ 3 \end{array} &
\begin{array}{ccc}
1 & 2 & 3 \\
\left[\begin{array}{ccc}
1 & b & a \\
0 & 1 & (c + d) \\
0 & 0 & 1
\end{array}\right]
\end{array}
\end{array}
$$

Note that a connection is supposed to exist only in the direction of the arrow.

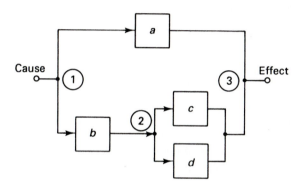

Figure 5.14   Figure 5.11 redrawn with nodes marked.

Node removal method.   In the node removal method, our objective is to eliminate all nodes except the input and output nodes. To remove a node k from a matrix, replace each element N_{ij} $(i, j \neq k)$ by a new element

$$N_{ij}^{0} = N_{ij} + (N_{ik} N_{kj}) \tag{5–44}$$

Thus, removal of node 2 is equivalent to eliminating the second row and second column from the original connection matrix. When we remove node 2, the entries in the resulting two-by-two matrix are

$$N_{11}^{0} = N_{11} + N_{12} N_{21} = 1 + b \cdot 0 = 1$$

$$N_{13}^{0} = N_{13} + N_{12} N_{23} = a + b(c + d) = (a + bc + bd)$$

$$N_{31}^{0} = N_{31} + N_{32} N_{21} = 0 + 0 \cdot 0 = 0$$

$$N_{33}^{0} = N_{33} + N_{32} N_{23} = 1 + 0 \cdot (c + d) = 1$$

The reduced matrix is

$$
\begin{array}{cc}
\begin{array}{c} 1 \\ 3 \end{array} &
\begin{array}{cc}
1 & 3 \\
\left[\begin{array}{cc}
1 & (a + bc + bd) \\
0 & 1
\end{array}\right]
\end{array}
\end{array}
$$

Since we are down to just two nodes, input and output, we stop at this point and focus on element N_{13}^0, which is the transmission from input to output. This transmission represents all possible paths that exist between the input and the output. In other words, these are the minimal tie sets of the system. Thus,

$$\text{System reliability} = R_s = P(a + bc + bd) \tag{5–45}$$

which is the same as that obtained earlier.

Matrix multiplication method. The matrix multiplication method gives the transmissions between all the nodes of a system instead of just those between the input and output nodes. The connection matrix **M** is multiplied by itself a number of times until, after the application of Boolean algebra to simplify the elements of the powers of **M**, the resulting matrix repeats itself. The entries in this repeating matrix give the transmissions between the various nodes of the system, including the input and output nodes. For the example from Figure 5.14,

$$\mathbf{M}^2 = \mathbf{MM} = \begin{bmatrix} 1 & b + b & a + a + b(c + d) \\ 0 & 1 & (c + d) + (c + d) \\ 0 & 0 & 1 \end{bmatrix}$$

$$= \begin{bmatrix} 1 & b & (a + bc + bd) \\ 0 & 1 & (c + d) \\ 0 & 0 & 1 \end{bmatrix}$$

It can easily be seen that $\mathbf{M}^3 = \mathbf{MM}^2$ is the same as the matrix $\mathbf{M}^2$. So we stop here and read off the transmissions between different nodes from the elements of $\mathbf{M}^2$. In particular, the transmission from the input (node 1) to the output (node 3) is $(a + bc + bd)$. The reliability of the system is therefore

$$R_s = P(a + bc + bd)$$

as before. The transmission from node 2 to node 3 is $(c + d)$, which is obvious from Figure 5.14. The advantages of matrix techniques are not obvious, because of the simple example considered. With complex systems involving many nodes and a large number of elements, these techniques enable us to develop a computer program to accomplish the requisite tasks.   ■■

5.10.8 Event Trees

The complete event-space, consisting of the events that are possible in a system, is represented pictorially in an event tree. The use of event trees becomes rather unwieldy if the number of components is greater than 5. For an n-component system, assuming that each component can be good or bad, there are 2^n possibilities (paths), and for $n = 6$, this number is 64! Moreover, if each component can reside in more than two states—say, three, UP, DERATED, and DOWN—then the number of paths become even greater (3^n instead of 2^n).

In the case of continuously operated systems, the components can be considered in any arbitrary order. But if the system involves standby units and sequential logic, the sequence of events must be considered in the chronological order in which they occur. The following examples illustrate the procedures involved.

■■ **Example 5–12 (Continuously Operated System)**
Consider yet again the four-component system of Figure 5.14, and assume that each of the components can be either good or bad with probabilities of R and Q with the proper subscript. Then the event-space will consist of 2^4, or 16, paths, as shown in Figure 5.15. After the event tree is completed, the outcome of each path is classified

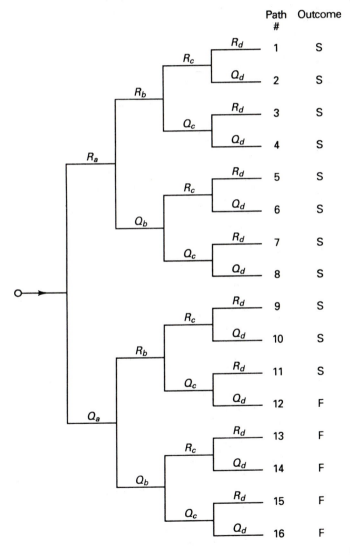

	Path #	Outcome
R_d	1	S
Q_d	2	S
R_d	3	S
Q_d	4	S
R_d	5	S
Q_d	6	S
R_d	7	S
Q_d	8	S
R_d	9	S
Q_d	10	S
R_d	11	S
Q_d	12	F
R_d	13	F
Q_d	14	F
R_d	15	F
Q_d	16	F

Component: *a*   *b*   *c*   *d*

Figure 5.15 Complete event tree for the system shown in Figure 5.14.

as a success or a failure. We note that all the paths are mutually exclusive, and the probability of occurrence of each path is obtained by multiplying the probabilities of occurrence of the events constituting the path. Then

$$\text{System reliability} = R_s = \sum \left(\begin{array}{c} \text{probabilities of occurrence} \\ \text{of all successful paths} \end{array} \right) \qquad (5\text{--}46)$$

For the complete event tree of Figure 5.15, there are 5 unsuccessful paths and 11 successful paths. Therefore, we will compute the system unreliability by summing the probability of occurrence of the 5 unsuccessful paths. This gives

$$Q_s = Q_a R_b Q_c Q_d + Q_a Q_b R_c (R_d + Q_d) + Q_a Q_b Q_c (R_d + Q_d)$$

$$= Q_a R_b Q_c Q_d + Q_a Q_b (R_c + Q_c)$$

$$= Q_a R_b Q_c Q_d + Q_a Q_b$$

If the components are identical and independent with a probability of success of p, then

$$Q_s = (1 - p)^3 p + (1 - p)^2$$

$$= 1 - p - 2p^2 + 3p^3 - p^4$$

and

$$R_s = (1 - Q_s) = (p^4 - 3p^3 + 2p^2 + p)$$

      To reduce the number of paths, we consider each component or event as before, but deduce the outcome before each new component or event is considered. Thus, if component a is successful, then irrespective of which state components b, c, and d are in, the system will succeed. Therefore, the path denoted by R_a (meaning that a is successful) need not be developed further. Similarly, the paths denoted by $Q_a R_b R_c$ and $Q_a Q_b$ need not be developed further. Following this line of reasoning, the event tree can be reduced to just 5 paths, instead of the 16 required by the complete event tree. Figure 5.16 shows the reduced event tree for this system.

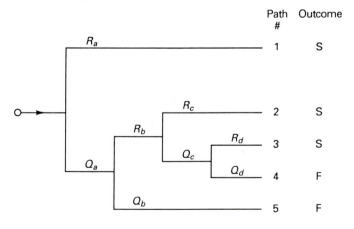

Figure 5.16   Reduced event tree for the system of Figure 5.14.

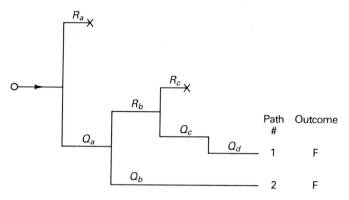

Figure 5.17   Failure event tree for the system shown in Figure 5.14.

A further reduction in the number of paths can be made by considering only those paths leading to success (or failure) and terminating the paths leading to failure (or success). In this case, the tree will be called a success (or failure) event tree. Figure 5.17 shows the failure event tree for the system of Figure 5.14. Note that this tree has only 2 paths.   ■■

■■ **Example 5–13 (System with Sequential Logic and Standby Units)**
Consider the problem of supplying power to a sensitive installation. Because of the nature of the installation, two standby emergency generators are provided (see Figure 5.18), any one of which is capable of handling the complete load. The loss of normal power is detected by the detector, and in the case of failure of normal power, a signal is sent to start the emergency generators, If both of these fail, then the installation experiences a power failure and fails.

In constructing the complete event tree, the logical sequence of events must be considered, starting with the failure of normal power, which is the initiating event. The resulting complete event tree is shown in Figure 5.19.

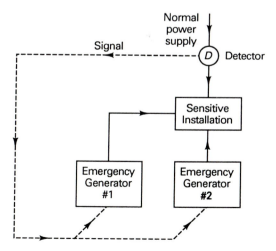

Figure 5.18   System involving sequential logic and standby units.

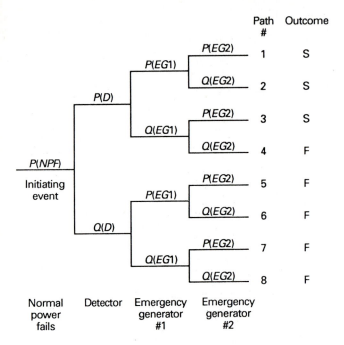

Figure 5.19 Complete event tree for the system of Figure 5.18.

The number of paths can be reduced as before by deducing the outcome before the next component or event is considered. The reduced event tree of Figure 5.20 has only four paths, instead of the eight paths needed in the complete event

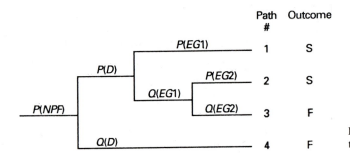

Figure 5.20 Reduced event tree for the system of Figure 5.18.

tree. We deduce the system reliability as

$$P(\text{system success}) = P(NPF)P(D)[P(EG1) + Q(EG1)P(EG2)]$$ ■■

5.10.9 Fault Trees

A fault tree symbolically represents the conditions that may cause a system to fail. It can pinpoint system weaknesses in a visible form. Thus, it acts as a visual tool in communicating and supporting decisions and in performing trade-off studies or determining the adequacy of the system design. Fault trees are used to analyze complex systems.

In constructing fault trees, we use a logic that is the reverse of the one used in constructing event trees. We start with a particular failure or undesired event and work backwards to explore all the combinations of events that may lead to this failure. The basic steps to be followed in developing fault trees are as follows:

a. Identify a particular undesired event or failure condition (called the *top event*) of the system under consideration.

b. Study and understand thoroughly the system and its intended use.

c. Determine the higher order functional events that can cause the undesired event identified in (a). Also, determine the logical relationships of lower order events that can cause the higher order functional events.

d. Construct a fault tree using a set of basic building blocks. This tree shows pictorially the different combinations and sequences of other events that lead to the top event. All of the input fault events must be defined in terms of basic, independent, and identifiable faults.

e. Evaluate the fault tree qualitatively and/or quantitatively as required.

The basic building blocks of fault trees are given in Figure 5.21. A failure event that is clearly identifiable, such as the power failing, a fuse blowing, a switch failing, or a valve failing, is called a *basic fault event*. An event that results from a logical combination of fault events is represented by a rectangle. Logical relationships are represented by logic gates, the chief ones of interest of which are AND, OR, EXCLUSIVE OR, PRIORITY AND, INHIBIT, and DELAY. In addition, special gates can be used to represent any legitimate combination of input events. Also, special symbols are used to represent special events, such as an incomplete event, a trigger event, and a conditional event. Some commonly used symbols are included in Figure 5.21.

Fault tree analysis is aided by classifying failures into three broad groups:

I. A *primary failure* is a failure that occurs while a component is functioning within the operating parameters for which it is designed.

II. A *secondary failure* is a failure that is due to excessive environmental or operational stress placed on a component.

III. *Command failure* is a failure that results from the proper operation of a component, but at the wrong time or place.

Figures 5.22, 5.23, and 5.24 illustrate fault trees involving primary, secondary, and command failures, respectively.

The following example illustrates the quantitative evaluation of a fault tree.

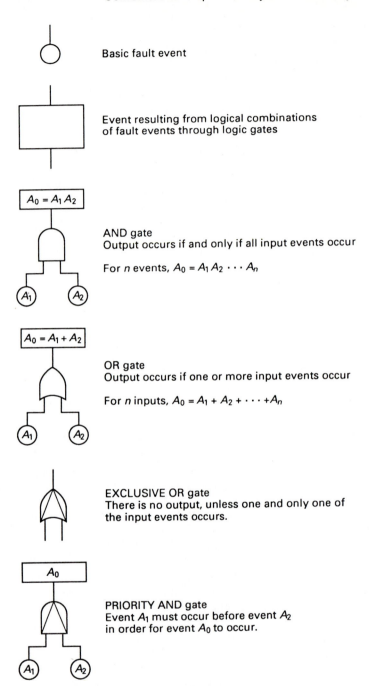

Basic fault event

Event resulting from logical combinations
of fault events through logic gates

$A_0 = A_1 A_2$

AND gate
Output occurs if and only if all input events occur

For n events, $A_0 = A_1 A_2 \cdots A_n$

$A_0 = A_1 + A_2$

OR gate
Output occurs if one or more input events occur

For n inputs, $A_0 = A_1 + A_2 + \cdots + A_n$

EXCLUSIVE OR gate
There is no output, unless one and only one of
the input events occurs.

A_0

PRIORITY AND gate
Event A_1 must occur before event A_2
in order for event A_0 to occur.

Figure 5.21 Basic building blocks of fault trees.

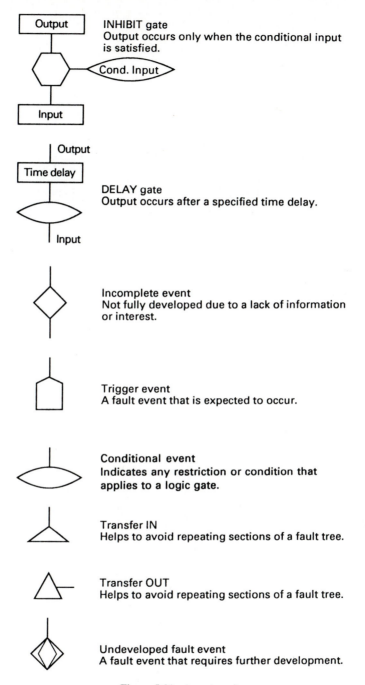

INHIBIT gate
Output occurs only when the conditional input is satisfied.

DELAY gate
Output occurs after a specified time delay.

Incomplete event
Not fully developed due to a lack of information or interest.

Trigger event
A fault event that is expected to occur.

Conditional event
Indicates any restriction or condition that applies to a logic gate.

Transfer IN
Helps to avoid repeating sections of a fault tree.

Transfer OUT
Helps to avoid repeating sections of a fault tree.

Undeveloped fault event
A fault event that requires further development.

Figure 5.21 (*continued*)

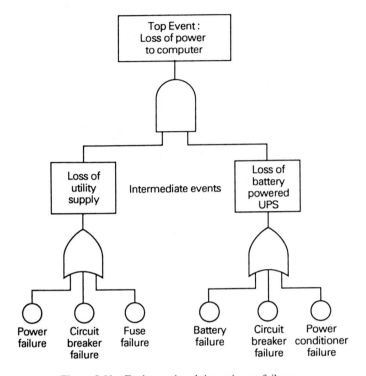

Figure 5.22   Fault tree involving primary failures.

■■   **Example 5–14**

In the fault tree shown in Figure 5.25, suppose all the basic events are known to be independent of each other with a probability of occurrence of $\frac{1}{4}$. Calculate the probability of the top event T.

By inspection,

$$T = T_1 C T_2$$

$$= (A + B)C(T_3 + D)$$

$$= (A + B)C(EFG + D)$$

Therefore,

$$P(T) = [P(A) + P(B) - P(AB)][P(C)][P(EFG) + P(D) - P(DEFG)]$$

$$= \left(\frac{1}{4} + \frac{1}{4} - \frac{1}{16}\right)\left(\frac{1}{4}\right)\left(\frac{1}{64} + \frac{1}{4} - \frac{1}{256}\right)$$

$$= \frac{469}{16,384}$$

If we assume that D is replaced by F, we have a repeated event, and the

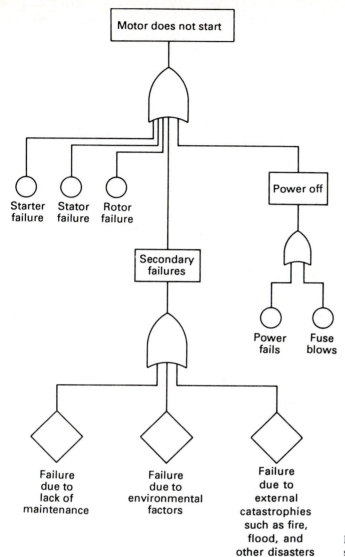

Figure 5.23 Fault tree involving secondary failures.

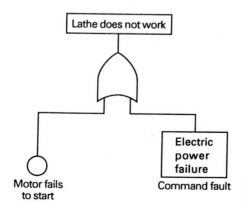

Figure 5.24 Fault tree with a command failure.

177

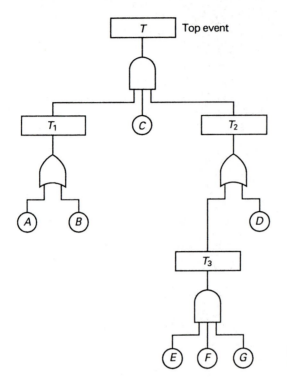

Figure 5.25 Fault tree for Example 5–14.

foregoing procedure becomes invalid. Now we have to remove all dependencies in a fault tree before calculating the probability of the top event.

With D replaced by F,

$$T = (A + B)C(T_3 + F)$$
$$= (A + B)C(EFG + F)$$
$$= (A + B)CF$$

and

$$P(T) = \left(\frac{7}{16}\right)\left(\frac{1}{4}\right)\left(\frac{1}{4}\right) = \frac{7}{256}$$

$$= \frac{448}{16,384}$$

which is different from the preceding analysis.

If the probabilities of occurrence of the basic events are small, then the difference between the results obtained for the two cases will be small. To illustrate this case, if the probabilities of the basic events are all $\frac{1}{8}$ instead of $\frac{1}{4}$, then the probabilities of the top event are (8,304/2,097,152) and (7,680/2,097,152) for the two cases, respectively. These two values are much closer to each other than the previous two values are. ∎∎

5.11 THREE-STATE DEVICES

Certain commonly used devices, such as semiconductor diodes and fluid flow valves, can fail either in open (open-circuit) or closed (short-circuit) mode. Such devices require three-state models, the states being normal operation, open-circuit failure, and short-circuit failure, with probabilities of p_n, q_o, and q_s, respectively. Moreover,

$$p_n + q_o + q_s = 1 \qquad (5\text{--}47)$$

If we have k components, the possible states can be enumerated and their probabilities found by expanding the product $(p_{n1} + q_{o1} + q_{s1})(p_{n2} + q_{o2} + q_{s2})$ $\dots (p_{nk} + q_{ok} + q_{sk})$. These states can then be identified and grouped as good (system success) or bad (system failure). Then the system reliability can be calculated by summing the probabilities of all states identified as good. If the components are all identical, the product can be replaced by $(p_n + q_o + q_s)^k$. Note that we must exercise care in identifying the failure and success states, since they depend on the way the components are put together in the system.

With three-state devices, the application of conditional probability techniques should be modified if the keystone component is a three-state device. Let the keystone component be the kth component. Then the modified equation is

$$R_s = P(\text{system success} \mid k \text{ is normal})P(k \text{ is normal})$$

$$+ \ P(\text{system success} \mid k \text{ is open})P(k \text{ is open}) \qquad (5\text{--}48)$$

$$+ \ P(\text{system success} \mid k \text{ is shorted})P(k \text{ is shorted})$$

5.11.1 Series Structure

Let us consider two identical three-state devices in series. There are 3^2 or 9 combinations of states possible, which can be obtained by expanding $(p_n + q_o + q_s)^2$:

$$(p_n + q_o + q_s)^2 = p_n^2 + q_o^2 + q_s^2 + 2p_nq_o + 2q_oq_s + 2q_sp_n \qquad (5\text{--}49)$$

Thus, the possible states of the system are NN, OO, SS, NO, OS, and SN.

The combinations that lead to system success are NN and SN. Therefore, the system reliability is

$$R_s = p_n^2 + 2p_nq_s = (1 - q_o - q_s)^2 + 2q_s(1 - q_o - q_s)$$

or

$$R_s = (1 - q_o)^2 - q_s^2 \qquad (5\text{--}50)$$

Open-mode failure corresponds to OO, NO, and OS. Hence, the open-mode failure probability is

$$Q_o = q_o^2 + 2p_n q_o + 2q_o q_s$$
$$= q_o^2 + 2(1 - q_o - q_s)q_o + 2q_o q_s$$

or

$$Q_o = 1 - (1 - q_o)^2 \tag{5-51}$$

Similarly, the short-mode failure probability is

$$Q_s = q_s^2 \tag{5-52}$$

Repeating this process for three identical three-state devices in series, we can show that

$$R_s = (1 - q_o)^3 - q_s^3 \tag{5-53}$$

$$Q_o = 1 - (1 - q_o)^3 \tag{5-54}$$

and

$$Q_s = q_s^3 \tag{5-55}$$

We can generalize this set of equations to k identical three-state devices in series. We obtain

$$R_s = (1 - q_o)^k - q_s^k \tag{5-56}$$

$$Q_o = 1 - (1 - q_o)^k \tag{5-57}$$

and

$$Q_s = q_s^k \tag{5-58}$$

If the components are not identical, the powers are replaced by products. For k nonidentical three-state devices in series, we have

$$R_s = \prod_{i=1}^{k} (1 - q_{oi}) - \prod_{i=1}^{k} q_{si} \tag{5-59}$$

$$Q_o = 1 - \prod_{i=1}^{k} (1 - q_{oi}) \tag{5-60}$$

and

$$Q_s = \prod_{i=1}^{k} q_{si} \tag{5-61}$$

5.11.2 Parallel Structure

For two identical three-state devices connected in parallel, the combinations that lead to system success are *NN* and *NO*. Therefore,

$$R_s = p_n^2 + 2p_n q_o$$
$$= (1 - q_o - q_s)^2 + 2q_o(1 - q_o - q_s)$$

or

$$R_s = (1 - q_s)^2 - q_o^2 \qquad (5\text{--}62)$$

Open-mode failure corresponds to only one combination, namely, *OO*. Thus,

$$Q_o = q_o^2 \qquad (5\text{--}63)$$

Short-mode failure corresponds to the combinations *SS*, *SO*, and *NS*. Thus,

$$Q_s = q_s^2 + 2q_o q_s + 2q_s p_n$$

or

$$Q_s = 1 - (1 - q_s)^2 \qquad (5\text{--}64)$$

Extending this line of reasoning to *k* identical three-state devices in parallel, we get

$$R_s = (1 - q_s)^k - q_o^k \qquad (5\text{--}65)$$
$$Q_o = q_o^k \qquad (5\text{--}66)$$

and

$$Q_s = 1 - (1 - q_s)^k \qquad (5\text{--}67)$$

If the components are not identical, the powers are replaced by products as before. The result is

$$R_s = \prod_{i=1}^{k} (1 - q_{si}) - \prod_{i=1}^{k} q_{oi} \qquad (5\text{--}68)$$

$$Q_o = \prod_{i=1}^{k} q_{oi} \qquad (5\text{--}69)$$

and

$$Q_s = 1 - \prod_{i=1}^{k} (1 - q_{si}) \qquad (5\text{--}70)$$

5.11.3 Series-Parallel Network

Let there be k identical independent units in series, each consisting of m non-identical independent components in parallel. Then, for each unit,

$$Q_o = \prod_{i=1}^{m} q_{oi} \tag{5-71}$$

$$Q_s = 1 - \prod_{i=1}^{m} (1 - q_{si}) \tag{5-72}$$

and for the series-parallel network, the system reliability can be expressed as

$$
\begin{aligned}
R_s &= (1 - Q_o)^k - Q_s^k \\
&= \left(1 - \prod_{i=1}^{m} q_{oi}\right)^k - \left(1 - \prod_{i=1}^{m} (1 - q_{si})\right)^k
\end{aligned} \tag{5-73}
$$

5.11.4 Parallel-Series Network

Let there be m identical independent branches in parallel, each consisting of k nonidentical independent components in series. Then, for each series branch,

$$Q_s = \prod_{i=1}^{k} q_{si} \tag{5-74}$$

$$Q_o = 1 - \prod_{i=1}^{k} (1 - q_{oi}) \tag{5-75}$$

and for the parallel-series network, the system reliability can be expressed as

$$
\begin{aligned}
R_s &= (1 - Q_s)^m - Q_o^m \\
&= \left(1 - \prod_{i=1}^{k} q_{si}\right)^m - \left(1 - \prod_{i=1}^{k} (1 - q_{oi})\right)^m
\end{aligned} \tag{5-76}
$$

■■   **Example 5–15**

Consider eight identical components with $q_o = 0.1$ and $q_s = 0.05$. Let us compute the system reliability for various series-parallel and parallel-series combinations using Equations (5–73) and (5–76). The results are as follows:

- Series-parallel with $k = 2$, $m = 4$:

$$R_s = [1 - (0.1)^4]^2 - [1 - (1 - 0.05)^4]^2 = 0.96539$$

- Series-parallel with $k = 4$, $m = 2$:

$$R_s = [1 - (0.1)^2]^4 - [1 - (1 - 0.05)^2]^4 = 0.96051$$

- Parallel-series with $k = 2$, $m = 4$:

$$R_s = [1 - (0.05)^2]^4 - [1 - (1 - 0.1)^2]^4 = 0.98873$$

- Parallel-series with $k = 4$, $m = 2$:

$$R_s = [1 - (0.05)^4]^2 - [1 - (1 - 0.1)^4]^2 = 0.88172$$

The parallel-series combination consisting of four branches in parallel with each branch consisting of two components in series results in the highest reliability.

Next, consider the same components, but with one change: Let $q_o = 0.05$ and $q_s = 0.1$. Then calculations similar to the preceding ones give the following values for the system reliability:

- Series-parallel with $k = 2$, $m = 4$:

$$R_s = 0.88172$$

- Series-parallel with $k = 4$, $m = 2$:

$$R_s = 0.98873$$

- Parallel-series with $k = 2$, $m = 4$:

$$R_s = 0.96051$$

- Parallel-series with $k = 4$, $m = 2$:

$$R_s = 0.96539$$

With the changes in the values of q_o and q_s, the series-parallel combination with four units in series, each of which consists of two components in parallel, results in the highest value for the system reliability.

In designing high-reliability systems involving three-state devices, the relative values of q_o and q_s must be considered carefully in finalizing the network configuration. ■■

■■ **Example 5–16**

The three-diode system shown in Figure 5.26 can be evaluated from the reliability point of view in several ways. We will discuss three approaches here. All three diodes will be assumed to be identical with probabilities of normal operation, open-circuit failure, and short-circuit failure of p_n, q_o, and q_s, respectively.

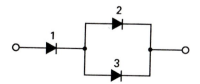

Figure 5.26 Three-diode system.

Approach No. 1:

Diodes 2 and 3 in parallel can be replaced by an equivalent diode with the following success and failure probabilities:

$$P(\text{normal operation}) = p_n^2 + 2q_o p_n$$

$$P(\text{open-circuit failure}) = q_o^2$$

$$P(\text{short-circuit failure}) = 2q_s - q_s^2$$

This equivalent diode is in series with diode 1, and the system will succeed for the following combinations of states:

| diode 1: | normal | normal | shorted |
| equivalent diode: | normal | shorted | normal |

All the other combinations will lead to system failure. The succesful combinations are mutually exclusive, and therefore, the system reliability can be found by summing the probabilities of occurrence of the three successful combinations of states. Thus,

$$R_s = p_n(p_n^2 + 2q_o p_n) + p_n(2q_s - q_s^2) + q_s(p_n^2 + 2q_o p_n)$$

Replacing p_n with $(1 - q_o - q_s)$ and simplifying we obtain

$$R_s = 1 - q_o - q_o^2 + q_o^3 - 2q_s^2 + q_s^3$$

Approach No. 2:

First, let us expand $(p_n + q_o + q_s)^2 (p_{ns} + q_{os} + q_{ss})$, in which we have added a second subscript s to denote diode 1, which is in series with the parallel combination of diodes 2 and 3. The expansion results in

$$(p_n^2 p_{ns}) + q_o^2 p_{ns} + (q_s^2 p_{ns}) + (2p_n q_o p_{ns}) + (2q_o q_s p_{ns}) + (2p_n q_s p_{ns})$$

$$+ p_n^2 q_{os} + q_o^2 q_{os} + q_s^2 q_{os} + 2p_n q_o q_{os} + 2q_o q_s q_{os} + 2p_n q_s q_{os} + (p_n^2 q_{ss})$$

$$+ q_o^2 q_{ss} + q_s^2 q_{ss} + (2p_n q_o q_{ss}) + 2q_o q_s q_{ss} + 2p_n q_s q_{ss}$$

The terms in parentheses correspond to successful combinations of system states. After determining all the success states, we can drop the additional subscript s and express the system reliability as

$$R_s = p_n^3 + q_s^2 p_n + 2p_n^2 q_o + 2q_o q_s p_n + 2p_n^2 q_s + p_n^2 q_s + 2p_n q_o q_s$$

Next, we replace p_n with $(1 - q_o - q_s)$, and after simplifying, we obtain

$$R_s = 1 - q_o - q_o^2 + q_o^3 - 2q_s^2 + q_s^3$$

as before.

Approach No. 3:

In this approach, we select diode 1 as the keystone component and employ the conditional probability method as follows:

$$R_s = P(\text{system success} \mid 1 \text{ is normal}) \, P(1 \text{ is normal})$$

$$+ P(\text{system success} \mid 1 \text{ is open}) \, P(1 \text{ is open})$$

$$+ P(\text{system success} \mid 1 \text{ is shorted}) \, P(1 \text{ is shorted})$$

Using the appropriate probabilities, we get

$$R_s = (1 - q_o^2)p_n + 0 + (p_n^2 + 2q_o p_n)q_s$$

Replacing p_n with $(1 - q_o - q_s)$ and simplifying, we once again obtain

$$R_s = 1 - q_o - q_o^2 + q_o^3 - 2q_s^2 + q_s^3$$

Clearly, this approach involves the least amount of work and algebraic manipulations. ■■

5.11.5 Star-Delta Structure

In Section 5.5, we discussed the star-delta equivalents for two-state (binary) components that can exist either in the good (success) or in the bad (failure) state. Although we did not specifically mention at that time, with binary components, we assume that failure means an open-mode failure. Thus, the equivalents derived in Section 5.5 are applicable to the case of open-mode failure if all of the components in the star and in the delta structure are assumed to be three-state devices.

However, with three-state devices, we have the possibility of a short-mode failure to consider. Thus, we have to have two sets of equivalents: one for the open-mode failure and the other for the short-mode failure. The open-mode failure equivalents are the same as the ones discussed in Section 5.5. Derivation of the equivalents for the short-mode failure is left as an exercise at the end of the chapter.

5.12 COMBINATORIAL ASPECTS OF TIME-DEPENDENT RELIABILITIES

The technique of decomposing a complex system into a set of subsystems or components whose reliabilities are known or can be computed or estimated and then combining the subsystem reliabilities to obtain the reliability of the complex system can be used even if the reliabilities are time dependent. The level of decomposition and the complexity (or simplicity) of the models of the individual subsystems or components depend on the nature of the system, the accuracy desired, how much knowledge one has of the behavior of the components, and engineering judgment. Models of individual components and subsystems should be kept simple enough to avoid complicating the calculations of system reliability, but detailed enough to preserve the behavior of the components and subsystems from the reliability point of view.

Time-dependent reliabilities can be obtained from a known or assumed hazard function $\lambda(t)$ using

$$R(t) = \exp\left[-\int_0^t \lambda(\xi) \, d\xi \right] \qquad (5\text{--}77)$$

The nature of $\lambda(t)$ depends on the distribution of the failure times involved. In Chapter 4, we discussed a number of distributions that have found applications in reliability studies. In this section, we will examine the effect of the introduction of time-dependent reliabilities (based on some of the well-known and widely used distributions) on the evaluation of system reliability for some of the commonly occurring system configurations.

5.12.1 Series Configuration

For n components in series, as shown in Figure 5.1, let $\lambda_i(t)$ be the hazard rate for the ith component. Then

$$R_i(t) = \exp\left[-\int_0^t \lambda_i(\xi) \, d\xi \right] \tag{5-78}$$

and the system reliability is

$$R_s(t) = \prod_{i=1}^{n} \exp\left[-\int_0^t \lambda_i(\xi) \, d\xi \right] \tag{5-79}$$

irrespective of the failure distribution or distributions involved, as long as the component failures are independent. Each of the n components can be represented by different distributions if necessary, and Equation (5–79) will still be valid. However, simplifications are possible for the case of some commonly used distributions. For example, if all the components have exponential failure distributions, we have

$$R_i(t) = \exp[-\lambda_i t] \tag{5-80}$$

where λ_i is the constant failure rate for the ith component. The series system reliability can now be expressed as

$$R_s(t) = \prod_{i=1}^{n} \exp[-\lambda_i t] \tag{5-81}$$

or

$$R_s(t) = \exp\left[\left(-\sum_{i=1}^{n} \lambda_i \right) t \right] \tag{5-82}$$

From Equation (5–82), it is clear that the series configuration of n constant-hazard components can be replaced by an equivalent component with a constant-hazard rate

$$\lambda_e = \sum_{i=1}^{n} \lambda_i \tag{5-83}$$

The effective failure rate of n constant-hazard components in series is equal to the sum of the constant failure rates of the individual components. However, such a simplification is not always possible with other distributions.

If each component has a linearly increasing hazard, let

$$\lambda_i(t) = K_i t \tag{5-84}$$

Then

$$R_i(t) = \exp\left[-\frac{(K_i t^2)}{2}\right] \tag{5-85}$$

and

$$R_s(t) = \prod_{i=1}^{n} \exp\left[-\frac{(K_i t^2)}{2}\right] \tag{5-86}$$

or

$$R_s(t) = \exp\left[\left(-\sum_{i=1}^{n} K_i\right)\frac{t^2}{2}\right] \tag{5-87}$$

In this case also, we can replace the n components in series by an equivalent component with a linearly increasing hazard

$$\lambda_e(t) = \left(\sum_{i=1}^{n} K_i\right) t \tag{5-88}$$

The effective mean time to failure for n identical components in series can easily be found. We have, for constant-hazard components,

$$(\text{MTTF})_e = \frac{1}{\lambda_e}$$

$$= \frac{1}{n\lambda} \tag{5-89}$$

$$= \left(\frac{1}{n}\right) [\text{MTTF of one component}]$$

and for components with linearly increasing hazards,

$$(\text{MTTF})_e = \sqrt{\frac{\pi}{2K_e}} \tag{5-90}$$

where

$$K_e = \sum_{i=1}^{n} K_i = nK \tag{5-91}$$

Therefore,

$$(\text{MTTF})_e = \left[\frac{1}{\sqrt{n}}\right] [\text{MTTF of one component}] \qquad (5\text{–}92)$$

Similar relationships can be derived for other simple failure distributions.

■■ **Example 5–17**

Suppose a complex system has 35 components, all of which are needed for system success. The types of components and their failure rates are as follows:

No. of identical components	Failure rate, λ
10	10^{-6} hr^{-1}
18	10^{-1} %/K
7	10 PPM/K

The overall failure rate λ_e for the system is obtained by summing all the individual failure rates:

$$\lambda_e = 10(10^{-6}) + 18(10^{-5} \times 10^{-1}) + 7(10^{-9} \times 10) \text{ hr}^{-1}$$

$$= 2.807 \times 10^{-5} \text{ hr}^{-1}$$

The system reliability for different mission times can now be calculated easily:

$$R(1 \text{ year}) = R(8{,}760) = \exp[-8{,}760 \times 2.807 \times 10^{-5}]$$

$$= 0.782$$

$$R(100 \text{ hrs}) = \exp[-100 \times 2.807 \times 10^{-5}]$$

$$= 0.997197$$

The system MTTF is the reciprocal of the equivalent failure rate, which is equal to λ_e^{-1}, or 35,625 hours.

If we want to maintain a minimum reliability of 0.9, the mission time must be limited to

$$\frac{-(\ln 0.9)}{(2.807 \times 10^{-5})}, \text{ or } 3{,}753.49 \text{ hours}$$ ■■

5.12.2 Parallel Configuration

For a truly parallel configuration of n components, as illustrated in Figure 5.2, the system reliability is given by

$$R_p(t) = 1 - \left[\prod_{i=1}^{n}\left\{1 - \exp\left(-\int_0^t \lambda_i(\xi)\, d\xi\right)\right\}\right] \qquad (5\text{–}93)$$

where $\lambda_i(t)$ is the time-dependent hazard rate for the ith component, with the tacit assumption that component failures are independent.

Assuming all the components have constant hazards, let λ_i be the constant-hazard rate for the ith component. Then

$$R_p(t) = 1 - \prod_{i=1}^{n} [1 - \exp(-\lambda_i t)] \tag{5-94}$$

In this case, we cannot replace the components in parallel by an equivalent component with a single equivalent failure rate. In other words, the failure times of parallel systems composed of constant-hazard components are not exponentially distributed, even though the failure times of each component are exponentially distributed.

■■ **Example 5–18**

For two dissimilar independent constant-hazard components in parallel,

$$R_p = 1 - [1 - \exp(-\lambda_1 t)][1 - \exp(-\lambda_2 t)]$$

$$= \exp(-\lambda_1 t) + \exp(-\lambda_2 t) - \exp[-(\lambda_1 + \lambda_2)t]$$

If $\lambda_1 = \lambda_2 = \lambda$, then

$$R_p(t) = 2\exp(-\lambda t) - \exp(-2\lambda t) \qquad ■■$$

If each component has a linearly increasing hazard as given by Equation (5–84), then the parallel system reliability is given by

$$R_p(t) = 1 - \prod_{i=1}^{n} \left[1 - \exp\left(-\frac{K_i t^2}{2} \right) \right] \tag{5-95}$$

Once again, we cannot come up with an equivalent component with an equivalent linearly increasing hazard.

The mean time to failure can always be found using Equation (3–28), repeated here as

$$\text{MTTF} = \int_0^{\infty} R(t)\, dt \tag{5-96}$$

For the case of n dissimilar independent constant-hazard components in parallel, we obtain, using Equation (5–94),

$$\text{MTTF} = \int_0^{\infty} \left[1 - \prod_{i=1}^{n} (1 - e^{-\lambda_i t}) \right] dt \tag{5-97}$$

or

$$\text{MTTF} = \sum_{i=1}^{n} \left(\frac{1}{\lambda_i} \right) - \sum_{i=1}^{n} \sum_{\substack{j=1 \\ i \neq j}}^{n} \left(\frac{1}{\lambda_i + \lambda_j} \right)$$

$$+ \sum_{i=1}^{n} \sum_{\substack{j=1 \\ i \neq j \neq k}}^{n} \sum_{k=1}^{n} \left(\frac{1}{\lambda_i + \lambda_j + \lambda_k} \right) - \cdots \qquad (5\text{--}98)$$

$$+ (-1)^{n+1} \frac{1}{\displaystyle\sum_{i=1}^{n} (\lambda_i)}$$

If the units are identical, $\lambda_1 = \lambda_2 = \cdots = \lambda_n \equiv \lambda$, and Equation (5–98) reduces to

$$\text{MTTF} = \frac{1}{\lambda} \left[\frac{\binom{n}{1}}{1} - \frac{\binom{n}{2}}{2} + \frac{\binom{n}{3}}{3} - \cdots + (-1)^{n+1} \frac{\binom{n}{n}}{n} \right] \qquad (5\text{--}99)$$

Alternatively,

$$\text{MTTF} = \frac{1}{\lambda} \left[\frac{n}{1} - \frac{n(n-1)}{2 \times 2!} + \frac{n(n-1)(n-2)}{3 \times 3!} - \cdots + (-1)^{n+1} \frac{1}{n} \right] \qquad (5\text{--}100)$$

The following are MTTF values for some commonly used values of n:

$$n = 2: \text{MTTF} = \frac{1}{\lambda} \left[2 - \frac{1}{2} \right] = \frac{1.5}{\lambda} \qquad (5\text{--}101)$$

$$n = 3: \text{MTTF} = \frac{1}{\lambda} \left[3 - \frac{3}{2!} + \frac{2}{3!} \right] = \frac{1.833}{\lambda} \qquad (5\text{--}102)$$

$$n = 4: \text{MTTF} = \frac{1}{\lambda} \left[4 - \frac{6}{2!} + \frac{8}{3!} - \frac{6}{4!} \right] = \frac{2.083}{\lambda} \qquad (5\text{--}103)$$

The next two examples illustrate system reduction and the calculation of the MTTF using series and parallel concepts.

■■ **Example 5–19**

The series-parallel system shown in Figure 5.27(a) consists of four constant-hazard components with hazard rates as indicated. It can be reduced in two steps. First, the components in series are replaced by equivalent components. Then, the two equivalent components in parallel are replaced by one overall equivalent.

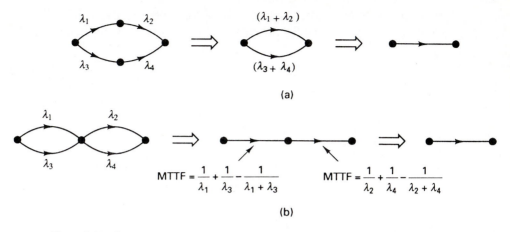

Figure 5.27   Some simple systems. (a) Series-parallel system. (b) Parallel-series system.

For the complete system, by inspection,

$$\text{MTTF} = \frac{1}{\lambda_1 + \lambda_2} + \frac{1}{\lambda_3 + \lambda_4} - \frac{1}{\lambda_1 + \lambda_2 + \lambda_3 + \lambda_4}$$ ■■

■■   **Example 5–20**

The MTTF for the parallel-series system illustrated in Figure 5.27(b) can be found by replacing the parallel elements by their equivalents and then considering the two equivalent elements in series. The equivalent elements, however, do not exhibit a constant hazard. Hence, we have to write down the complete expression for the system reliability and integrate it from zero to infinity to find the system MTTF. The system reliability is

$$R(t) = [e^{-\lambda_1 t} + e^{-\lambda_3 t} - e^{-(\lambda_1 + \lambda_3)t}][e^{-\lambda_2 t} + e^{-\lambda_4 t} - e^{-(\lambda_2 + \lambda_4)t}]$$

Integrating this expression from zero to infinity, we obtain

$$\text{MTTF} = \frac{1}{\lambda_1 + \lambda_2} + \frac{1}{\lambda_1 + \lambda_4} + \frac{1}{\lambda_2 + \lambda_3} + \frac{1}{\lambda_3 + \lambda_4} - \frac{1}{\lambda_1 + \lambda_2 + \lambda_4}$$

$$- \frac{1}{\lambda_2 + \lambda_3 + \lambda_4} - \frac{1}{\lambda_1 + \lambda_2 + \lambda_3} - \frac{1}{\lambda_1 + \lambda_3 + \lambda_4} + \frac{1}{\lambda_1 + \lambda_2 + \lambda_3 + \lambda_4}$$

which is rather difficult to write down by inspection.   ■■

5.12.3 An r-out-of-n Structure

Systems having an r-out-of-n structure are known as *partially redundant* (or *majority-vote*) systems. Equation (5–8) for system reliability is valid even when

p is time dependent. For example, if all the components are identical, are independent, and have a constant hazard of λ, then the system reliability R_s is

$$R_s(t) = \sum_{k=r}^{n} {}_nC_k e^{-k\lambda t} (1 - e^{-\lambda t})^{n-k} \qquad (5\text{--}104)$$

If the components are not identical, but have constant hazards of λ_1, λ_2, ... , λ_n, we must first establish which combinations lead to system success. The probabilities of those states that correspond to system success can be deduced from the expansion of

$$\prod_{i=1}^{n} [R_i(t) + Q_i(t)]$$

where

$$R_i(t) = e^{-\lambda_i t} \qquad (5\text{--}105)$$

and

$$Q_i(t) = 1 - e^{-\lambda_i t} \qquad (5\text{--}106)$$

for $i = 1, 2, ... , n$.

The sum of the probabilities of all the states that have been identified as leading to system success will then be equal to the system reliability.

■■ **Example 5–21**

Let us consider a system consisting of three components, a, b, and c, with constant failure rates of 0.05, 0.1, and 0.15 failure per year, respectively. Suppose we know that, in order for the system to succeed, a must be good and one of the other two components must also be good. We will develop an expression for the time-dependent reliability of the system by following two different approaches.

Approach No. 1: First expand

$$(R_a + Q_a)(R_b + Q_b)(R_c + Q_c)$$

and identify all the terms that correspond to system success. The product is

$$= R_a R_b R_c + R_a R_b Q_c + R_a Q_b R_c + R_a Q_b Q_c$$

$$+ Q_a R_b R_c + Q_a R_b Q_c + Q_a Q_b R_c + Q_a Q_b Q_c$$

The system success terms are $R_a R_b R_c$, $R_a R_b Q_c$, and $R_a Q_b R_c$.

Moreover,

$$R_a = e^{-0.05t}$$

$$R_b = e^{-0.1t}$$

$$R_c = e^{-0.15t}$$

$$Q_a = 1 - e^{-0.05t}$$

$$Q_b = 1 - e^{-0.1t}$$

$$Q_c = 1 - e^{-0.15t}$$

Therefore,

$$\text{System reliability} = R_a R_b R_c + R_a R_b Q_c + R_a Q_b R_c$$

$$= R_a R_b (R_c + Q_c) + R_a Q_b R_c$$

$$= R_a R_b + R_a Q_b R_c$$

$$= e^{-0.15t} + e^{-0.2t} [1 - e^{-0.1t}]$$

$$= e^{-0.15t} + e^{-0.2t} - e^{-0.3t}$$

Approach No. 2: Logically speaking, we have component a in series with components b and c in a truly parallel (redundant) configuration. Using Equation (5–94), we find that the reliability of the parallel subsystem is $\{1 - (1 - e^{-0.1t})(1 - e^{-0.15t})\}$. Therefore,

$$\text{System reliability} = e^{-0.05t}[1 - (1 - e^{-0.1t})(1 - e^{-0.15t})]$$

$$= e^{-0.15t} + e^{-0.2t} - e^{-0.3t}, \text{ as before}$$

The numerical value of the system reliability can be easily calculated once the mission time is known. For a mission time of one month, $t = \frac{1}{12}$ and

$$R(\tfrac{1}{12}) = 0.9957394$$

If the mission time is one year, then $t = 1$ and

$$R(1) = 0.93862 \qquad \blacksquare\blacksquare$$

■■ **Example 5–22**

Subsystem A consists of two constant-hazard components in series, with failure rates of 0.001 and 0.002 per hour. Subsystem B consists of two components in series also, but they have linearly increasing hazards as given by $0.00015t$ and $0.00025t$ per hour. One of these two subsystems must be good for the mission to succeed. We want to calculate the system reliability for a mission time of 2 days.

For A, we have Equivalent failure rate $= 0.001 + 0.002 = 0.003 \text{ hr}^{-1}$

and

$$R(48 \text{ hrs}) = \exp(-48 \times 0.003)$$

$$= 0.865888$$

For B, Equivalent hazard function $= (0.00015 + 0.00025)t \text{ hr}^{-1}$

$$= 0.0004t \text{ hr}^{-1}$$

and

$$R(48 \text{ hrs}) = \exp\left[-\frac{(0.0004)48^2}{2} \right]$$

$$= 0.630779$$

Since the two subsystems are in a parallel configuration, we get

$$\text{System reliability} = R_s(48) = 1 - (1 - 0.865888)(1 - 0.630779)$$

$$= 0.950483 \qquad \blacksquare\blacksquare$$

5.13 STANDBY SYSTEMS MODELING BASED ON PROBABILITY DISTRIBUTIONS

In Section 5.9, we introduced the basics of standby redundancy and its influence on improving system reliability. In the system shown in Figure 5.5, the main operating unit A is backed by a standby unit B, and if A fails, the changeover

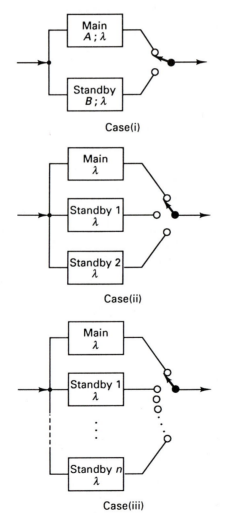

Figure 5.28   Standby system cases.

switch brings *B* into service. This arrangement is only one of several options available to us. For example, we can have more than one standby unit with appropriate switching configurations, the units may or may not be identical, and we may even want to include failures in the standby mode. The procedure for considering all of these factors is discussed in this section, based on the assumption that all the components have constant hazards, meaning that their failure times are exponentially distributed. The following cases are examined:

Case (*i*)   Perfect switching and two identical components—one main unit and one standby unit.

Case (*ii*)   Perfect switching and three identical components—one main unit and two standby units.

Case (*iii*)   Perfect switching, one main unit, and *n* standby units.

Case (*iv*)   Imperfect switching, one main unit, and an identical standby unit.

Case (*v*)   Perfect switching and two nonidentical units—one main unit and one standby.

Case (*vi*)   Same as case (*v*), except that failures in the standby mode are also considered.

Schematics of these six cases are shown in Figure 5.28.

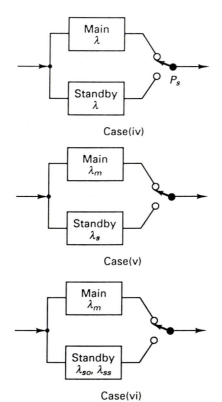

Case(iv)

Case(v)

Case(vi)

Figure 5.28   Standby system cases (*continued*)

Case (*i*).   Perfect switching implies that when the main unit A fails, the sensing and changeover switch brings the standby unit B into operation definitively and in zero time. Thus, for the system to fail, A must fail and then B must fail. In other words, the system succeeds if there are no failures or only one failure. With the assumption of constant-hazard components, we can employ the Poisson distribution to find the probabilities of zero failures and one failure:

$$P_0(t) = P(\text{zero failures}) = \frac{(\lambda t)^0 e^{-\lambda t}}{0!} = e^{-\lambda t} \tag{5-107}$$

$$P_1(t) = P(\text{one failure}) = \frac{(\lambda t)^1 e^{-\lambda t}}{1!} = \lambda t e^{-\lambda t} \tag{5-108}$$

Therefore,

$$\text{System reliability} = R(t) = P_0(t) + P_1(t)$$
$$= e^{-\lambda t}[1 + \lambda t] \tag{5-109}$$

In applying the Poission distribution, we note that since λ is the failure rate of either component, and since only one component is operating at any given time, the expected number of failures in time t is equal to λt.

Case (*ii*).   In this case, the system will succeed if there are zero, one, or two failures. Proceeding as in case (*i*), we get

$$P_2(t) = P(\text{two failures}) = \frac{(\lambda t)^2 e^{-\lambda t}}{2!} \tag{5-110}$$

and

$$\text{System reliability} = R(t) = P_0(t) + P_1(t) + P_2(t)$$
$$= e^{-\lambda t}\left[1 + \lambda t + \frac{(\lambda t)^2}{2!}\right] \tag{5-111}$$

Case (*iii*).   Since we can tolerate n failures in this case, the system reliability $R(t)$ is given as the sum of the first n terms of the Poisson distribution. Thus, proceeding as in the previous two cases, we have

$$R(t) = e^{-\lambda t}\left[1 + \lambda t + \frac{(\lambda t)^2}{2!} + \frac{(\lambda t)^3}{3!} + \ldots + \frac{(\lambda t)^n}{n!}\right] \tag{5-112}$$

Also,

$$\text{MTTF} = \int_0^\infty R(t)\, dt$$
$$= \int_0^\infty \left(\sum_{i=0}^{n} \frac{(\lambda t)^i e^{-\lambda t}}{i!}\right) dt \tag{5-113}$$

Interchanging the summation and integration processes, we get

$$\mathrm{MTTF} = \sum_{i=0}^{n} \frac{\lambda^i}{i!} \int_0^{\infty} t^i e^{-\lambda t} \, dt$$

$$= \sum_{i=0}^{n} \frac{\lambda^i}{i!} \frac{\Gamma(i+1)}{\lambda^{i+1}}$$

$$= \sum_{i=0}^{n} \frac{1}{\lambda} \tag{5-114}$$

$$= \frac{n+1}{\lambda}$$

Intuitively, we could have guessed the foregoing result. As one unit fails, another one instantaneously takes its place, and we have $(n + 1)$ such failures before the system fails. The mean time taken for each failure is $1/\lambda$, and therefore, the system MTTF is equal to the sum of $(n + 1)$ such time intervals, which is $(n + 1)/\lambda$.

■■   **Example 5–23**

Let us suppose that a component has a constant failure rate of 1 in 40 hours of operation during its normal lifetime. The reliability of this component for a one-day mission is then

$$R(24 \text{ hours}) = \exp\left(\frac{-24}{40}\right) = 0.5488$$

We want to explore the different options available to improve the reliability of systems using this component. We have:

- Two identical components operated in parallel:

$$R(24 \text{ hours}) = 1 - (1 - 0.5488)^2 = 0.79642$$

- One component in standby mode with perfect switching:

$$R(24 \text{ hours}) = \left[\exp\left(\frac{-24}{40}\right)\right]\left[1 + \frac{24}{40}\right] = 0.87808$$

  The additional expense incurred in providing a sensing and changeover switch has increased the system reliability by 10.25%, as compared to the parallel redundant arrangement.

- Two components in standby mode with perfect switching:

$$R(24 \text{ hours}) = \left[\exp\left(\frac{-24}{40}\right)\right]\left[1 + \frac{24}{40} + 0.5\left(\frac{24}{40}\right)^2\right] = 0.97686$$

- Three components operated in parallel:

$$R(24 \text{ hours}) = 1 - (1 - 0.5488)^3 = 0.90814$$

Note that, for the same number of components, the standby arrangement always results in a higher reliability than the parallel redundant case, as long as we have perfect switching.

- Three components in standby mode with perfect switching:

$$R(24 \text{ hours}) = e^{-0.6} \left[1 + 0.6 + \frac{1}{2} (0.6)^2 + \frac{1}{6} (0.6)^3 \right] = 0.99662$$

If we want to improve the system reliability to 0.99 or higher, we must have at least three identical spares backing the normally operating unit, as well as a perfect sensing and changeover device.

The MTTFs for the six cases considered are calculated next:

- For just one component in operation, MTTF = 40 hours.
- For two identical components in parallel,

$$\text{MTTF} = 40 \left[2 - \frac{1}{2} \right] = 60 \text{ hours}$$

- For one standby unit and one normally operating unit,

$$\text{MTTF} = 2 \times 40 = 80 \text{ hours}$$

- For two components in standby mode,

$$\text{MTTF} = 3 \times 40 = 120 \text{ hours}$$

- For three components in parallel,

$$\text{MTTF} = 40 \left[3 - \frac{3}{2} + \frac{1}{3} \right] = 73.33 \text{ hours}$$

- For three standby components backing one component in operation,

$$\text{MTTF} = 4 \times 40 = 160 \text{ hours}$$

Even though the component, by itself, does not have a high reliability for the one-day mission, by employing the concepts of parallel and standby redundancies, we can achieve any desired level of reliability and MTTF values. Of course, no practical switch is perfect, and as we increase the number of components required, the overall system cost increases. Once again, final design decisions will be based on economic considerations. ■■

Case (*iv*). In this case, the inclusion of imperfections in the sensing and switchover process requires another parameter, namely, P_s, the probability of successful operation of the sensing and changeover switch. Typically, the value of P_s is estimated from historical data as the ratio of the number of successful operations to the total number of times the device was expected to operate successfully. The value of P_s can be kept high by periodic testing and maintenance. If this sensing and switching device consists of several components with known failure rates in a series-parallel or any other configuration, the combinatorial techniques discussed in this chapter can be used to arrive at a proper value for P_s.

The probability of zero failures given by Equation (5–107) is unaffected by the quality of the switch; it depends only on the failure rate of the main (or normally operating) unit. On the other hand, the probability of one failure is indeed influenced by the quality of the switch because, unless we have a successful changeover, we cannot have successful operation with one failure. Therefore,

$$P_1(t) = \lambda t e^{-\lambda t} P_s \tag{5–115}$$

and

$$R(t) = e^{-\lambda t}[1 + P_s \lambda t] \tag{5–116}$$

The MTTF is easily calculated to be

$$\begin{aligned} \text{MTTF} &= \int_0^\infty e^{-\lambda t}[1 + P_s \lambda t]\, dt \\ &= \left(\frac{1 + P_s}{\lambda}\right) \end{aligned} \tag{5–117}$$

By comparing the expression for the reliability of two identical constant-hazard components in a truly parallel configuration with that for the reliability of a standby arrangement with an imperfect switch, we can arrive at a lower bound for P_s that is necessary to realize an improvement in system reliability in the standby case. Since

$$\text{Reliability of truly parallel system} = 2e^{-\lambda t} - e^{-2\lambda t}$$

and

$$\left.\begin{array}{r}\text{Reliability of standby arrangement} \\ \text{with an imperfect switch}\end{array}\right\} = e^{-\lambda t}[1 + P_s \lambda t]$$

it follows that for

$$e^{-\lambda t}[1 + P_s \lambda t] > 2e^{-\lambda t} - e^{-2\lambda t}$$

we must have

$$P_s > \frac{1 - e^{-\lambda t}}{\lambda t} \tag{5–118}$$

Case (v).   In practice, we often encounter situations in which the standby unit is not identical to the main unit. A spare tire in an automobile that is smaller than a regular tire, an inverter-battery system backing up a remote generator, and a reconditioned motor used as a spare for a normally operating motor are some examples that fall into this category.

We will assume that both units have constant hazards—λ_m for the main unit and λ_s for the standby unit.

Let

$$t_m = \text{time to failure of the main unit}$$

and

$$t_s = \text{time to failure of the standby unit}$$

Since we assume perfect switching, the time to failure of the system is simply

$$t = t_m + t_s \tag{5-119}$$

In this equation, all three time variables—t, t_m, and t_s—are continuous random variables. Also, the density functions of t_m and t_s are, respectively,

$$f_m(t_m) = \lambda_m \exp[-\lambda_m t_m] \tag{5-120}$$

and

$$f_s(t_s) = \lambda_s \exp[-\lambda_s t_s] \tag{5-121}$$

What we want is the density function $f(t)$ for the random variable $t = t_m + t_s$. Since both t_m and t_s can assume only positive values, we have

$$f(t) = \int_0^t f_m(\tau) f_s(t - \tau) \, d\tau \tag{5-122}$$

Using Equations (5–120) and (5–121) in Equation (5–122), we get

$$f(t) = \int_0^t \lambda_m \exp(-\lambda_m \tau) \lambda_s \exp[-\lambda_s(t - \tau)] \, d\tau$$
$$= \left(\frac{\lambda_m \lambda_s}{\lambda_m - \lambda_s}\right) \left\{ \exp(-\lambda_s t) - \exp(-\lambda_m t) \right\} \tag{5-123}$$

The system reliability is obtained by integrating the density function from t to infinity:

$$R(t) = \int_t^\infty f(\xi) \, d\xi$$
$$= \left(\frac{\lambda_m \lambda_s}{\lambda_m - \lambda_s}\right) \int_t^\infty [\exp(-\lambda_s \xi) - \exp(-\lambda_m \xi)] \, d\xi$$

or

$$R(t) = \exp(-\lambda_m t) + \left(\frac{\lambda_m}{\lambda_m - \lambda_s}\right)[\exp(-\lambda_s t) - \exp(-\lambda_m t)] \tag{5-124}$$

Moreover,

$$\text{MTTF} = \int_0^\infty R(t) \, dt = \frac{1}{\lambda_m} + \frac{1}{\lambda_s} \tag{5-125}$$

We can include the possibility of imperfect switching by a slight modification of Equation (5–124). With an imperfect switch,

$$R(t) = \exp(-\lambda_m t) + \left(\frac{P_s \lambda_m}{\lambda_m - \lambda_s}\right)[\exp(-\lambda_s t) - \exp(-\lambda_m t)] \quad (5\text{–}126)$$

The density function in this case is known as the *joint density function* and is given by

$$
\begin{aligned}
f(t) &= f_m(t_m)\, f_s(t_s) \\
&= \lambda_m \exp(-\lambda_m t_m)\lambda_s \exp[-\lambda_s(t - t_m)]
\end{aligned}
\quad (5\text{–}127)
$$

The right-hand side of this equation has two time functions, t_m and t. To express $f(t)$ strictly in terms of t, we integrate the right-hand side over t_m from 0 to t, since t_m can assume any value between 0 and t, and obtain

$$f(t) = \int_0^t \lambda_m \exp(-\lambda_m t_m)\lambda_s \exp[-\lambda_s(t - t_m)]\, dt_m$$

which results in Equation (5–123).

An extension of this approach for the case of one main unit with a failure density function $f_m(t_m)$ and n standby units with failure density functions $f_{s1}(t_{s1})$, $f_{s2}(t_{s2})$, ... , $f_{sn}(t_{sn})$ yields

$$f(t) = \int_{t_{sn}=0}^{t} \cdots \int_{t_{s1}=0}^{t_{s2}} \int_{t_m=0}^{t_{s1}} f_m(t_m)f_{s1}(t_{s1}) \cdots f_{sn}(t_{sn})\, dt_m dt_{s1} \cdots dt_{sn} \quad (5\text{–}128)$$

Alternative approach.   An expression for system reliability can also be obtained by listing all the events leading to system success, grouping them into a set of mutually exclusive events, and then summing the probabilities of these mutually exclusive events. For case (*v*), any one of the following mutually exclusive events will lead to system success at time t:

a. The main unit does not fail in the time interval $(0, t)$.
b. The main unit fails at $\tau < t$, and the standby unit does not fail in the time interval (τ, t).

Thus, we have

$$P(\text{event a}) = \exp(-\lambda_m t) \quad (5\text{–}129)$$

Also, the probability of main unit failure in $(\tau, \tau + d\tau)$ and no failure of the standby unit during (τ, t) is

$$[\lambda_m \exp(-\lambda_m \tau)\, d\tau][\exp(-\lambda_s(t - \tau))]$$

Since τ can have any value between 0 and t, we see that

$$P(\text{event b}) = \int_0^t \lambda_m \exp(-\lambda_m\tau) \exp[-\lambda_s(t - \tau)] \, d\tau \tag{5-130}$$

$$= \left(\frac{\lambda_m}{\lambda_m - \lambda_s}\right) [\exp(-\lambda_s t) - \exp(-\lambda_m t)]$$

By summing the probabilities of event a and event b, we obtain the system reliability expression given in Equation (5–124).

■■ **Example 5–24**

A small diesel generator with a failure rate of 0.001 per hour is backed by another generator in standby mode. The standby generator has a failure rate of 0.0015 per hour. The probability of success of the sensing and changeover unit is estimated to be 0.98. Then, for a mission time of 100 hours, the system reliability is, using Equation (5–126),

$$R(100) = e^{-0.1} + \frac{0.98 \times 0.001}{-0.0005}\left(e^{-0.15} - e^{-0.1}\right)$$

$$= 0.99133$$

Next, let us calculate the system reliability for a truly parallel configuration. We have, for the main unit,

$$R(100) = e^{-0.1} = 0.904837$$

For the standby unit,

$$R(100) = e^{-0.15} = 0.860708$$

and

$$\text{Parallel system reliability} = 1 - (1 - 0.904837)(1 - 0.860708)$$

$$= 0.986745$$

We can also calculate the minimum reliability required of the changeover switch for the system reliability to be no less than that of the parallel arrangement. Since we want

$$0.986745 \leq e^{-0.1} + \frac{P_s\,0.001}{-0.0005}[e^{-0.15} - e^{-0.1}]$$

using the equality sign, we find that $P_s = 0.928$. Therefore, the changeover switch reliability should be no less than 92.8%.   ■■

Case (vi).   In this case, we include the possibility of the standby unit failing in the standby mode. Let λ_{so} and λ_{ss} be the failure rates of the standby unit in operating and standby modes, respectively.

Using the alternative approach discussed under case (v), we list all the mutually exclusive events leading to system success at time t. They are:

a. The main unit does not fail in the interval $(0, t)$

b. The main unit fails at $\tau < t$, the standby unit has not failed at τ, and the standby unit does not fail in the interval (τ, t)

Accordingly,

$$P(\text{event a}) = \exp[-\lambda_m t] \tag{5-131}$$

$$P(\text{event b}) = \int_{\tau=0}^{t} \lambda_m \exp(-\lambda_m \tau) \exp(-\lambda_{ss} \tau) \exp[-\lambda_{so}(t - \tau)] \, d\tau$$

$$= \lambda_m \exp(-\lambda_{so} t) \int_{0}^{t} \exp[-(\lambda_m + \lambda_{ss} - \lambda_{so})\tau] \, d\tau \tag{5-132}$$

$$= \frac{\lambda_m}{\lambda_m + \lambda_{ss} - \lambda_{so}} [\exp(-\lambda_{so} t) - \exp[-(\lambda_m + \lambda_{ss})t]]$$

Adding the reliabilities of events a and b, we obtain the system reliability:

$$R(t) = \exp(-\lambda_m t) + \left(\frac{\lambda_m}{\lambda_m + \lambda_{ss} - \lambda_{so}}\right) [\exp(-\lambda_{so} t) - \exp[-(\lambda_m + \lambda_{ss})t]]$$

$$\tag{5-133}$$

The six cases considered here are by no means exhaustive as far as standby systems are concerned. However, we have systematically developed a methodology that can be expanded and applied to other cases as well. Some of these other cases are as follows:

(*i*) Two nonidentical units operating in parallel, backed by a standby unit that is different from them.

(*ii*) An r-out-of-n system backed by one or more standby units.

(*iii*) Three or more nonidentical units operating in parallel, backed by a different standby unit.

(*iv*) One main unit backed by two smaller nonidentical standby units operated in parallel.

The list can be expanded as the situation warrants. Also, in each of these cases, we may want to include failures of the standby unit in the standby mode, and, in addition, we may want to consider the imperfections of the sensing and changeover device when on standby as well as after switching. The methodology discussed is capable of handling all such cases.

5.14 MODELING SPARES WITH INSTANT REPLACEMENT

There are many practical situations where a number N of identical components are operated with a certain number n of identical spares. Assuming that all the N operating components are needed for system success, when one of them fails, it is replaced by a spare in a very short amount of time.

For modeling purposes, we assume that replacement of the component is instantaneous and that the system continues to operate with no breaks as long as there are spares available for replacement. The failed components are discarded (meaning that no repairs are performed on them). In addition, we assume that the failure times of each and every component are exponentially distributed. Let λ represent the constant hazard of a component. Since all of the N components are needed for system success, they are logically in series, and the overall failure rate [see Equation (5–83)] is equal to $N\lambda$. Moreover,

$$P(\text{zero failures in time } t) = e^{-N\lambda t} \tag{5–134}$$

$$P(\text{exactly 1 failure in time } t) = (N\lambda t)e^{-N\lambda t} \tag{5–135}$$

$$P(\text{exactly 2 failures in time } t) = \frac{(N\lambda t)^2}{2!} e^{-N\lambda t} \tag{5–136}$$

and so on.

A system with n spares can tolerate up to n component failures before it fails. Therefore, the system reliability is equal to the probability of 0 or 1 or 2 or … or n failures in time t. This reasoning leads to the equation

$$R(t) = e^{-N\lambda t}\left(1 + N\lambda t + \frac{(N\lambda t)^2}{2!} + \dots + \frac{(N\lambda t)^n}{n!}\right) \tag{5–137}$$

The MTTF is given by

$$\text{MTTF} = \int_0^\infty R(t)\, dt = \frac{n+1}{N\lambda} \tag{5–138}$$

[see equation (5–114) with λ replaced by $N\lambda$].

For the special case of $N = 1$ and $n = \infty$,

$$R(t) = e^{-\lambda t} \sum_{k=0}^\infty \frac{(\lambda t)^k}{k!}$$

$$= e^{-\lambda t}e^{\lambda t}$$

$$= 1$$

and the MTTF $= \infty$. The system never fails, since we have an infinite number of spares and we can keep on instantaneously replacing the failed component forever!

Since

$$\text{System reliability with no spares } = e^{-N\lambda t} \qquad (5\text{--}139)$$

with n spares, we can define a reliability improvement ratio as

$$\text{RIR} \equiv \frac{R(t) \text{ with } n \text{ spares}}{R(t) \text{ with no spares}} \qquad (5\text{--}140)$$

or

$$\text{RIR} = \sum_{k=0}^{n} \frac{(N\lambda t)^k}{k!} \qquad (5\text{--}141)$$

Plots of RIR versus the number of spares, n, for different values of N are presented in Figure 5.29 for $\lambda t = 0.1$—that is, a mission time equal to 10% of the MTTF of the individual components. It can be seen that the first spare results in the greatest improvement in system reliability. Also, the greater the number of components, N, in the system, the greater is the improvement realized with the provision of the first few spares. We can use these results and work backwards to find the number of spares needed to achieve a certain specified minimum level of system reliability.

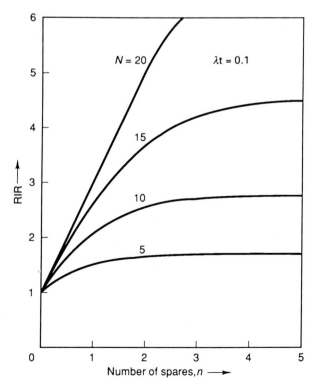

Figure 5.29 Dependence of reliability improvement ratio (RIR) on number of spares.

The improvement in system MTTF is another useful quantity for evaluation. With no spares, the system MTTF is equal to $1/(N\lambda)$. With the provision of n spares, the system MTTF is $(n + 1)/(N\lambda)$. Therefore,

$$\text{MTTF improvement ratio} \equiv \frac{\text{System MTTF with spares}}{\text{System MTTF with no spares}} \quad (5\text{--}142)$$
$$= n + 1$$

The system MTTF increases linearly with the number of spares, as long as the assumption of constant-hazard components is valid. In reality, wearout may set in before some of the higher MTTF values are realized.

Improvements in system reliability and MTTF realized with the provision of spares come with an increase in overall system cost, as expected. Eventually, the choice of the number of spares becomes a strictly economic decision based on the costs of not improving system performance and the additional cost incurred in maintaining a stock of spares.

■■ **Example 5–25**

For a system consisting of eight identical constant-hazard components, each with a failure rate of 10^{-4} hr^{-1}, calculate the number of spares required to achieve a minimum system reliability of 0.999 for a mission time of 3 days. Assume that all eight components are needed for system success and that replacement is instantaneous.

Without spares, the system reliability is

$$R_s(72 \text{ hours}) = \exp[-8 \times 10^{-4} \times 72] = 0.944027$$

The required RIR must be greater than or equal to

$$\frac{0.999}{0.944027} = 1.05823$$

Using Equation (5–141), we have

$$1.05823 \le \sum_{k=0}^{n} \frac{(0.0576)^k}{k!}$$

The terms of the summation are 1, 0.0576, 0.0016589, ... It is obvious, then, that $n = 2$ will accomplish our objective. Therefore, two spares are required.   ■■

■■ **Example 5–26**

A system consists of 32 identical components, each with an MTTF of 960 hours. How many spares are required to achieve a system MTTF of 100 hours?

Without spares, the system MTTF is equal to 960/32, or 30, hours. Therefore, we want an MTTF improvement ratio of 100/30, or 3.33. Equating 3.3 to $n + 1$, where n is the number of spares, we see that $n = 2.3$ and conclude that three spares are required.   ■■

5.15 SUMMARY

Combinatorial aspects of system reliability are extremely useful in evaluating the reliability of complex systems. We decompose the complex system into functional entities connected together in series, parallel, series-parallel, an *r*-out-of-*n* structure, star-delta, etc., and employ network modeling techniques to arrive at the overall system reliability.

Among the general techniques available to evaluate the reliability of complex systems that have been presented are inspection, the event-space method, the path-tracing method, the decomposition method, the minimal cut set method, the minimal tie set method, connection matrix techniques, event trees, and fault trees. Topological aspects of system reliability and the use of computer algorithms to evaluate system reliability are vast areas of study in themselves. It is hoped that the brief introduction given in this chapter will kindle the interest of some readers to delve into the literature on these topics.

Systems composed of three-state devices and their combinatorial aspects have been introduced. Procedures to handle time-dependent reliabilities were discussed for the case of some of the commonly used distributions, primarily the exponential distribution. Extensions of these techniques to several special cases of standby systems were presented systematically. Finally, modeling spares with instant replacement was considered. This chapter and the two preceding ones contain the bulk of the information necessary to model catastrophic failures and to evaluate the reliability of simple as well as complex systems composed of components that undergo failures with no repair.

A consideration of repair entails the use of Markov models, which is the topic of the next two chapters.

PROBLEMS

5.1 What is the minimum allowable component reliability to achieve a system reliability of 0.90 for a series system consisting of 2, 4, 6, 8, and 10 identical components?

5.2 For *n* identical independent components in series, each with a reliability of *R*, what is the reliability deterioration ratio over that of a single component? Plot this ratio as a function of *R* for *n* = 2, 4, 6, 8, and 10. The ratio is given as

$$\text{Ratio} = \frac{\text{Reliability of a single component}}{\text{Reliability of the series structure}}$$

5.3 For a fully redundant configuration with two components in parallel, find the reliability improvement ratio over that of a single component.

5.4 It is required that a system have a reliability of 99%. How many components are required in parallel if each component has a reliability of 70%? Can you achieve the required level of system reliability by putting these components in series?

5.5 An *m/N parallel system* is a parallel system in which a minimum of *m* units should be good out of a total of *N* units in order for the system to be successful. If all the components are identical and independent with a reliability of *R*, show that the system reliability can be expressed as

$$R_s = \sum_{k=0}^{N-m} {}_NC_k(1 - R)^k R^{N-k}$$

or

$$R_s = 1 - \sum_{k=N-m+1}^{N} {}_NC_k(1 - R)^k R^{N-k}$$

5.6 Consider a star-delta structure.
   (a) If $R_\Delta = 0.8$, what is the corresponding R_Y?
   (b) If $R_Y = 0.8$, what is the corresponding R_Δ?

5.7 By employing delta-star techniques, derive an expression for the reliability of the system shown in Figure P 5.7. Assume that all the units are identical, are independent, and have a probability of success of *p*. Evaluate the system reliability for $p = 0.9$.

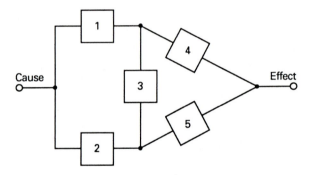

Figure P 5.7

5.8 Repeat Problem 5.7 if the reliabilities of units 1 and 2 are 0.95 and 0.85, respectively, and the reliabilities of units 3, 4, and 5 are identical with $p = 0.9$.

5.9 Calculate the reliabilities of the two system configurations shown in Figure P 5.9.

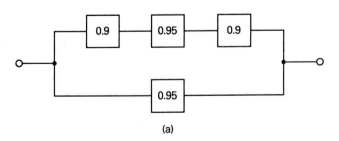

(a)

Figure P 5.9

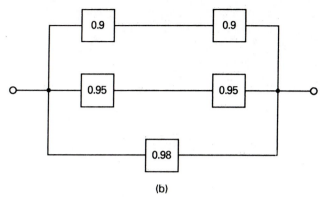

(b)

Figure P 5.9   (*continued*)

5.10 In Figure P 5.10, the reliabilities of the components shown are indicated inside the boxes. Derive an expression for the system reliability R_s in terms of p_1, p_2, and p_3. Evaluate the system reliability for $p_1 = p_2 = p_3 = 0.9$.

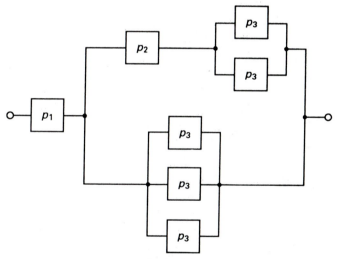

Figure P 5.10

5.11 A system consists of four identical components arranged as shown in Figure P 5.11. If a system reliability of at least 0.95 is desired, calculate the minimum value of component reliability.

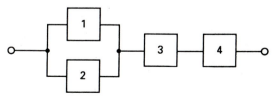

Figure P 5.11

5.12 Two arrangements for four identical components are shown in Figure P 5.12. Let p be the probability of success of each component. Derive expressions and compare the reliabilities of the two arrangements by plotting their values against p for $0 \leq p \leq 1$. Which arrangement is better?

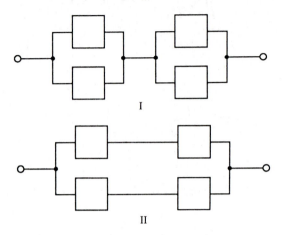

Figure P 5.12

5.13 What is the reliability of a TMR system with unit reliabilities of 0.9 and a voter reliability of 0.95?

5.14 Consider a three-out-of-five NMR system. Calculate its reliability with a perfect voter and unit reliabilities of 0.9.

5.15 A normally operating component has a reliability of 0.95 and is backed by another component with a reliability of 0.9, given that the normally operating component has failed.
  (*i*) What is the system reliability if the changeover from the normally operating component to the standby component is fault free?
  (*ii*) If the changeover system has a reliability of 0.9, what is the overall system reliability?
  (*iii*) How bad could the changeover system be and still maintain an overall system reliability of 0.98?

5.16 Consider a component with a reliability of 0.9, backed by an identical component. The failure sensing and changeover mechanism has a reliability of 0.95, and the switch has an operating reliability of 0.96. What is the overall system reliability?

5.17 Derive an expression for the reliability of the system shown in Figure P 5.17. All the units are identical, are independent, and have a reliability of p.

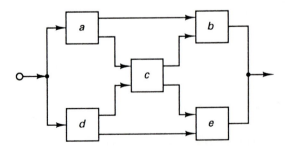

Figure P 5.17

5.18 Derive an expression (in terms of p) for the reliability of the bridge network shown in Figure P 5.18 using the decomposition method. All the components are identical, independent, and binary and have a success probability of p.

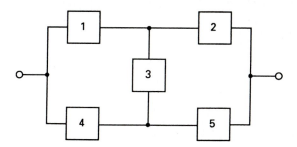

Figure P 5.18

5.19 Apply the event-space method to the bridge network of Problem 5.18, and derive an expression for the system reliability in terms of q, the unreliability of a component.

5.20 For the bridge network of Problem 5.18, (a) set up the connection matrix, (b) find the minimum tie sets by the matrix multiplication method, and (c) find an expression for the system unreliability Q in terms of component unreliability q.

5.21 The bridge network shown in Figure P 5.21 is made up of different units. Draw a complete event tree, and derive expressions for the system reliability and unreliability in terms of R_i and Q_i, where $i = 1, 2, \ldots, 5$.

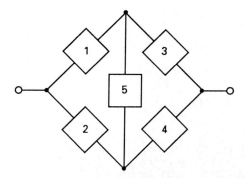

Figure P 5.21

5.22 For the simple linked configuration shown in Figure P 5.22, derive an expression for the system reliability using the event-tree method. If all the components are identical, are independent, and have a probability of success of p, what is the system reliability?

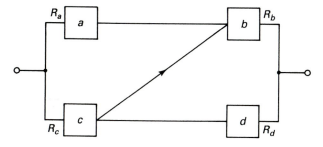

Figure P 5.22

5.23 Repeat Problem 5.22 by using (*i*) the decomposition method, (*ii*) the minimal cut set method, (*iii*) the path-tracing method, and (*iv*) the event-space method.

5.24 Evaluate the reliability of the system shown in Figure P 5.24 using the path-tracing method. The reliabilities of the individual components are shown outside the blocks.

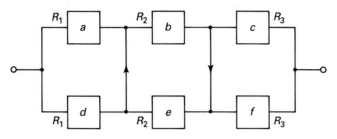

Figure P 5.24

5.25 Repeat Problem 5.24 using the conditional probability method. Choose *b* as the keystone component.

5.26 Derive an expression for the reliability of the system shown in Figure P 5.26 by employing (*i*) the path-tracing method and (*ii*) the conditional probability method.

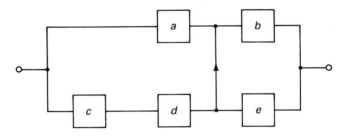

Figure P 5.26

5.27 For the system shown in Figure P 5.27, set up the connection matrix *M*. Eliminate nodes 2 and 3, and obtain an expression for the transmission from node 1 to node 4.

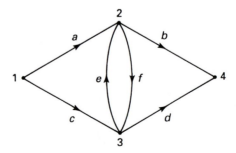

Figure P 5.27

5.28 For the system configuration of Problem 5.27, deduce the minimal tie sets by the matrix multiplication technique.

5.29 For the fault tree shown in Figure P 5.29,
   (a) Develop a Boolean expression.
   (b) Reduce the expression to remove all dependencies, and redraw the fault tree with no repeated events.
   (c) If the probabilities of all the basic events are $\frac{1}{4}$, find the probability of occurrence of the top event T_0.

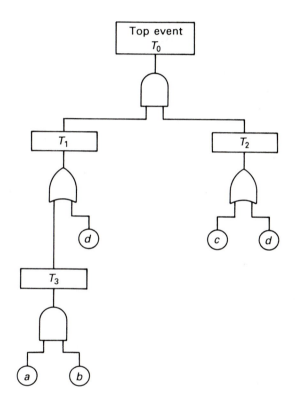

Figure P 5.29

5.30 Consider the fault tree illustrated in Figure P 5.30.
   (*i*) Draw the reduced fault tree with no repeated events.
   (*ii*) Calculate $P(T_0)$ if the probabilities of all the basic events are 0.25.
   (*iii*) What is the difference in the value of $P(T_0)$ obtained using the original and the reduced fault trees?

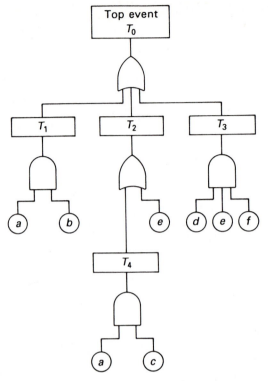

Figure P 5.30

5.31 Develop a fault tree for the reliability block diagram given in Figure P 5.31.

5.32 **(a)** Construct the fault tree corresponding to the logic expression

$$T = (abc + f)[(a + d)f](a + be)$$

(b) Simplify the expression to remove all repetitions, and construct an equivalent reliability block diagram.

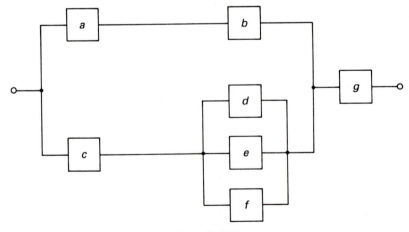

Figure P 5.31

5.33 Draw a fault tree corresponding to the reliability block diagram given in Figure P 5.33.

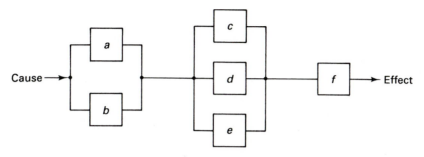

Figure P 5.33

5.34 Rework Example 5–14 if (*i*) *C* is replaced by *D*, (*ii*) *C* is replaced by *B*, and (*iii*) *A* is replaced by *D*.

5.35 Evaluate the reliabilities of series-parallel arrays of six identical fluid flow valves with
    (*i*) three units in series, each unit consisting of two in parallel, and
    (*ii*) two branches in parallel, each branch consisting of three units in series.
Take $q_o = 0.01$ and $q_s = 0.05$.

5.36 Consider two three-state devices with normal operation, open-circuit failure, and short-circuit failure probabilities of p_n, q_o, and q_s, respectively. Under what conditions will the series configuration do better than the parallel configuration?

5.37 Evaluate the reliability of the quad-configuration of diodes shown in Figure P 5.37. For each diode, $p_n = 0.96$ and $q_o = 0.015$. Comparing this quad with just one diode, (*i*) by what factor is the reliability improved, and (*ii*) what is the unreliability improvement ratio?

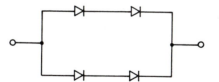

Figure P 5.37

5.38 Figure P 5.38 shows two arrangements of four identical three-state devices. Indicate your preferred arrangement, based on numerical values of system reliabilities. For each device, $p_n = 0.95$ and $q_s = 0.03$. Repeat the problem for $q_o = 0.03$ and $q_s = 0.02$.

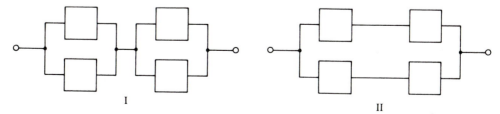

Figure P 5.38

5.39 Show that the ratio of the hazard function of a single component to that of two identical components in parallel in a fully redundant configuration is $[1 + Q(t)]/[2Q(t)]$.

5.40 An electronic system consists of four components in parallel. The MTTFs of the components are 200, 300, 400, and 500 hours. Evaluate the system reliability for 800 hours of operation.

5.41 A communication subsystem is designed using two 8086 MPUs in a fully redundant configuration. Each MPU has a failure rate of 1.25 per million hours.
    (*i*) Calculate the gain in reliability achieved over using only one MPU for a mission time of 2,000 hours.
    (*ii*) Recalculate the gain if a third MPU is added in parallel.

5.42 Let $R(t)$ and $Q(t)$ be the probabilities of success and failure, respectively, for a particular component. For a fully redundant configuration with n such components in parallel, derive an expression for the reliability improvement ratio over that of a single component. Calculate this ratio for $R(t) = 0.6$ and $n = 2, 3,$ and 4.

5.43 (**a**) Repeat Problem 5.39 for the case of three identical components in parallel in a fully redundant configuration.
    (**b**) Generalize your result for the case of n components in parallel.

5.44 A system consists of three subsystems, identified as I, II, and III, and all three subsystems are necessary for system success. Subsystem I consists of a critical component with a constant hazard of λ_1. Subsystem II consists of two identical components, each with a constant hazard of λ_2, and only one of the two need be good for this subsystem to succeed. Subsystem III consists of three identical units, each with a linearly increasing hazard of kt. Two of these three units must be good for success. Derive an expression for the system reliability.

5.45 For the system shown in Figure P 5.45, derive an expression for the system MTTF. All the units have constant hazards as indicated and are independent.

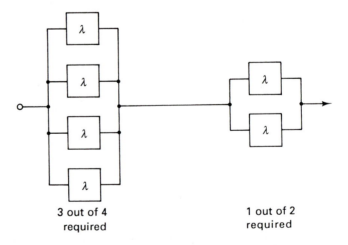

3 out of 4 required   1 out of 2 required

Figure P 5.45

5.46 It is desired to set the design life T_d of a system such that its reliability is no less than a preselected value p_d.

(*i*) For a constant-hazard component, obtain the relationship between design life and failure rate.

(*ii*) Repeat (*i*) for two identical components in parallel.

(*iii*) By what factor did the design life increase in (*ii*) as compared to (*i*)? Does this factor depend on the individual component failure rate? Calculate this factor for $p_d = 0.95, 0.9, 0.85$, and 0.8.

5.47 Show that the MTTF for a system consisting of n identical constant-hazard components in parallel is given by

$$\text{MTTF} = \sum_{k=1}^{n} (-1)^{k-1} \left[\frac{{}_nC_k}{k\lambda} \right]$$

where λ is the failure rate for one component.

5.48 Consider n constant-hazard components in parallel, of which r must be good for the system to succeed. Show that if $\lambda t \ll 1$, the system unreliability can be approximated as

$$Q_s \cong {}_nC_{n-r+1}(\lambda t)^{n-r+1}$$

This technique is called a *rare-event approximation*.

5.49 A class of components is found to have a linearly increasing hazard

$$\lambda(t) = 10^{-6.2}t \text{ hr}^{-1}$$

How many such components can be operated in series if the system MTTF is not to go below 500 hours?

5.50 Compute the reliability and MTTF of the system shown in Figure P 5.50. Elements a and b form an ordinary parallel circuit. However, since all three elements dissipate substantial amounts of heat and are confined to a small space, their failure rates are dependent. The units are identical, and with 1, 2, and 3 units operating, the constant-hazard rates are λ, 2λ, and 3λ, respectively, for the elements as a group.

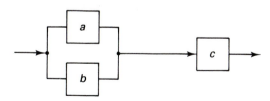

Figure P 5.50

5.51 By employing the minimal cut set method, derive an expression for the reliability of the system of Figure P 5.51. If each component has a constant hazard of λ, what is the system MTTF?

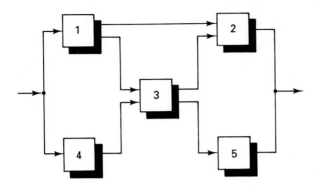

Figure P 5.51 Identical components, each with a reliability of p.

5.52 For the system shown in Figure P 5.52, what is the maximum allowable failure rate for each of the four identical units such that a system MTTF of 1,000 hours can be achieved? Assume that the units are independent and have constant hazards.

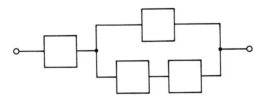

Figure P 5.52

5.53 In the system of Figure P 5.53, each block represents a constant-hazard element with a failure rate of λ. Find the system MTTF.

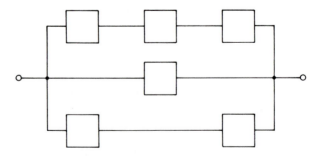

Figure P 5.53

5.54 For the simple parallel system shown in Figure P 5.54, derive an expression for the system MTTF.

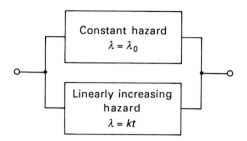

Figure P 5.54

5.55 A system has four different parts, each with a constant hazard of λ_i, $i = 1, \ldots, 4$. Parts 1 and 2 must work for the system to perform its intended function. Also, at least one of the other two parts must be good for the system to succeed. Derive an expression for the reliability of this system.

5.56 The probability of a relief valve being unable to operate on demand is 2%. Five such valves are used in conjunction with a pressure vessel, and at least two of them must work to control the possible pressure transients. Assuming that the valves operate independently, calculate the system reliability using the rare-event approximation of Problem 5.48.

5.57 A device has a constant hazard of λ failures per hour. An identical unit is used as a standby. The failure sensing and switching unit has a reliability of P_{fss}. Derive an expression for the system reliability as a function of time.

5.58 A standby system consisting of one main unit and one standby unit with a sensing and changeover switch has the following parameters:

$$\text{Failure rate of the main unit} = 10^{-4} \text{ hr}^{-1}$$

$$\text{Failure rate of the standby unit} = 15 \times 10^{-5} \text{ hr}^{-1}$$

It is desired that the system have a reliability of 0.998 for a mission time of 100 hours. Calculate the minimum reliability required of the sensing and changeover switch.

5.59 Show that as λ_s tends towards λ_m, Equation (5–126) becomes Equation (5–116).

5.60 Develop an expression for the reliability of the standby system shown in Figure P 5.60. Components 1 and 2 are operated in a truly parallel manner. When both fail, the changeover switch brings component 3 into service. The sensing and switchover mechanism can be assumed to be perfect. $\lambda_1, \lambda_2, \lambda_3$ are the failure rates of components 1, 2, and 3, respectively, under normal operating conditions. λ_{3s} is the failure rate of component 3 when on standby.

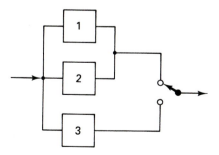

Figure P 5.60

5.61 For a system consisting of N identical components backed by a set of n identical spares, what should be the number of spares to make the system MTTF equal to the component MTTF? All of the N components must be good for the system to be good.

5.62 If there are 10 identical subsystems, each with an MTTF of 1,000 hours, how many spares should be available to make the system MTTF equal to 55% of the individual subsystem MTTF? What is the MTTF improvement ratio? The system succeeds only if every subsystem succeeds.

5.63 A component has a constant failure rate of 4%/K and a mean wearout time of 2,000 hours with a standard deviation of 500 hours. Calculate the reliability for a 400-hr mission starting at $T = 1,500$ hr in the life of the system for (*i*) two identical components in series, (*ii*) two identical components in parallel, and (*iii*) a two-out-of-three configuration.

5.64 The components described in Problem 5.63 are used in the system configurations shown in Figure P 5.64. Evaluate the system reliability for a 200-hr mission starting at $T = 1,900$ hr in the life of the components.

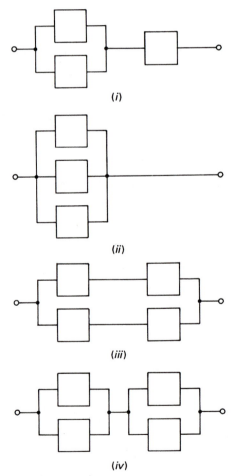

(*i*)

(*ii*)

(*iii*)

(*iv*)   **Figure P 5.64**

5.65 A system consists of four identical components, two of which are required for system success. There is one spare backing all four components. Assuming a perfect sensing and changeover arrangement, develop an expression for the reliability of this system. All the components can be assumed to have a constant hazard of λ.

5.66 Let q_{sA}, q_{sB}, and q_{sC} be the short-mode failure probabilities of components connected in a star configuration. Also, let q_{sAB}, q_{sBC}, and q_{sCA} be the short-mode failure probabilities of components connected in a delta configuration. Derive the unreliability expressions for the star components in terms of the delta values in order for the two arrangements to be equivalent.

6

Markov Models

6.1 INTRODUCTION

Many physical happenings and measurements of practical interest come under the general term *stochastic processes*. Such processes are concerned with observations that cannot be predicted precisely beforehand, but the probabilities of the different possible states at a particular time can be specified. Thus, a stochastic process is a family of random variables observed at different times t (or, to be general, indexed by the parameter t) and defined on a specified probability space. A *state* is a value assumed by a random variable, and the *state space* of the stochastic process is the set of all possible values that the random variable can assume.

In reliability work, we are interested in the evolution of physical systems with respect to time. Invariably, the state space of the associated stochastic process is discrete and finite. Discrete-state processes are also called *chains*. The index set, or the time of observation, could be discrete or continuous.

In 1907, the Russian mathematician A. A. Markov (1856–1922) introduced a special type of stochastic process whose future probabilistic behavior is uniquely determined by its present state. It is obvious that this type of behavior is non-hereditary, or memoryless. With the inclusion of the Markovian property, problems are simplified considerably, since knowledge of the present decouples the past from the future. The behavior of a variety of physical systems falls into this category; hence our interest in Markov models.

   A Markovian stochastic process with a discrete state space and discrete time space is referred to as a *Markov chain*. If the time (index parameter) space is continuous, then we refer to the process as a *Markov process*. There are two other possibilities—continuous-state discrete-time models and continuous-space continuous-time models—that are not of much interest in the field of reliability engineering.

   We will discuss Markov chains first, with the objective of developing an understanding of the basic concepts and techniques. This discussion will be followed by a study of Markov processes, which play an important role in the reliability evaluation of engineering systems.

6.2 DEFINITIONS

A Markov model is defined in terms of a set of transition probabilities p_{ij}; $i, j = 1, \ldots, n$; which denote the probabilities of transition from state i to state j in one step or in a specified time interval. For an n-state system, these probabilities can be arranged in the following matrix form:

$$\mathbf{P} = \begin{bmatrix} p_{11} & p_{12} & \cdots & p_{1n} \\ p_{21} & p_{22} & \cdots & p_{2n} \\ \vdots & \vdots & \vdots & \vdots \\ p_{n1} & p_{n2} & \cdots & p_{nn} \end{bmatrix} \tag{6-1}$$

The matrix $\mathbf{P}$ is called the *stochastic transitional probability matrix*. The diagonal entry p_{ii} is the probability that the system will remain in state i during one step (or transition) or specified time interval.

   Representation of a stochastic process by a Markov model implies a "lack of memory." In other words, the probability of transition p_{ij} depends only on the states i and j and the duration involved. It is completely independent of all past states, except the one it is in at the present time. Thus, the future behavior of the process depends only on the present and not on past history. Certain special types of processes are of interest:

1. *Homogeneous (stationary) process.* All the p_{ij}'s are independent of time; in other words, the probability of transition from one state to another is the same at all times in the past and in the future.
2. *Ergodic process.* In addition to being homogeneous, the final value of the probability of being in any state is independent of the initial conditions.

   The combination of a lack of memory and being homogeneous means that the overall behavior of the system does not change with time (the component does not "age"). In other words, the system has a constant hazard rate and the times involved have exponential distributions.

We will be concerned only with finite-state Markov models (models in which n is finite). A state is called *absorbing* if it is not possible to reach any outside state from that state. Once the system enters an absorbing state, it stays there until a new mission is started. If the kth state is absorbing, then $p_{kk} = 1$ and $p_{ki} = 0$ for all i except k. It can be shown that a finite-state homogeneous process is ergodic if every state can be reached from any other state with a nonzero probability. This condition implies that there are no zeros or ones in the stochastic transitional probability matrix **P**.

Each of the entries in **P** is a probability, and therefore, all the elements of the matrix should be between 0 and 1. Also, since each row consists of probabilities of a set of exhaustive events, the sum of the entries in each row of **P** must be unity. In other words, each row is a probability vector, and **P** is a stochastic matrix. Therefore, we have, for $i = 1, 2, \ldots, n$,

$$\sum_{j=1}^{n} p_{ij} = 1 \tag{6-2}$$

Let $S_1, S_2, \ldots, S_n$ be the possible states of the system, and let Δt be the specified time interval. Then the probability that the system is in state S_j at time $(t + \Delta t)$ is

$$p_{s_j}(t + \Delta t) = p_{1j}p_{s_1}(t) + p_{2j}p_{s_2}(t) + \ldots + p_{nj}p_{s_n}(t) \tag{6-3}$$

$$= \sum_{i=1}^{n} p_{ij}p_{s_i}(t)$$

6.3 DISCRETE MARKOV CHAINS

Let us consider a system with a state space consisting of three discrete states, S_1, S_2, and S_3. At each step in the evolution of the system behavior with time, let p_{ij} be the probability of transition from S_i to S_j. All this information can be exhibited in a state-space diagram, such as that shown in Figure 6.1. Also,

$$\mathbf{P} = \begin{bmatrix} p_{11} & p_{12} & p_{13} \\ p_{21} & p_{22} & p_{23} \\ p_{31} & p_{32} & p_{33} \end{bmatrix} \tag{6-4}$$

If $\mathbf{p}(0) = [p_1(0) \quad p_2(0) \quad p_3(0)]$ is the initial probability distribution vector, then, after one step, the probability distribution vector $\mathbf{p}(1)$ is

$$\mathbf{p}(1) = \mathbf{p}(0)\mathbf{P} = [p_1(1) \quad p_2(1) \quad p_3(1)] \tag{6-5}$$

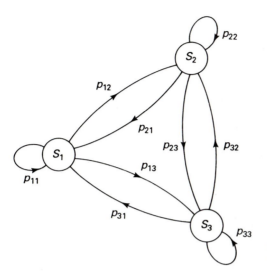

Figure 6.1 A three-state discrete Markov chain.

Similarly,

$$\mathbf{p}(2) = \mathbf{p}(1)\mathbf{P} = \mathbf{p}(0)\mathbf{P}\mathbf{P} = \mathbf{p}(0)\mathbf{P}^2 \tag{6-6}$$

and after k steps,

$$\mathbf{p}(k) = \mathbf{p}(0)\mathbf{P}^k \tag{6-7}$$

The matrix $\mathbf{P}^k = \overbrace{\mathbf{P}\mathbf{P}... \mathbf{P}}^{k \text{ times}}$ is the kth-step transition matrix. The typical entry in this matrix is $p_{ij}^{(k)}$, which is the probability of transition from S_i to S_j in exactly k steps. All the vectors, $\mathbf{p}(0)$, $\mathbf{p}(1)$, ... , $\mathbf{p}(k)$, ... , are probability vectors, and therefore, their elements are between 0 and 1 and they add up to 1. If any one entry is 1, then all the rest of the entries must be zero.

A stochastic matrix $\mathbf{P}$ is said to be *regular* if all of the entries of *some* power $\mathbf{P}^m$ are positive. If $\mathbf{P}$ is a regular stochastic matrix, then the sequence $\mathbf{P}$, $\mathbf{P}^2$, $\mathbf{P}^3$, ... , of the powers of $\mathbf{P}$ approaches the matrix $\mathbf{T}$ whose rows are each the fixed or limiting state probability vector $\boldsymbol{\alpha}$. In the long run, the probability that state S_j occurs is equal to the jth component α_j of $\boldsymbol{\alpha}$. For this reason, $\boldsymbol{\alpha}$ is also known as the *unique fixed probability vector* of $\mathbf{P}$, or the *stationary distribution* of the Markov chain. The effect of the initial state or the initial probability distribution wears off as the number of steps of the process increases. As mentioned earlier, such systems are said to be ergodic.

The limiting state vector $\boldsymbol{\alpha}$ is found by simultaneously solving

$$\boldsymbol{\alpha}\mathbf{P} = \boldsymbol{\alpha} \tag{6-8}$$

and

$$\sum_{j=1}^{n} \alpha_j = 1 \tag{6-9}$$

where

$$\boldsymbol{\alpha} = [\alpha_1 \quad \alpha_2 \quad ... \quad \alpha_n] \tag{6-10}$$

■■ **Example 6–1**

For the stochastic transitional probability matrix

$$\mathbf{P} = \begin{bmatrix} a & 1-a \\ 1-b & b \end{bmatrix}$$

with $0 < a < 1$ and $0 < b < 1$, the second-step transition matrix is

$$\mathbf{P}^2 = \mathbf{PP} = \begin{bmatrix} a^2 + (1-a)(1-b) & (1-a)(a+b) \\ (1-b)(a+b) & b^2 + (1-a)(1-b) \end{bmatrix}$$

We can easily verify that the sum of the entries in each row is equal to unity.
To find the limiting state vector $\boldsymbol{\alpha}$, let

$$\boldsymbol{\alpha} = [\alpha_1 \quad \alpha_2]$$

The requirement that $\boldsymbol{\alpha}\mathbf{P} = \boldsymbol{\alpha}$ results in the equations

$$\alpha_1 = a\alpha_1 + (1-b)\alpha_2$$

and

$$\alpha_2 = (1-a)\alpha_1 + b\alpha_2$$

Moreover,

$$\alpha_1 + \alpha_2 = 1$$

Solving these equations simultaneously, we obtain

$$\alpha_1 = \frac{1-b}{2-a-b}$$

$$\alpha_2 = \frac{1-a}{2-a-b}$$

For $a = \frac{1}{3}$ and $b = \frac{1}{4}$, the limiting state vector is

$$\boldsymbol{\alpha} = \begin{bmatrix} \dfrac{9}{17} & \dfrac{8}{17} \end{bmatrix}$$

As we compute higher and higher powers of $\mathbf{P}$, we approach the matrix

$$\mathbf{T} = \begin{bmatrix} \dfrac{9}{17} & \dfrac{8}{17} \\ \dfrac{9}{17} & \dfrac{8}{17} \end{bmatrix}$$

The determinant of $\mathbf{P}$ is

$$|\mathbf{P}| = ab - (1-a)(1-b)$$

$$= (a+b-1)$$

Since the determinant of a product of matrices is equal to the product of the determinants, we conclude that

$$| \mathbf{P}^n | = (a + b - 1)^n$$

We know that $\mathbf{P}^n$ is also a stochastic matrix. So let

$$\mathbf{P}^n = \begin{bmatrix} p_{11}^{(n)} & p_{12}^{(n)} \\ p_{21}^{(n)} & p_{22}^{(n)} \end{bmatrix}$$

or

$$\mathbf{P}^n = \begin{bmatrix} p_{11}^{(n)} & 1 - p_{11}^{(n)} \\ p_{21}^{(n)} & 1 - p_{21}^{(n)} \end{bmatrix}$$

from which it follows that

$$| \mathbf{P}^n | = p_{11}^{(n)} - p_{21}^{(n)}$$

Equating the two expressions derived for $| \mathbf{P}^n |$, we get

$$p_{21}^{(n)} = p_{11}^{(n)} - (a + b - 1)^n$$

We can now express all the entries in the nth-step transition matrix $\mathbf{P}^n$ in terms of $p_{11}^{(n)}$ as

$$\mathbf{P}^n = \begin{bmatrix} p_{11}^{(n)} & 1 - p_{11}^{(n)} \\ p_{11}^{(n)} - (a + b - 1)^n & 1 - p_{11}^{(n)} + (a + b - 1)^n \end{bmatrix}$$

A special case that occurs is when a is equal to unity. Then

$$\mathbf{P} = \begin{bmatrix} 1 & 0 \\ (1 - b) & b \end{bmatrix}$$

Since $\mathbf{P}$ is not regular now, we cannot talk about a limiting state probability vector $\boldsymbol{\alpha}$. As the process evolves, it will ultimately enter the first state and stay there for the rest of the time. This phenomenon is called *absorption*, and we will discuss it further in the next section.

Similarly, if b is equal to unity, then

$$\mathbf{P} = \begin{bmatrix} a & (1 - a) \\ 0 & 1 \end{bmatrix}$$

which is also not regular. In this case, the process will eventually be absorbed into the second state.   ■■

■■ **Example 6–2**

If the stochastic transitional probability matrix has all ones along the diagonal, then the rest of the entries must be zero, and $\mathbf{P}$ becomes the identity matrix $\mathbf{I}$. That is,

$$\mathbf{P} = \begin{bmatrix} 1 & 0 & 0 & \dots & 0 \\ 0 & 1 & 0 & \dots & 0 \\ \vdots & \vdots & \vdots & \vdots & \vdots \\ 0 & 0 & 0 & \dots & 1 \end{bmatrix} = \mathbf{I}$$

All the powers of **P** are also equal to **I**. In other words, once the initial probability distribution of finding the system in various states is known, we know the distribution forever! There are no transitions taking place, no matter how long we wait. Another way of saying the same thing is that every state is an absorbing state.   ■■

■■ **Example 6–3**

A person who is struggling hard not to put on too much weight has the following lunch habits: If that person has a heavy lunch on one day, then there is a 60% probability that he or she will have a light lunch the next day; on the other hand, there is an even chance that a light lunch on one day is followed by a light lunch on the next day. Examine this person's lunch habits in the long run.

Let us use HL and LL to denote heavy lunch and light lunch, respectively. Based on the available information, we can derive the transition matrix

$$\mathbf{P} = \begin{matrix} \text{HL} \\ \text{LL} \end{matrix} \begin{matrix} \text{HL} & \text{LL} \\ \begin{bmatrix} 0.4 & 0.6 \\ 0.5 & 0.5 \end{bmatrix} \end{matrix}$$

The long-run probabilities of having heavy and light lunches are given by the corresponding stationary distribution

$$[\alpha_1 \quad \alpha_2] = [\alpha_1 \quad \alpha_2]\begin{bmatrix} 0.4 & 0.6 \\ 0.5 & 0.5 \end{bmatrix}$$

where

$$\alpha_1 + \alpha_2 = 1$$

Solving these equations simultaneously, we obtain

$$\alpha = \begin{bmatrix} \dfrac{5}{11} & \dfrac{6}{11} \end{bmatrix}$$

Thus, in the long run, the probability of the person having a light lunch is $\frac{6}{11}$, or 54.545%, so there is still some hope for him or her to lose weight!

If there is a 60% chance of this person participating in a big luncheon meeting tomorrow, then, assuming

$$\mathbf{p}(0) = [0.6 \quad 0.4]$$

we get

$$\mathbf{p}(1) = p(0)\mathbf{P} = [0.44 \quad 0.56]$$

Hence, there is a 44% probability of having a heavy lunch and a 56% probability of having a light lunch the day after tomorrow.   ■■

6.4 ABSORBING MARKOV CHAINS

An *n-state absorbing Markov chain* has one or more (say, *r*) absorbing states and (*n − r*) nonabsorbing or transient states. Corresponding to each of the absorbing states, there will be a one in the main diagonal and a zero everywhere else in that row. We can always rearrange the stochastic transitional probability matrix so as to put all the absorbing states first and obtain

$$P = \begin{bmatrix} \mathbf{I} & \vdots & \mathbf{0} \\ \cdots\cdots\cdots & \vdots & \cdots\cdots\cdots \\ \mathbf{R} & \vdots & \mathbf{Q} \end{bmatrix} \begin{array}{l} r \text{ rows} \\ \\ (n-r) \text{ rows} \end{array}$$

$$\begin{array}{cc} r & (n-r) \\ \text{columns} & \text{columns} \end{array}$$

(6–11)

In this equation, $\mathbf{I}$ is an identity matrix of size r, and $\mathbf{Q}$ is called the *truncated matrix associated with* $\mathbf{P}$.

As the number of steps increases, eventually the process will enter one of the absorbing states. Often, it is useful to evaluate the average number of steps (or time intervals) the system resides in each of the transient states before entering an absorbing state.

Extending the procedure for finding the expected value in the case of discrete random variables, we can say that the matrix whose entries are the expected number of steps (or time intervals or visits) is

$$\mathbf{N} = \mathbf{1I} + \mathbf{1Q} + \mathbf{1Q}^2 + \mathbf{1Q}^3 + \dots \tag{6–12}$$

in which $\mathbf{I}$ is an identity matrix of size $(n - r)$. The rationale for this analysis lies in the fact that, as we consider higher and higher powers of $\mathbf{P}$, the region $\mathbf{I}$ remains $\mathbf{I}$, and powers of $\mathbf{Q}$ tend to zero since the entries are less than unity. To compute $\mathbf{N}$, we consider the identity

$$[\mathbf{I} - \mathbf{Q}][\mathbf{I} + \mathbf{Q} + \mathbf{Q}^2 + \dots + \mathbf{Q}^{n-1}] = \mathbf{I} - \mathbf{Q}^n \tag{6–13}$$

which can easily be verified by multiplying out the left-hand side. As n becomes large, $\mathbf{Q}^n \to \mathbf{0}$, and therefore, $[\mathbf{I} - \mathbf{Q}^n] \to \mathbf{I}$. In other words, for sufficiently large n, $[\mathbf{I} - \mathbf{Q}^n]$ must have a nonzero determinant, and therefore, $[\mathbf{I} - \mathbf{Q}^n]^{-1}$ exists. Since the determinant of a product of two matrices is equal to the product of the determinants, we conclude that $[\mathbf{I} - \mathbf{Q}]$ must have a nonzero determinant, and therefore, $[\mathbf{I} - \mathbf{Q}]^{-1}$ exists. Premultiplying both sides of the identity by $[\mathbf{I} - \mathbf{Q}]^{-1}$, we obtain

$$[\mathbf{I} - \mathbf{Q}]^{-1}[\mathbf{I} - \mathbf{Q}^n] = \sum_{k=0}^{n-1} \mathbf{Q}^k \tag{6–14}$$

As $n \to \infty$, we get

$$[\mathbf{I} - \mathbf{Q}]^{-1} = \sum_{k=0}^{\infty} \mathbf{Q}^k \equiv \mathbf{N} \tag{6–15}$$

The matrix $\mathbf{N}$ is called the *fundamental matrix of the absorbing Markov chain.*

The mean of the total number of times the process is in a given transient state is always finite, and such means are given by the elements of **N**. In general,

$$
\mathbf{N} = \begin{array}{c} S_{r+1} \\ S_{r+2} \\ \vdots \\ S_n \end{array}
\begin{bmatrix}
n_{11} & n_{12} & \cdots & n_{1,n-r} \\
n_{21} & n_{22} & \cdots & n_{2,n-r} \\
\vdots & \vdots & \vdots & \vdots \\
n_{n-r,1} & n_{n-r,2} & \cdots & n_{n-r,n-r}
\end{bmatrix}
\tag{6-16}
$$

If the process starts in the transient state S_{r+1}, then it will be in the same state an average of n_{11} times, it will be in the state S_{r+2} an average of n_{12} times, and so on. Therefore,

$$
\left. \begin{array}{r} \text{mean number of steps} \\ \text{before absorption} \end{array} \right\} = n_{11} + n_{12} + \cdots + n_{1,n-r}
\tag{6-17}
$$

$$
= \sum \text{elements in the row}
$$
$$
\text{corresponding to the}
$$
$$
\text{starting transient state}
$$

Finally, let b_{ij} be the probability that if the Markov chain starts in transient state S_i, it will be absorbed by entering the absorbing state S_j. Then the $(n - r)$-by-r matrix $[b_{ij}]$ is

$$
[b_{ij}] = \mathbf{B} = \mathbf{N}\mathbf{R}
\tag{6-18}
$$
$$
(n - r) \times (n - r) \qquad (n - r) \times r
$$

In reliability analysis, a system or component might have several operating states and a (catastrophic) failure state. Once the system enters the failure state, it cannot come out (it cannot be repaired). In such a case, the failure state will be the absorbing state, and the mean number of steps (or time intervals) the system takes before entering the absorbing state is of interest and can be found from the fundamental matrix **N**.

Even if we do not have a real-life absorbing state such as that mentioned in the previous paragraph, we can find the mean number of steps the system will operate in other states before entering a particular state—say, S_i—by designating S_i as an absorbing state and computing the corresponding fundamental matrix **N**. An example of such a system is a repairable system, in which case we are interested in the mean number of time intervals the system operates before failing.

■■ Example 6–4

Let us consider the simple two-state Markov chain of Example 6–3:

$$
\mathbf{P} = \begin{array}{c} \\ HL \\ LL \end{array}
\begin{array}{c} \overset{\displaystyle HL \quad LL}{} \\ \begin{bmatrix} 0.4 & 0.6 \\ 0.5 & 0.5 \end{bmatrix} \end{array}
$$

If we designate HL as an absorbing state, then the modified state transition matrix
is

$$\mathbf{P} = \begin{bmatrix} 1 & \vdots & 0 \\ \text{-- -- --} & \text{--} & \text{-- -- --} \\ 0.5 & \vdots & 0.5 \end{bmatrix}$$

and the truncated matrix $\mathbf{Q}$ is simply a one-by-one matrix. The fundamental matrix
is also one by one and is

$$\mathbf{N} = [\mathbf{I} - \mathbf{Q}]^{-1} = [(1 - 0.5)]^{-1} = [2]$$

With the modified matrix $\mathbf{P}$, HL becomes an absorbing state, and we are left with
only one transient state—namely, LL. Using the meaning of $\mathbf{N}$, we conclude that,
on average, if the person has a light lunch today, it will be two days before he or
she has a heavy lunch.

Next, we repeat the procedure with a slight modification: We designate LL
as the absorbing state and rearrange $\mathbf{P}$ to get

$$\mathbf{P} = \begin{array}{c} \\ LL \\ HL \end{array} \begin{array}{c} \overset{\displaystyle LL \quad\ HL}{\begin{bmatrix} 1 & \vdots & 0 \\ \text{-- -- --} & + & \text{-- -- --} \\ 0.6 & \vdots & 0.4 \end{bmatrix}} \end{array}$$

Now $\mathbf{N} = [(1 - 0.4)]^{-1} = [1.667]$. Thus, if the person has a heavy lunch today,
then, on average, it will be 1.667 days before he or she will have a light lunch.
Fractional days creep into consideration because of the long-term averaging that is
inherent in the procedure used.

Since in both cases we have only one absorbing state, the process has to be
absorbed by entering that state only. As such, the computation of $\mathbf{B}$ in both cases
using $\mathbf{B} = \mathbf{NR}$ results in [1], as expected.   ■■

■■   **Example 6–5**

In this example, we focus on a three-state Markov chain with the transition matrix

$$\mathbf{P} = \begin{array}{c} \\ S_1 \\ S_2 \\ S_3 \end{array} \begin{array}{c} \overset{\displaystyle S_1 \quad\ S_2 \quad\ S_3}{\begin{bmatrix} 0 & 0.5 & 0.5 \\ 0.3 & 0.7 & 0 \\ 0.2 & 0 & 0.8 \end{bmatrix}} \end{array}$$

Given that the system is in state S_3, we want to calculate the mean number of steps
or time intervals taken before it enters state S_2. To find this number, we first designate
S_2 as an absorbing state and rearrange $\mathbf{P}$ as

$$\mathbf{P} = \begin{array}{c} \\ S_2 \\ \\ S_1 \\ S_3 \end{array} \begin{array}{c} \overset{\displaystyle S_2 \qquad S_1 \qquad\ S_3}{\begin{bmatrix} 1 & \vdots & 0 & 0 \\ \text{-- -- --} & + & \text{-- -- -- -- --} \\ 0.5 & \vdots & 0 & 0.5 \\ 0 & \vdots & 0.2 & 0.8 \end{bmatrix}} \end{array}$$

Then

$$\mathbf{Q} = \begin{bmatrix} 0 & 0.5 \\ 0.2 & 0.8 \end{bmatrix}$$

and

$$\mathbf{R} = \begin{bmatrix} 0.5 \\ 0 \end{bmatrix}$$

The fundamental matrix is

$$\mathbf{N} = [\mathbf{I} - \mathbf{Q}]^{-1}$$

$$= \begin{bmatrix} 1 & -0.5 \\ -0.2 & 0.2 \end{bmatrix}^{-1}$$

$$= \begin{array}{c} \\ S_1 \\ S_3 \end{array} \begin{array}{c} S_1 \quad S_3 \\ \begin{bmatrix} 2 & 5 \\ 2 & 10 \end{bmatrix} \end{array}$$

Now we are ready to exploit the significance of the entries in the fundamental matrix. If the process starts in state S_3, then, on average, it will visit state S_1 2 times and will stay in state S_3 10 times before it is absorbed (makes a transition to state S_2). Therefore, the mean number of steps before the process enters state S_2 is equal to the sum of the entries in the second row of $\mathbf{N}$—that is, $(2 + 10)$, or 12. Also, if the process starts in state S_1, then the mean number of steps (or time intervals) before it enters state S_2 is the sum of the entries in the first row of $\mathbf{N}$—namely, $(2 + 5)$, or 7. Moreover,

$$\mathbf{B} = \mathbf{NR} = \begin{bmatrix} 2 & 5 \\ 2 & 10 \end{bmatrix} \begin{bmatrix} 0.5 \\ 0 \end{bmatrix}$$

$$= \begin{bmatrix} 1 \\ 1 \end{bmatrix}$$

This equation indicates that, whether the process starts in S_1 or S_3, it will eventually be absorbed by entering state S_2, the only absorbing state there is. This behavior should come as no surprise to us. ■■

■■ **Example 6–6**

The general mood of Pistol Pete, the mascot of Oklahoma State University, varies between cheerful, so-so, and glum, and his mood can change only overnight. If he is cheerful today, then there is a 60% chance that he will continue to be cheerful the next day, and the chances of his being so-so or glum are equally likely. If he is so-so today, then there is a 40% chance that he will continue to be so-so the next day and only a 30% chance that he will become glum. If he is glum today, then he is never cheerful the next day, and in fact, there is a 70% chance that he will continue to be glum the next day.

Let S_1, S_2, and S_3 denote the cheerful, so-so, and glum moods of Pistol. Based on the information available on the probabilities of his mood changes, we can con-

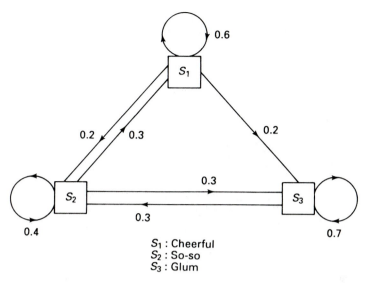

S_1 : Cheerful
S_2 : So-so
S_3 : Glum

Figure 6.2   State space diagram for Example 6–6.

struct the state-space diagram shown in Figure 6.2 and arrive at the corresponding stochastic transitional probability matrix

$$\mathbf{P} = \begin{array}{c} \\ S_1 \\ S_2 \\ S_3 \end{array} \begin{array}{ccc} S_1 & S_2 & S_3 \\ \left[\begin{array}{ccc} 0.6 & 0.2 & 0.2 \\ 0.3 & 0.4 & 0.3 \\ 0 & 0.3 & 0.7 \end{array} \right] \end{array}$$

Given that Pistol is so-so today, the initial probability distribution vector is $\mathbf{p}(0) = [0 \ \ 1 \ \ 0]$ and the probability distribution vector for tomorrow is

$$\mathbf{p}(1) = \mathbf{p}(0)\mathbf{P} = [0.3 \ \ 0.4 \ \ 0.3]$$

Thus, there is a 30% chance that Pistol will be cheerful tomorrow, a 40% chance that he will continue to be so-so, and a 30% chance that he will become glum. Also, given that a transition occurs, the conditional probability that he becomes glum before becoming cheerful is (0.3)/(0.3 + 0.3), or 50%.

The probability distribution vector for the day after tomorrow is $\mathbf{p}(2) = \mathbf{p}(1)\mathbf{P} = [0.3 \ \ 0.31 \ \ 0.39]$. Proceeding in a similar fashion, we obtain the probability distribution vectors for successive days as

$$\mathbf{p}(3) = [0.2730 \ \ 0.3010 \ \ 0.4260]$$

$$\mathbf{p}(4) = [0.2541 \ \ 0.3028 \ \ 0.4431]$$

$$\mathbf{p}(5) = [0.2433 \ \ 0.3049 \ \ 0.4518]$$

We readily observe that the probabilities of finding Pistol in various moods are converging to some steady-state values.

In the long run, the probabilities of finding Pistol in different moods are given by the elements of the unique fixed probability vector α obtained by simultaneously solving

$$\alpha P = \alpha$$

and

$$\sum_{i=1}^{3} \alpha_i = 1$$

We thus have the following equations (after renaming α_i as P_i) to solve simultaneously:

$$0.6P_1 + 0.3P_2 + 0P_3 \quad = P_1$$

$$0.2P_1 + 0.4P_2 + 0.3P_3 = P_2$$

$$0.2P_1 + 0.3P_2 + 0.7P_3 = P_3$$

$$P_1 + \quad P_2 + \quad P_3 = 1$$

Solution of these equations yields $P_1 = \dfrac{3}{13}$, $P_2 = \dfrac{4}{13}$, and $P_3 = \dfrac{6}{13}$. Hence, in the long run, Pistol will be cheerful for a fraction $\dfrac{3}{13}$, or 23.1%, of the time, so-so a fraction $\dfrac{4}{13}$, or 30.8%, of the time, and glum for a fraction $\dfrac{6}{13}$, or 46.1%, of the time.

Next, suppose Pistol is glum today, and we would like to find out how long, on average, it will be before he is cheerful. To do this, we will designate S_1 as an absorbing state and write $\mathbf{P}$ in partitioned form as

$$\mathbf{P} = \left[\begin{array}{c:cc} 1.0 & 0.0 & 0.0 \\ \hdashline 0.3 & 0.4 & 0.3 \\ 0.0 & 0.3 & 0.7 \end{array}\right]$$

The corresponding truncated matrix is extracted from $\mathbf{P}$ and is

$$\mathbf{Q} = \begin{bmatrix} 0.4 & 0.3 \\ 0.3 & 0.7 \end{bmatrix}$$

The fundamental matrix is

$$\mathbf{N} = [\mathbf{I} - \mathbf{Q}]^{-1} = \begin{bmatrix} 0.6 & -0.3 \\ -0.3 & 0.3 \end{bmatrix}^{-1} = \left(\frac{1}{0.09}\right) \begin{bmatrix} 0.3 & 0.3 \\ 0.3 & 0.6 \end{bmatrix}$$

$$= \begin{array}{c} \\ S_2 \\ S_3 \end{array} \begin{array}{c} \begin{array}{cc} S_2 & S_3 \end{array} \\ \left[\begin{array}{cc} \dfrac{10}{3} & \dfrac{10}{3} \\ \dfrac{10}{3} & \dfrac{20}{3} \end{array}\right] \end{array}$$

Consequently, if Pistol is glum today, then

$$\left.\begin{array}{r}\text{mean number of days before he is}\\\text{cheerful}\end{array}\right\} = \Sigma \text{ entries in the 2nd row}$$

$$= \frac{10}{3} + \frac{20}{3} = 10 \text{ days}$$

If Pistol is so-so today, then

$$\left.\begin{array}{r}\text{mean number of days before he is}\\\text{cheerful}\end{array}\right\} = \Sigma \text{ entries in the 1st row}$$

$$= \frac{10}{3} + \frac{10}{3} = 6.67 \text{ days}$$

The mean number of days before Pistol is glum can be obtained by designating S_3 as the absorbing state. Rearranging the entries in **P**, we have

$$\mathbf{P} = \begin{array}{c} \\ S_3 \\ S_1 \\ S_2 \end{array} \begin{array}{c} \begin{array}{ccc} S_3 & S_1 & S_2 \end{array} \\ \left[\begin{array}{ccc} 1 & 0 & 0 \\ 0.2 & 0.6 & 0.2 \\ 0.3 & 0.3 & 0.4 \end{array}\right]\end{array}$$

The corresponding truncated matrix is

$$\mathbf{Q} = \begin{bmatrix} 0.6 & 0.2 \\ 0.3 & 0.4 \end{bmatrix}$$

and the fundamental matrix is calculated as before. Thus, we get

$$\mathbf{N} = [\mathbf{I} - \mathbf{Q}]^{-1}$$

$$= \begin{array}{c} \\ S_1 \\ S_2 \end{array} \begin{array}{c} \begin{array}{cc} S_1 & S_2 \end{array} \\ \left[\begin{array}{cc} \dfrac{10}{3} & \dfrac{10}{9} \\ \dfrac{5}{3} & \dfrac{20}{9} \end{array}\right]\end{array}$$

Hence, if Pistol is cheerful today, then

$$\text{mean number of days before he is glum} = \Sigma \text{ entries in 1st row}$$

$$= 4.44 \text{ days}$$

If Pistol is so-so today, then

$$\text{mean number of days before he is glum} = \Sigma \text{ entries in 2nd row}$$

$$= 3.89 \text{ days}$$

A similar procedure can be used to obtain the mean number of days before Pistol is so-so by designating S_2 as the absorbing state.   ■■

■■ **Example 6–7**

A single-chain birth-and-death process has probabilities of transition that satisfy the condition

$$p_{ij} = 0 \text{ for } j \neq (i - 1),\ i,\ \text{and } (i + 1)$$

The corresponding stochastic transitional probability matrix is

$$\mathbf{P} = \begin{bmatrix}
p_{11} & p_{12} & 0 & 0 & 0 & \cdots & 0 & 0 & 0 \\
p_{21} & p_{22} & p_{23} & 0 & 0 & \cdots & 0 & 0 & 0 \\
0 & p_{32} & p_{33} & p_{34} & 0 & \cdots & 0 & 0 & 0 \\
0 & 0 & p_{43} & p_{44} & p_{45} & \cdots & 0 & 0 & 0 \\
\vdots & \vdots & \vdots & \vdots & \vdots & \vdots & \vdots & \vdots & \vdots \\
0 & 0 & 0 & 0 & 0 & \cdots & p_{n-1,n-2} & p_{n-1,n-1} & p_{n-1,n} \\
0 & 0 & 0 & 0 & 0 & \cdots & 0 & p_{n,n-1} & p_{nn}
\end{bmatrix}$$

and the state-space diagram has the simple structure shown in Figure 6.3. In the figure, the birth-death notation is used to denote the transition probabilities thus:

$$p_{i,i+1} = b_i$$

$$p_{i,i-1} = d_i$$

Moreover, **P** is a stochastic matrix, and all the entries in a row add up to unity. The diagonal entry can be expressed in terms of the off-diagonal values as

$$p_{ii} = 1 - p_{i,i-1} - p_{i,i+1}$$

$$= 1 - d_i - b_i$$

The matrix **P** can be rewritten in terms of the birth-and-death notation as

$$\mathbf{P} = \begin{bmatrix}
1 - b_1 & b_1 & 0 & 0 & \cdots & 0 & 0 \\
d_2 & 1 - (b_2 + d_2) & b_2 & 0 & \cdots & 0 & 0 \\
0 & d_3 & 1 - (b_3 + d_3) & b_3 & \cdots & 0 & 0 \\
\vdots & \vdots & \vdots & \vdots & \vdots & \vdots & \vdots \\
0 & 0 & 0 & 0 & \cdots & d_n & 1 - d_n
\end{bmatrix}$$

The elements of the unique fixed probability vector **α** are obtained by solving

$$\alpha \mathbf{P} = \alpha$$

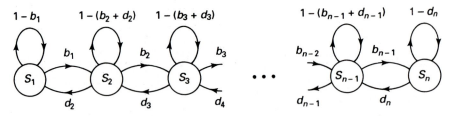

Figure 6.3 State-space diagram for a single-chain birth-and-death process.

and

$$\sum_{i=1}^{n} \alpha_i = 1$$

Expanding these equations, we have

$$\alpha_1(1 - b_1) + \alpha_2 d_2 = \alpha_1$$

$$\alpha_1 b_1 + \alpha_2[1 - b_2 - d_2] + \alpha_3 d_3 = \alpha_2$$

$$\alpha_2 b_2 + \alpha_3[1 - b_3 - d_3] + \alpha_4 d_4 = \alpha_3$$

$$\vdots$$

$$\alpha_{n-1} b_{n-1} + \alpha_n(1 - d_n) = \alpha_n$$

$$\alpha_1 + \alpha_2 + \ldots + \alpha_n = 1$$

Arranging these in a compact form, we get

$$\alpha_{i+1} = \frac{\alpha_i b_i}{d_{i+1}}, \quad i = 1, 2, \ldots, (n-1)$$

or, alternatively,

$$\alpha_{i+1} = \frac{\prod\limits_{k=1}^{i} b_k}{\prod\limits_{k=2}^{i+1} d_k} \alpha_1 \text{ for } i = 1, 2, 3, \ldots, (n-1)$$

By using the last equation above to express all the elements of $\boldsymbol{\alpha}$ in terms of α_1 in

$$\alpha_1 + \alpha_2 + \ldots + \alpha_n = 1$$

we can find $\boldsymbol{\alpha}$. The condition

$$\alpha_i b_i = \alpha_{i+1} d_{i+1}$$

implies that, in the long run, the probability of a transition from state S_i to state S_{i+1} is equal to the probability of transition from state S_{i+1} to state S_i during the next trial. Many practical problems fall into this category. A simple numerical example is used to illustrate this point next. ■■

■■   **Example 6–8**

The great running back Joe Teedee usually makes several touchdowns for his team. If he makes a touchdown in one quarter, the probability of his making another touchdown in the next quarter is 0.7, and if he does not make a touchdown in one quarter, then there is an even chance of his not making a touchdown in the next quarter also.

The state-space diagram illustrating Joe's exploits is shown in Figure 6.4. The associated stochastic transitional probability matrix is

$$\mathbf{P} = \begin{array}{c} \\ S_1 \\ S_2 \end{array} \begin{array}{cc} S_1 & S_2 \\ \left[\begin{array}{cc} 0.7 & 0.3 \\ 0.5 & 0.5 \end{array} \right. & \left. \begin{array}{l} \end{array} \right] \end{array} \begin{array}{l} \textit{makes touchdown} \\ \textit{does not make touchdown} \end{array}$$

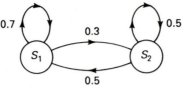

Makes
touchdown

Does not make
touchdown

Figure 6.4   State-space diagram for
Example 6–8.

If Joe makes a touchdown in the first quarter of the football season, then the probabilities of his making and of his not making a touchdown during the fourth quarter are obtained using

$$[1 \quad 0]\begin{bmatrix} 0.7 & 0.3 \\ 0.5 & 0.5 \end{bmatrix}^3 = [1 \quad 0]\begin{bmatrix} 0.628 & 0.372 \\ 0.620 & 0.380 \end{bmatrix}$$

$$= [0.628 \quad 0.372]$$

Thus, there is a 62.8% probability that Joe will make a touchdown during the fourth quarter and a 37.2% probability that he will not.

The probability of Joe's making a touchdown in the 30th quarter of the season is given by the long-term probability of state S_1, which is α_1 in the unique fixed probability vector $\boldsymbol{\alpha} = [\alpha_1 \quad \alpha_2]$. We have

$$\alpha_2 = \alpha_1 \begin{bmatrix} 0.3 \\ \overline{0.5} \end{bmatrix}$$

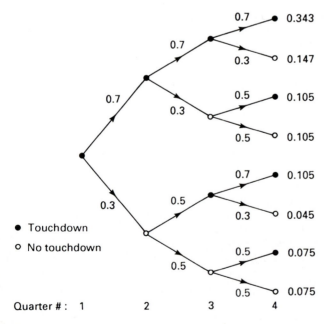

● Touchdown

○ No touchdown

Quarter # :   1   2   3   4

Figure 6.5   Tree diagram for Example 6–8.

and

$$\alpha_1 + \alpha_2 = 1$$

Solving these equations simultaneously, we obtain $\alpha_1 = \frac{5}{8}$ and $\alpha_2 = \frac{3}{8}$. Therefore, the probability of Joe's making a touchdown in the 30th quarter is $\frac{5}{8}$. We have made the tacit assumption that Joe does not get hurt during this period and that the "process" of his scoring touchdowns has "settled down" by the 30th quarter, allowing us to use the fixed probability vector.

The probability of Joe's making touchdowns both in the 30th and in the 31st quarters is $(\frac{5}{8})(0.7)$. It is not equal to $(\frac{5}{8})(\frac{5}{8})$, because if Joe makes a touchdown in one quarter, the probability of his making another touchdown in the next quarter is 0.7.

The probability of Joe's making touchdowns in the 3rd and 33rd quarters is $(0.628)(\frac{5}{8})$. These two quarters are sufficiently far apart that events that occur during them can be considered independent.

The tree diagram of Figure 6.5 helps us to answer questions involving conditional probability. For example, if Joe did not make a touchdown during the third quarter, then the probability that he made a touchdown in the second quarter is obtained as follows. Let TD_i denote the event *making a touchdown in the ith quarter* and NTD_i denote the event *not making a touchdown in the ith quarter*. Then

$$P(TD_2 \mid NTD_3) = \frac{P(TD_2 \text{ and } NTD_3)}{P(NTD_3)}$$

$$= \frac{0.7 \times 0.3}{0.7 \times 0.3 + 0.3 \times 0.5}$$

$$= 0.5833$$

In the same way, if Joe made a touchdown during the third quarter, then the probability that he also made a touchdown in the second quarter is calculated as follows:

$$P(TD_2 \mid TD_3) = \frac{P(TD_2 \text{ and } TD_3)}{P(TD_3)}$$

$$= \frac{0.7 \times 0.7}{0.7 \times 0.7 + 0.3 \times 0.5}$$

$$= 0.7656 \qquad\blacksquare\blacksquare$$

6.5 CONTINUOUS MARKOV PROCESSES

Markov processes are stochastic processes because they develop in time in a manner controlled by probabilistic laws. They can be used in a wide variety of reliability problems. In particular, we will be interested in homogeneous discrete-state continuous-time Markov processes and assume that the system involved can exist in one of n discrete states, $S_1, S_2, \ldots, S_n$. Transitions occur from one state to another as determined by a set of numbers ρ_{ij} (called rates of departure), and the system continuously stays in one state between transitions. Let

$$\rho_{ij} = \text{rate of departure from state } S_i \text{ to state } S_j$$

Then

$$\rho_{ij}\Delta t = P_{ij} = \text{probability of transition from } S_i \text{ to } S_j \text{ in the time interval } \Delta t;$$

(note the change in notation from lower case to upper case as compared to the one used in discussing discrete Markov chains).

If $P_i(t)$ is the probability of finding the system in S_i at time t, then the probability of finding the system in S_i at time $(t + \Delta t)$ is

$$P_i(t + \Delta t) = \sum_{\substack{j=1 \\ j \neq i}}^{n} \rho_{ji}\,\Delta t P_j(t) + [1 - \sum_{\substack{j=1 \\ j \neq i}}^{n} \rho_{ij}\Delta t]P_i(t) \qquad (6\text{--}19)$$

Rearranging and dividing by Δt, we get

$$\frac{P_i(t + \Delta t) - P_i(t)}{\Delta t} = \sum_{\substack{j=1 \\ j \neq i}}^{n} \rho_{ji}P_j(t) - [P_i(t)] \sum_{\substack{j=1 \\ j \neq i}}^{n} \rho_{ij} \qquad (6\text{--}20)$$

Taking the limit as $\Delta t \to 0$ leads to

$$P_i'(t) = \sum_{\substack{j=1 \\ j \neq i}}^{n} \rho_{ji}P_j(t) - [P_i(t)] \sum_{\substack{j=1 \\ j \neq i}}^{n} \rho_{ij} \qquad (6\text{--}21)$$

For $i = 1, 2, \ldots, n$, we have n such first-order differential equations that can be written in standard vector matrix form as

$$
\begin{bmatrix} P_1'(t) \\ P_2'(t) \\ \vdots \\ P_n'(t) \end{bmatrix}
=
\begin{bmatrix}
-\sum_{j=2}^{n} \rho_{1j} & \rho_{21} & \cdots & \rho_{n1} \\
\rho_{12} & -\sum_{\substack{j=1 \\ j \neq 2}}^{n} \rho_{2j} & \cdots & \rho_{n2} \\
\vdots & \vdots & \vdots & \vdots \\
\rho_{1n} & \rho_{2n} & \cdots & -\sum_{j=1}^{n-1} \rho_{nj}
\end{bmatrix}
\begin{bmatrix} P_1(t) \\ P_2(t) \\ \vdots \\ P_n(t) \end{bmatrix} \qquad (6\text{--}22)
$$

The time-dependent system state probabilities can be obtained by solving this set of differential equations and by using the appropriate initial conditions. When n is large, numerical techniques are used to solve the equations.

It is useful to compare the matrix of rates of departures with the coefficient matrix of the Markov differential equations. For convenience, the two matrices are shown side by side for a four-state system:

$$
\begin{array}{cccc}
 & S_1 & S_2 & S_3 & S_4 \\
S_1 & \rho_{11} & \rho_{12} & \rho_{13} & \rho_{14} \\
S_2 & \rho_{21} & \rho_{22} & \rho_{23} & \rho_{24} \\
S_3 & \rho_{31} & \rho_{32} & \rho_{33} & \rho_{34} \\
S_4 & \rho_{41} & \rho_{42} & \rho_{43} & \rho_{44}
\end{array} \quad ; \quad
\begin{bmatrix}
-\sum_{j=2}^{4} \rho_{1j} & \rho_{21} & \rho_{31} & \rho_{41} \\[2em]
\rho_{12} & -\sum_{\substack{j=1 \\ j\neq 2}}^{4} \rho_{2j} & \rho_{32} & \rho_{42} \\[2em]
\rho_{13} & \rho_{23} & -\sum_{\substack{j=1 \\ j\neq 3}}^{4} \rho_{3j} & \rho_{43} \\[2em]
\rho_{14} & \rho_{24} & \rho_{34} & -\sum_{j=1}^{3} \rho_{4j}
\end{bmatrix}
$$

Plainly, the coefficient matrix of the Markov differential equations is obtained by transposing the ρ-matrix and replacing the diagonal entries by the negative of the sum of all the rest of the entries in the corresponding columns.

The stochastic transitional probability matrix **P** for an n-state Markov process can be obtained by considering a small time interval Δt that is small enough that the probability of more than one transition occurring is zero.

For an n-state process, let the ρ-matrix be

$$
\begin{bmatrix}
\rho_{11} & \rho_{12} & \cdots & \rho_{1n} \\
\rho_{21} & \rho_{22} & \cdots & \rho_{2n} \\
\rho_{n1} & \rho_{n2} & \cdots & \rho_{nn}
\end{bmatrix}
$$

Then the corresponding matrix **P** is

$$
\begin{bmatrix}
\left(1 - \sum_{j=2}^{n} \rho_{1j}\,\Delta t\right) & \rho_{12}\,\Delta t & \cdots & \rho_{1n}\,\Delta t \\[2em]
\rho_{21}\,\Delta t & \left(1 - \sum_{\substack{j=1 \\ j\neq 2}}^{n} \rho_{2j}\,\Delta t\right) & \cdots & \rho_{2n}\,\Delta t \\[1.5em]
\vdots & \vdots & \vdots & \vdots \\
\rho_{n1}\,\Delta t & \rho_{n2}\,\Delta t & \cdots & \left(1 - \sum_{j=1}^{n-1} \rho_{nj}\,\Delta t\right)
\end{bmatrix}
$$

The matrix **P** is truly a stochastic matrix in that all the entries are between zero and one and all the entries in each row add up to unity. Thus, each row is a probability vector.

In many cases, in working with **P**, the term Δt drops out. Therefore, for the sake of convenience, in writing **P**, all the Δt's are often omitted. In this form, we cannot claim that **P** is a stochastic matrix, since the entries are not probabilities but are transition rates. However, written without the Δt's, the matrix **P** has many uses and is employed frequently. This form is as follows:

$$\begin{bmatrix} \left(1 - \sum_{j=2}^{n} \rho_{1j}\right) & \rho_{12} & \cdots & \rho_{1n} \\ \rho_{21} & \left(1 - \sum_{\substack{j=1 \\ j\neq 2}}^{n} \rho_{2j}\right) & \cdots & \rho_{2n} \\ \vdots & \vdots & \vdots & \vdots \\ \rho_{n1} & \rho_{n2} & \cdots & \left(1 - \sum_{j=1}^{n-1} \rho_{nj}\right) \end{bmatrix}$$

■■ Example 6–9

A monkey named Ms. M is involved in a psychological experiment. She is confined to two rooms with a passageway in between. Whenever Ms. M is in room 1, the probability that she will enter room 2 during the next time interval Δt is found to be $\lambda_{12}\Delta t$. Similarly, $\lambda_{21}\Delta t$ is the probability that she will go from room 2 to room 1 during the time interval Δt.

With just these two pieces of data, we can deduce a lot of information on the behavior of Ms. M. The state-space diagram (or the state transition diagram) can be drawn either in terms of the transition probabilities or in terms of the rates of departures, as shown in Figure 6.6.

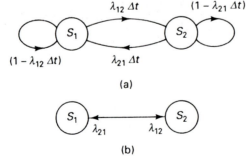

Figure 6.6   Discrete-state continuous-transition Markov process representation for the case of two discrete states (a) in terms of transition probabilities and (b) in terms of rates of departures.

Let S_1 denote the state corresponding to Ms. M's being in room 1 and S_2 denote the state corresponding to Ms. M's being in room 2. Then the matrix of rates of departures—that is, the ρ-matrix—is

$$\begin{array}{cc} & \begin{array}{cc} S_1 & S_2 \end{array} \\ \begin{array}{c} S_1 \\ S_2 \end{array} & \begin{bmatrix} 0 & \lambda_{12} \\ \lambda_{21} & 0 \end{bmatrix} \end{array}$$

The coefficient matrix of the corresponding Markov differential equations is

$$\begin{array}{cc} & \begin{array}{cc} S_1 & S_2 \end{array} \\ \begin{array}{c} S_1 \\ S_2 \end{array} & \left[\begin{array}{cc} -\lambda_{12} & \lambda_{21} \\ \lambda_{12} & -\lambda_{21} \end{array}\right] \end{array}$$

The Markov differential equations in vector-matrix notation are

$$\begin{bmatrix} P_1'(t) \\ P_2'(t) \end{bmatrix} = \begin{bmatrix} -\lambda_{12} & \lambda_{21} \\ \lambda_{12} & -\lambda_{21} \end{bmatrix} \begin{bmatrix} P_1(t) \\ P_2(t) \end{bmatrix}$$

Once the initial conditions are known, these two first-order linear differential equations can be solved to find $P_1(t)$ and $P_2(t)$.

In the long run (meaning under steady-state conditions), all time derivatives disappear, and we have

$$\begin{bmatrix} 0 \\ 0 \end{bmatrix} = \begin{bmatrix} -\lambda_{12} & \lambda_{21} \\ \lambda_{12} & -\lambda_{21} \end{bmatrix} \begin{bmatrix} P_{10} \\ P_{20} \end{bmatrix}$$

where P_{10} and P_{20} are the long-run probabilities of finding Ms. M in rooms 1 and 2, respectively. To find P_{10} and P_{20}, we have to include

$$P_{10} + P_{20} = 1$$

in our set of equations. Solution of these equations gives

$$P_{10} = \frac{\lambda_{21}}{\lambda_{12} + \lambda_{21}}$$

$$P_{20} = \frac{\lambda_{12}}{\lambda_{12} + \lambda_{21}}$$

The use of models such as this in system reliability evaluation will be discussed in detail in the next chapter. ■■

■■ **Example 6–10**

A discrete-state continuous-transition birth-and-death process is shown in Figure 6.7 for the case of three discrete states. This is a special case of an n-state process for which $\lambda_{ij} = 0$ if $j \neq (i - 1)$ or $(i + 1)$.

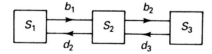

Figure 6.7 Discrete-state continuous-time birth-and-death process for three discrete states.

The matrix of rates of departures is assembled first. It is

$$\begin{array}{c} \begin{array}{ccc} & 1 & & 2 & & 3 \end{array} \\ \begin{array}{c} 1 \\ 2 \\ 3 \end{array} \left[\begin{array}{ccc} 0 & b_1 & 0 \\ d_2 & 0 & b_2 \\ 0 & d_3 & 0 \end{array}\right] \end{array}$$

The corresponding Markov differential equations can be written down immediately as

$$\begin{bmatrix} P_1'(t) \\ P_2'(t) \\ P_3'(t) \end{bmatrix} = \begin{bmatrix} -b_1 & d_2 & 0 \\ b_1 & -(d_2 + b_2) & d_3 \\ 0 & b_2 & -d_3 \end{bmatrix} \begin{bmatrix} P_1(t) \\ P_2(t) \\ P_3(t) \end{bmatrix}$$

These three first-order differential equations can be solved simultaneously to find the state probabilities as functions of time for prescribed initial conditions. Obviously, the three state probabilities must add up to unity at all times.

By definition, under steady-state conditions, all the derivatives of the state probabilities must vanish. Therefore, we have

$$\begin{bmatrix} 0 \\ 0 \\ 0 \end{bmatrix} = \begin{bmatrix} -b_1 & d_2 & 0 \\ b_1 & -(d_2 + b_2) & d_3 \\ 0 & b_2 & -d_3 \end{bmatrix} \begin{bmatrix} P_{10} \\ P_{20} \\ P_{30} \end{bmatrix}$$

in which P_{i0} is the steady-state probability of finding the system in state P_i. A quick check reveals that the determinant of the coefficient matrix is zero. Therefore, these three equations are not independent and cannot be solved to find the steady-state probabilities. Fortunately, we can add to them the condition

$$P_{10} + P_{20} + P_{30} = 1$$

Now we can pick any two of the three steady-state equations and include the equation added to solve for the steady-state probabilities. Picking the first and the third, we have to solve simultaneously

$$-b_1 P_{10} + d_2 P_{20} = 0$$

$$b_2 P_{20} - d_3 P_{30} = 0$$

$$P_{10} + P_{20} + P_{30} = 1$$

The results are

$$P_{10} = \frac{d_2 d_3}{d_2 d_3 + b_1 d_3 + b_2 b_1}$$

$$P_{20} = \frac{b_1 d_3}{d_2 d_3 + b_1 d_3 + b_2 b_1}$$

$$P_{30} = \frac{b_1 b_2}{d_2 d_3 + b_1 d_3 + b_2 b_1}$$

The steady-state equations also tell us that, under steady-state conditions,

$$P_{10} b_1 = P_{20} d_2$$

$$P_{20} b_2 = P_{30} d_3$$

$$P_{20}(d_2 + b_2) = P_{10} b_1 + P_{30} d_3$$

The implications of these equations are very interesting and useful. In the next chapter, we will see that the frequency of encountering a state is equal to the steady-state probability of being in that state times the total rate of departure from that state. What the foregoing equations tell us is that, under steady-state conditions, the number of transitions out of a state should be equal to the number of transitions into that state.   ■■

■■ **Example 6–11**

People arrive at a service facility that can handle only one customer at a time. If a person arrives and finds that the facility is occupied, then he or she leaves and does not become a customer. The average arrival rate of people is 4 per day, and the mean time taken to service a customer is 1 hour.

Let S_0 and S_1 be the events that there are no customers and that there is one customer at the facility. We want to find the probabilities of having no customers or one customer as functions of time.

Using hours as time units, we see that the arrival rate is $\frac{1}{6}$ customer per hour. If the facility is occupied continuously, with no time gaps, then it can service 24 customers per day, or 1 customer per hour. Figure 6.8 shows the state-space diagram for the facility, with transition probabilities for an incremental time interval Δt included. Often, the state-space diagram is drawn with only rates of departures indicated. This situation is shown in Figure 6.9 for this example.

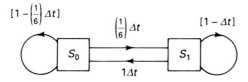

Figure 6.8   State-space diagram for Example 6–11, with transition probabilities for an incremental time Δt.

Figure 6.9   State-space diagram for the system in Figure 6.8, with only transition rates indicated.

The matrix of rates of departures is

$$
\begin{array}{c}
\\
S_0 \\
S_1
\end{array}
\begin{array}{cc}
S_0 & S_1 \\
\left[\begin{array}{cc}
0 & \dfrac{1}{6} \\
1 & 0
\end{array}\right]
\end{array}
$$

The stochastic transitional probability matrix is

$$
\mathbf{P} = \left[\begin{array}{cc}
1 - \left(\dfrac{1}{6}\right)\Delta t & \left(\dfrac{1}{6}\right)\Delta t \\
\Delta t & 1 - \Delta t
\end{array}\right]
$$

The Markov differential equations in vector-matrix notation are given as

$$\begin{bmatrix} P_0'(t) \\ P_1'(t) \end{bmatrix} = \begin{bmatrix} -\left(\dfrac{1}{6}\right) & 1 \\ \dfrac{1}{6} & -1 \end{bmatrix} \begin{bmatrix} P_0(t) \\ P_1(t) \end{bmatrix}$$

Also, the probabilities of the two states must add up to 1 at all times. That is,

$$P_0(t) + P_1(t) = 1$$

Next, we will solve the Markov differential equations with initial conditions $P_0(0) = p$ and $P_1(0) = 1 - p$. Replacing $P_1(t)$ by $[1 - P_0(t)]$ in the first differential equation, we obtain

$$P_0'(t) + \left(\frac{7}{6}\right) P_0(t) = 1$$

The solution of this first-order differential equation with constant coefficients for

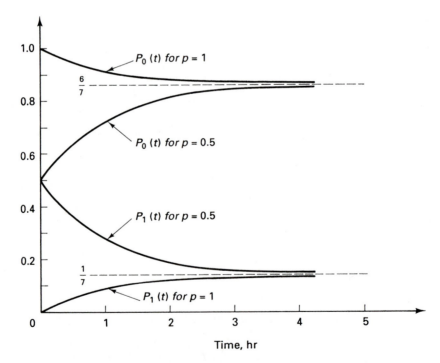

Figure 6.10   Variation of $P_0(t)$ and $P_1(t)$ with time for Example 6–11.

the initial condition $P_0(0) = p$ is

$$P_0(t) = \left[p - \frac{6}{7} \right] e^{-(7/6)t} + \frac{6}{7} \text{ for } t \geq 0$$

By subtracting $P_0(t)$ from 1, we obtain

$$P_1(t) = \frac{1}{7} - \left[p - \frac{6}{7} \right] e^{-(7/6)t} \text{ for } t \geq 0$$

Figure 6.10 sketches $P_0(t)$ and $P_1(t)$ for two different values (1 and 0.5) of the initial condition p. By letting t tend to infinity, steady-state probabilities of finding the facility in states S_0 and S_1 can be found. They are

$$P_0 = \lim_{t \to \infty} P_0(t) = \frac{6}{7}$$

and

$$P_1 = \lim_{t \to \infty} P_1(t) = \frac{1}{7}$$

In the long run, then the probability of having no customers is $\frac{6}{7}$, or 85.7%, and the probability of having one customer is $\frac{1}{7}$, or 14.3%. Obviously, this statement assumes that the facility is open continuously (24 hours a day, 7 days a week).

Since the rate of departure from S_1 to S_0 is six times the rate of departure from S_0 to S_1, we are six times as likely to see the facility with no customers as with one customer. Later, we will see that the mean residence time in a state of a Markov process is equal to the reciprocal of the total rate of departure from that state. Applying this relationship to the example at hand, we find that the mean residence time in state S_0 is 6 hours and in state S_1 is 1 hour.   ■■

If the ith state of an n-state Markov process is absorbing, then the corresponding matrix of departure rates will have a zero ith row:

$$i\text{th row} \rightarrow \begin{bmatrix} p_{11} & p_{12} & \cdots & p_{1,i-1} & p_{1i} & p_{1,i+1} & \cdots & p_{1n} \\ p_{21} & p_{22} & \cdots & p_{2,i-1} & p_{2i} & p_{2,i+1} & \cdots & p_{2n} \\ \vdots & \vdots & \vdots & \vdots & \vdots & \vdots & \vdots \\ p_{i-1,1} & p_{i-1,2} & \cdots & p_{i-1,i-1} & p_{i-1,i} & p_{i-1,i+1} & \cdots & p_{i-1,n} \\ 0 & 0 & \cdots & 0 & 0 & 0 & \cdots & 0 \\ p_{i+1,1} & p_{i+1,2} & \cdots & p_{i+1,i-1} & p_{i+1,i} & p_{i+1,i+1} & \cdots & p_{i+1,n} \\ \vdots & \vdots & \vdots & \vdots & \vdots & \vdots \\ p_{n1} & p_{n2} & & p_{n,i-1} & p_{ni} & p_{n,i+1} & \cdots & p_{nn} \end{bmatrix}$$

The associated Markov differential equations will have the following coefficient matrix, with a zero in the ith column.

$$
\begin{bmatrix}
-\sum_{j=2}^{n}\rho_{1j} & \rho_{21} & \cdots & \rho_{i-1,1} & 0 & \rho_{i+1,1} & \cdots & \rho_{n1} \\[2ex]
\rho_{12} & -\sum_{\substack{j=1\\j\neq2}}^{n}\rho_{2j} & \cdots & \rho_{i-1,2} & 0 & \rho_{i+1,2} & \cdots & \rho_{n2} \\[2ex]
\vdots & \vdots & & \vdots & \vdots & \vdots & \vdots & \vdots \\[1ex]
\rho_{1,i-1} & \rho_{2,i-1} & \cdots & -\sum_{\substack{j=1\\j\neq(i-1)}}^{n}\rho_{i-1,j} & 0 & \rho_{i+1,i-1} & \cdots & \rho_{n,i-1} \\[2ex]
\rho_{1i} & \rho_{2i} & \cdots & \rho_{i-1,i} & 0 & \rho_{i+1,i} & \cdots & \rho_{ni} \\[2ex]
\rho_{1,i+1} & \rho_{2,i+1} & \cdots & \rho_{i-1,i+1} & 0 & -\sum_{\substack{j=1\\j\neq(i+1)}}^{n}\rho_{i+1,j} & \cdots & \rho_{n,i+1} \\[2ex]
\vdots & \vdots & & \vdots & \vdots & \vdots & & \vdots \\[1ex]
\rho_{1n} & \rho_{2n} & \cdots & \rho_{i-1,n} & 0 & \rho_{i+1,n} & \cdots & -\sum_{j=1}^{n-1}\rho_{nj}
\end{bmatrix}
$$

↑
ith column   (6–23)

Thus, it is easy to identify the existence of absorbing states by a visual inspection of either the ρ-matrix or the coefficient matrix of the Markov differential equations.

■■ Example 6–12

The Markov process illustrated in the state-space diagram of Figure 6.11 has two absorbing states (S_1 and S_3) and three transient states (S_2, S_4, and S_5). The associated matrices are the matrix of rates of departure,

$$
\begin{array}{c}
\\ S_1 \\ S_2 \\ S_3 \\ S_4 \\ S_5
\end{array}
\begin{array}{c}
\begin{array}{ccccc} S_1 & S_2 & S_3 & S_4 & S_5 \end{array} \\
\begin{bmatrix}
0 & 0 & 0 & 0 & 0 \\
\lambda_1 & 0 & \lambda_2 & 0 & \lambda_3 \\
0 & 0 & 0 & 0 & 0 \\
\mu_2 & 0 & \lambda_5 & 0 & \lambda_4 \\
0 & \mu_3 & \lambda_6 & \mu_4 & 0
\end{bmatrix}
\end{array}
$$

and the coefficient matrix of Markov differential equations,

$$
\begin{bmatrix}
0 & \lambda_1 & 0 & \mu_2 & 0 \\
0 & -(\lambda_1 + \lambda_2 + \lambda_3) & 0 & 0 & \mu_3 \\
0 & \lambda_2 & 0 & \lambda_5 & \lambda_6 \\
0 & 0 & 0 & -(\lambda_4 + \lambda_5 + \mu_2) & \mu_4 \\
0 & \lambda_3 & 0 & \lambda_4 & -(\mu_3 + \mu_4 + \lambda_6)
\end{bmatrix}
$$

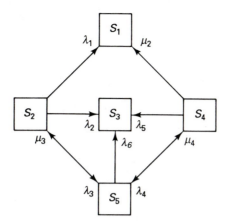

Figure 6.11   State-space diagram for Example 6–12.

If we rearrange the entries such that all the absorbing states are collected toward the top, then the matrix of rates of departure becomes

$$
\begin{array}{c c}
 & \begin{array}{c c c c c} S_1 & S_3 & S_2 & S_4 & S_5 \end{array} \\
\begin{array}{c} S_1 \\ S_3 \\ S_2 \\ S_4 \\ S_5 \end{array} &
\left[\begin{array}{c c c c c}
0 & 0 & 0 & 0 & 0 \\
0 & 0 & 0 & 0 & 0 \\
\lambda_1 & \lambda_2 & 0 & 0 & \lambda_3 \\
\mu_2 & \lambda_5 & 0 & 0 & \lambda_4 \\
0 & \lambda_6 & \mu_3 & \mu_4 & 0
\end{array}\right]
\end{array}
$$

The coefficient matrix of the rearranged Markov differential equations can easily be written down.   ■■

6.6 SUMMARY

The field of applied probability is full of a variety of powerful and useful stochastic models. Markov chains and Markov processes belong to this family of models, and they have found important applications in the field of system reliability evaluation—especially in modeling repairable components. No attempt has been made to bring out the mathematical finesse of these models. A simple introduction has been given to them, along with some examples to provide a glimpse of their versatility and utility. Many fine books are available to the reader who is interested in learning more about the power and subtleties of these and other similar mathematical models.

PROBLEMS

6.1 The state transition matrix for a discrete process is

$$\mathbf{P} = \begin{bmatrix} 0 & \frac{1}{4} & \frac{3}{4} \\ \frac{1}{3} & \frac{1}{3} & \frac{1}{3} \\ \frac{1}{2} & 0 & \frac{1}{2} \end{bmatrix}$$

    (a) Draw the corresponding state-space diagram.
    (b) What is the probability of transition from state S_2 to S_3 in exactly three steps?
    (c) Find the unique fixed probability vector associated with this transition matrix.

6.2 Calculate the powers of **P** for

$$\mathbf{P} = \begin{bmatrix} 0 & 0 & 1 \\ 0 & 0 & 1 \\ \frac{1}{2} & \frac{1}{2} & 0 \end{bmatrix}$$

What conclusions can you draw about the nature of the discrete process associated with **P**?

6.3 Draw a state transition diagram corresponding to the transition matrix

$$\begin{matrix} S_1 \\ S_2 \\ S_3 \end{matrix} \begin{bmatrix} \frac{1}{3} & \frac{1}{3} & \frac{1}{3} \\ \frac{1}{2} & \frac{1}{4} & \frac{1}{4} \\ \frac{1}{2} & 0 & \frac{1}{2} \end{bmatrix}$$

and find the associated unique fixed probability vector.

6.4 Show by the process of induction that, for the stochastic transitional probability matrix

$$\mathbf{P} = \begin{bmatrix} 1 - a & a \\ b & 1 - b \end{bmatrix}$$

the nth-step transition probability matrix is

$$\mathbf{P}^n = \begin{bmatrix} \dfrac{b + a(1 - a - b)^n}{a + b} & \dfrac{a - a(1 - a - b)^n}{a + b} \\ \dfrac{b - b(1 - a - b)^n}{a + b} & \dfrac{a + b(1 - a - b)^n}{a + b} \end{bmatrix}$$

6.5 A discrete process has the state-space diagram shown in Figure P 6.5. Find the state

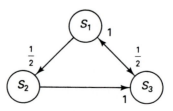

Figure P 6.5

transition matrix **P**, and compute its powers up to the fifth power. Also, find the corresponding stationary distribution.

6.6 A computer system can be either busy (S_1), idle (S_2), or under repair (S_3). The system is observed on a daily basis at 10:00 A.M., and its behavior appears to follow a Markov chain with the transition probability matrix

$$\mathbf{P} = \begin{bmatrix} 0.5 & 0.2 & 0.3 \\ 0.2 & 0.7 & 0.1 \\ 0.5 & 0.0 & 0.5 \end{bmatrix}$$

What are the long-term probabilities of finding the computer system in the three different states?

6.7 A *doubly stochastic matrix* is a stochastic matrix in which all the elements are non-negative and lie between 0 and 1, all the rows individually add up to unity, *and* all the columns individually add up to unity. Consider the 3×3 doubly stochastic matrix

$$\begin{bmatrix} 0.5 & 0.2 & 0.3 \\ 0.2 & 0.7 & 0.1 \\ 0.3 & 0.1 & 0.6 \end{bmatrix}$$

Show that the steady-state probabilities of the three states associated with this matrix are all equal to $\frac{1}{n}$ with $n = 3$, the number of states. (*Note:* This result is valid for all finite, irreducible, aperiodic, doubly stochastic matrices. For a simple explanation of these terms, see Trivedi, 1982).

6.8 The transition diagram for a Markov chain is given in Figure P 6.8. If the chain starts in state S_3, find the mean number of steps before it transitions to state S_2.

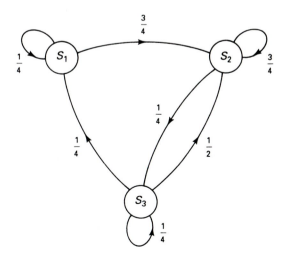

Figure P 6.8

6.9 Consider a Markov chain having the transition matrix

$$
\mathbf{P} = \begin{array}{c} \\ S_1 \\ S_2 \\ S_3 \\ S_4 \end{array}
\begin{array}{cccc}
S_1 & S_2 & S_3 & S_4 \\
\end{array}
\left[\begin{array}{cccc}
\frac{1}{4} & \frac{1}{4} & \frac{1}{4} & \frac{1}{4} \\
0 & 1 & 0 & 0 \\
\frac{1}{8} & \frac{1}{2} & \frac{1}{8} & \frac{1}{4} \\
0 & 0 & 0 & 1
\end{array}\right]
$$

(*i*) Draw the corresponding state transition diagram.
(*ii*) If the system starts in S_1, what is the mean number of steps before absorption?
(*iii*) If the system starts in S_3, what is the mean number of steps before absorption?
(*iv*) If the system starts in S_1, find the probabilities of absorption through states S_2 and S_4.
(*v*) If the system starts in S_3, find the probabilities of absorption through states S_2 and S_4.

6.10 A family has three cars—one very old (car *A*), one moderately old (car *B*), and one brand new (car *C*). A member of the family takes one of these cars to work every day of the week. If car *A* is taken today, then cars *A* and *B* are equally likely to be taken the next day. Car *B* is never taken two days in a row. Car *C* is twice as likely as car *A* to be taken the day after car *B* is taken. Whenever car *C* is taken on a particular day, car *B* is always taken the next day.

(a) Set up the state transition matrix for this set of circumstances.
(b) Given that car *A* is taken today, what is the probability of taking car *A* again the day after tomorrow?
(c) What are the long-run probabilities of taking the various cars to work?

6.11 Under normal conditions, the economy of a state can be modeled as a three-state Markov chain. The three states are:

The economy increases 5% or more compared to the preceding year (*A*)

The economy changes (increases or decreases) by less than 5% compared to the preceding year (*B*)

The economy decreases 5% or more compared to the preceding year (*C*)

Based on an analysis of data collected over a number of years, the following transition probability matrix is obtained:

	This Year		
	A	*B*	*C*
Last Year *A*	0.8	0.2	0.0
B	0.35	0.3	0.35
C	0.0	0.4	0.6

Compute the long-term probabilities of the state's economy being in the three states *A*, *B*, and *C*.

6.12 Each year, I trade my old car for a new car. If I have a Buick, I trade it for a Chrysler. If I have a Chrysler, I trade it for a Ford. However, if I have a Ford, I am just as likely to trade it for a new Ford as to trade it for a Buick or a Chrysler. In 1986, I bought a new Chrysler.

(*i*) In the long run, how often will I have (a) a Ford, (b) a Buick, and (c) a Chrysler?

(*ii*) What is the mean number of years before I buy a Buick?

6.13 There are four green marbles in box *A* and two blue marbles in box *B*. An experiment is conducted in which each step consists of selecting a marble from each box and interchanging them. Let S_i be the state wherein there are *i* green marbles in box *B*.

   (*i*) Find the transition matrix for this experiment.

   (*ii*) What is the probability that there is one green marble in box *B* after three exchanges?

   (*iii*) Calculate the long-run probability of finding two green marbles in box *B*.

6.14 A three-state generator model consists of an *up* state, a *down* state, and a *derated* state. The coefficient matrix of the corresponding Markov differential equations is as follows:

$$
\begin{array}{c}
Up \\
Down \\
Derated
\end{array}
\begin{bmatrix}
-(\lambda_1 + \lambda_2) & h_1 & h_2 \\
\lambda_1 & -(h_1 + h_3) & \lambda_3 \\
\lambda_2 & h_3 & -(\lambda_3 + h_2)
\end{bmatrix}
$$

Draw a state-space diagram for this system, and identify the states clearly.

6.15 Draw the state-space diagram and set up the corresponding Markov differential equations for a system that has the following matrix of rates of departures from various states:

$$
\begin{array}{c}
\\
S_1 \\
S_2 \\
S_3
\end{array}
\begin{array}{c}
\begin{array}{ccc} S_1 & S_2 & S_3 \end{array} \\
\begin{bmatrix}
0 & \lambda_1 & \lambda_2 \\
0 & 0 & 0 \\
\mu_2 & \mu_1 & 0
\end{bmatrix}
\end{array}
$$

6.16 A peaking generation unit in a power plant is modeled using four discrete states, as shown in Figure P 6.16. Set up the associated Markov differential equations.

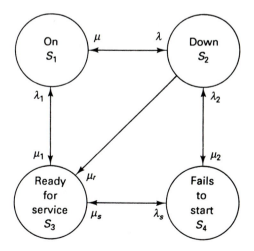

Figure P 6.16

6.17 A multicomponent system is represented by a state-space diagram with five discrete states, as illustrated in Figure P 6.17. Find the coefficient matrix of the corresponding Markov differential equations.

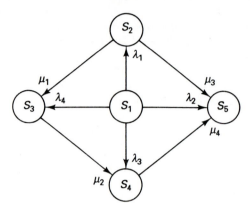

Figure P 6.17

6.18 The coefficient matrix of Markov differential equations (sometimes also known as the *Markov transition matrix*) for a system is

$$\begin{bmatrix} -(\lambda_1 + \lambda_2) & \mu & \mu & 0 \\ (1 - p)\lambda_1 & -(\lambda_3 + \mu) & 0 & \mu \\ \lambda_2 & 0 & -(\lambda_1 + \mu) & \mu \\ p\lambda_1 & \lambda_3 & \lambda_1 & -2\mu \end{bmatrix}$$

Draw the associated state-space diagram, mark all the important values, and identify the states.

6.19 Two generators operating in parallel supply electric power to an industry. In this mode of operation, their failure rates are λ_1 and λ_2, respectively. If one of the two units fails, the other unit becomes overloaded, and its failure rate increases. The overloaded failure rates are denoted λ_{1o} and λ_{2o}. Assuming no repair and neglecting the possibility of the simultaneous failure of both units, derive the time-dependent probability expressions for the different states of the system. Draw a neat state-space diagram.

6.20 Set up the state-space diagram for a two-unit nonrepairable system with failure rates of λ_1 and λ_2, respectively. Develop expressions for the time-dependent probabilities of occurrence of the various states. Neglect the possibility of the simultaneous failure of both units.

6.21 Let λ_o and λ_s be the constant failure rates for open-circuit and short-circuit failures, respectively, of a three-state device. Show that the time-dependent probability expressions for the two failure states are

$$q_o(t) = \left(\frac{\lambda_o}{\lambda_o + \lambda_s}\right)[1 - \exp\{-(\lambda_o + \lambda_s)t\}]$$

and

$$q_s(t) = \left(\frac{\lambda_s}{\lambda_o + \lambda_s}\right)[1 - \exp\{-(\lambda_o + \lambda_s)t\}]$$

6.22 A nonrepairable component has a good, a fair, and a bad state. The transitional probabilities of failure are:

From good to fair: $\lambda_{gf} \, \Delta t$
From good to bad: $\lambda_{gb} \, \Delta t$
From fair to bad: $\lambda_{fb} \, \Delta t$

Formulate a Markov model and derive expressions for the probabilities of being in the various states. Assume that at time $t = 0$, the component is in the good state.

6.23 Find $P_1(t)$ and $P_2(t)$ in Example 6–9 if $\lambda_{12} = 0.6$, $\lambda_{21} = 0.2$, and $P_1(0) = 1.0$. What are the steady-state probabilities?

6.24 A mom-and-pop grocery store can handle only one customer at any one time. The average time taken to serve a customer is 5 minutes. During the hours the store is open, people come in at an average rate of 8 per hour, and if there is already a customer in the store, they leave and go elsewhere. Set up the associated Markov differential equations, and find the long-run probabilities of having no customers and one customer in the store. How will your answers change if 15 people arrive per hour on the average?

6.25 Four subscribers who never call each other share a four-line switchboard. The durations of all phone calls can be assumed to be independent random variables with identical exponential distributions and an expected value of $1/d$. For each subscriber, the interval between the end of any call and the time of placing the next call can also be assumed to be an independent exponentially distributed random variable, but with an expected value of $1/w$. Draw a state-space diagram for this system, set up the Markov differential equations in matrix form, and find the limiting state probabilities for the number of input lines in use at any one time.

7

Reliability Evaluation of Engineering Systems Using Markov Models

7.1 INTRODUCTION

Many practical engineering systems operate in one of set of discrete states, with changeovers from one state to another being possible at any time. The discrete-state continuous-time Markov model (continuous Markov process) can be used to study such systems. Repairable systems, units that can operate in one or more derated states, and flow-type systems with different outputs instead of a simple success or failure are amenable for analysis by Markov techniques.

7.2 SOLUTION OF $\dot{\mathbf{X}} = \mathbf{AX}$

The time-dependent state probabilities of an n-state Markov process can be found by solving the Markov differential equations for that process. These equations have the general format of the well-known state model used in the analysis of dynamic systems. The state model can be written in the compact form

$$\dot{\mathbf{X}} = \mathbf{AX} \qquad (7-1)$$

256

where the dot refers to the time derivative, the state vector

$$\mathbf{X} = \begin{bmatrix} x_1(t) \\ x_2(t) \\ \vdots \\ x_n(t) \end{bmatrix} \equiv \begin{bmatrix} P_1(t) \\ P_2(t) \\ \vdots \\ P_n(t) \end{bmatrix} \tag{7-2}$$

and the constant coefficient matrix

$$\mathbf{A} = \begin{bmatrix} -\displaystyle\sum_{j=2}^{n} p_{1j} & p_{21} & \cdots & p_{n1} \\[2em] p_{12} & -\displaystyle\sum_{\substack{j=1 \\ j\neq 2}}^{n} p_{2j} & \cdots & p_{n2} \\[2em] \vdots & \vdots & \vdots & \vdots \\[1em] p_{1n} & p_{2n} & \cdots & -\displaystyle\sum_{j=1}^{n-1} p_{nj} \end{bmatrix} \tag{7-3}$$

The solution of Equation (7–1) with known initial conditions $\mathbf{X}(0)$ at $t = 0$ is

$$\mathbf{X}(t) = e^{\mathbf{A}t}\mathbf{X}(0) \tag{7-4}$$

The matrix $e^{\mathbf{A}t}$ is called the *state transition matrix*. Computation of the state transition matrix can be done in many ways. We will only enumerate the techniques here. Detailed theories and more information on these techniques can be found in books on systems analysis.

1. *Series method:*

$$e^{\mathbf{A}t} = \mathbf{I} + \mathbf{A}t + \frac{\mathbf{A}^2 t^2}{2!} + \cdots + \frac{\mathbf{A}^n t^n}{n!} + \cdots \tag{7-5}$$

2. *Resolvent matrix method:*

$$e^{\mathbf{A}t} = \mathcal{L}^{-1}[\mathbf{R}(s)] \tag{7-6}$$

where

$$\mathbf{R}(s) = [s\mathbf{I} - \mathbf{A}]^{-1} \tag{7-7}$$

can be found by using Leverrier's algorithm, which is given in terms of a ratio of polynomials in s.

3. *Eigenvalue method:*
Case (*i*): **A** has distinct eigenvalues $\lambda_1, \lambda_2, \ldots, \lambda_n$. Then

$$e^{\mathbf{A}t} = \mathbf{M}e^{\mathbf{\Lambda}t}\mathbf{M}^{-1} \tag{7-8}$$

where **M**, the modal matrix of **A**, is given by

$$\mathbf{M} = [\mathbf{e}_1 \quad \mathbf{e}_2 \quad \mathbf{e}_3 \quad \ldots \quad \mathbf{e}_n] \tag{7-9}$$

in which $\mathbf{e}_i$ is the eigenvector corresponding to the eigenvalue λ_i, $i = 1, 2, \ldots, n$, and

$$e^{\mathbf{\Lambda}t} = \begin{bmatrix} e^{\lambda_1 t} & & & \\ & e^{\lambda_2 t} & & \\ & & \ddots & \\ & & & e^{\lambda_n t} \end{bmatrix} \tag{7-10}$$

Case (*ii*): **A** has repeated eigenvalues. Then

$$e^{\mathbf{A}t} = \mathbf{T}e^{\mathbf{J}t}\mathbf{T}^{-1} \tag{7-11}$$

in which **J** is the Jordan canonical form and is given by

$$\mathbf{J} = \mathbf{T}^{-1}\mathbf{A}\mathbf{T} \tag{7-12}$$

4. *Other techniques:*
   a. Cayley-Hamilton method
   b. Application of Sylvester's theorem

■■ **Example 7–1**
We will illustrate the computation of $e^{\mathbf{A}t}$ for

$$\mathbf{A} = \begin{bmatrix} -0.1 & 10 \\ 0.1 & -10 \end{bmatrix}$$

by each of the first three techniques mentioned.

Series method:

$$\mathbf{A}^2 = \mathbf{A}\mathbf{A} = \begin{bmatrix} 1.01 & -101 \\ -1.01 & 101 \end{bmatrix}$$

$$\mathbf{A}^3 = \mathbf{A}^2\mathbf{A} = \begin{bmatrix} -10.201 & 1{,}020.1 \\ 10.201 & -1{,}020.1 \end{bmatrix}$$

and so on.
Substituting all these into Equation (7–5) and rearranging the terms, we get

$$e^{\mathbf{A}t} = \begin{bmatrix} 1 - 0.1t + 1.01\left(\frac{t^2}{2!}\right) - 10.201\left(\frac{t^3}{3!}\right) + \ldots & 10t - 101\left(\frac{t^2}{2!}\right) + 1{,}020.1\left(\frac{t^3}{3!}\right) - \ldots \\ 0.1t - 1.01\left(\frac{t^2}{2!}\right) + 10.201\left(\frac{t^3}{3!}\right) - \ldots & 1 - 10t + 101\left(\frac{t^2}{2!}\right) - 1{,}020.1\left(\frac{t^3}{3!}\right) + \ldots \end{bmatrix}$$

Using the series expansions of exponential functions, we can express the preceding as

$$e^{\mathbf{A}t} = \left(\frac{1}{101}\right) \begin{bmatrix} 100 + \exp(-10.1t) & 100 - 100\exp(-10.1t) \\ 1 - \exp(-10.1t) & 1 + 100\exp(-10.1t) \end{bmatrix}$$

Resolvent matrix method:

$$[s\mathbf{I} - \mathbf{A}] = \begin{bmatrix} s + 0.1 & -10 \\ -0.1 & s + 10 \end{bmatrix}$$

$$[s\mathbf{I} - \mathbf{A}]^{-1} = \frac{1}{s(s + 10.1)} \begin{bmatrix} s + 10 & 10 \\ 0.1 & s + 0.1 \end{bmatrix}$$

Taking the inverse Laplace transform of the right-hand side, we get the result obtained earlier for the state transition matrix.

Eigenvalue method:

The eigenvalues of $\mathbf{A}$ are found by solving the characteristic equation

$$|\mathbf{A} - \lambda\mathbf{I}| = 0$$

which is

$$\begin{vmatrix} -0.1 - \lambda & 10 \\ 0.1 & -10 - \lambda \end{vmatrix} = 0$$

or

$$\lambda^2 + 10.1\lambda = 0$$

The eigenvalues are thus $\lambda_1 = 0$ and $\lambda_2 = -10.1$. The associated eigenvectors, found by using standard techniques, are $[100 \quad 1]^T$ and $[1 \quad -1]^T$, respectively.

The modal matrix is

$$\mathbf{M} = \begin{bmatrix} 100 & 1 \\ 1 & -1 \end{bmatrix}$$

and its inverse is

$$\mathbf{M}^{-1} = -\left(\frac{1}{101}\right) \begin{bmatrix} -1 & -1 \\ -1 & 100 \end{bmatrix}$$

Since the two eigenvalues are distinct, we have

$$e^{\mathbf{A}t} = \begin{bmatrix} e^{0t} & 0 \\ 0 & e^{-10.1t} \end{bmatrix}$$

Using these three matrices in Equation (7–8), we obtain the same state transition matrix as before. The other techniques are left for self-study and self-assignment.

■■

7.3 STEADY-STATE OR LIMITING PROBABILITIES

Once the complete solution of Markov differential equations is found, the steady-state or limiting probabilities can easily be derived by letting the time t tend to infinity. However, if we are interested only in the steady-state probabilities, they can be calculated without solving the Markov differential equations. To calculate them, recall that, under steady-state conditions,

$$P_i'(t) = \frac{d}{dt} P_i(t) = 0 \text{ for } i = 1, 2, \ldots, n \qquad (7\text{--}13)$$

and the Markov differential equations become

$$
\begin{bmatrix} 0 \\ 0 \\ \vdots \\ 0 \end{bmatrix}
=
\begin{bmatrix}
-\sum\limits_{j=2}^{n} \rho_{1j} & \rho_{21} & \cdots & \rho_{n1} \\
\rho_{12} & -\sum\limits_{\substack{j=1 \\ j \neq 2}}^{n} \rho_{2j} & \cdots & \rho_{n2} \\
\vdots & \vdots & \vdots & \vdots \\
\rho_{1n} & \rho_{2n} & \cdots & -\sum\limits_{j=1}^{n-1} \rho_{nj}
\end{bmatrix}
\begin{bmatrix} P_1 \\ P_2 \\ \vdots \\ P_n \end{bmatrix}
\qquad (7\text{--}14)
$$

These equations, along with $\sum\limits_{i=1}^{n} P_i = 1$, can be solved simultaneously to obtain the limiting state probabilities. Any $(n - 1)$ of the n equations from the Markov set and $\sum\limits_{i=1}^{n} P_i = 1$ are sufficient to find the solution. Taking the first $(n - 1)$ equations and the summation equation, we get

$$
\begin{bmatrix} 0 \\ 0 \\ \vdots \\ 0 \\ 1 \end{bmatrix}
=
\begin{bmatrix}
-\sum\limits_{j=2}^{n} \rho_{1j} & \rho_{21} & \cdots & \rho_{n-1,1} & \rho_{n1} \\
\rho_{12} & -\sum\limits_{\substack{j=1 \\ j \neq 2}}^{n} \rho_{2j} & \cdots & \rho_{n-1,2} & \rho_{n2} \\
\vdots & \vdots & \vdots & \vdots & \vdots \\
\rho_{1,n-1} & \rho_{2,n-1} & \cdots & -\sum\limits_{\substack{j=1 \\ j \neq (n-1)}}^{n} \rho_{n-1,j} & \rho_{n,n-1} \\
1 & 1 & \cdots & 1 & 1
\end{bmatrix}
\begin{bmatrix} P_1 \\ P_2 \\ \vdots \\ P_{n-1} \\ P_n \end{bmatrix}
\qquad (7\text{--}15)
$$

These algebraic equations can be solved by one of many well-known methods to find the steady-state probabilities.

■■ **Example 7–2**

Let us consider the Markov differential equations for a three-state Markov process with the coefficient matrix

$$\mathbf{A} = \begin{bmatrix} -0.2 & 10 & 0 \\ 0.2 & -10.1 & 20 \\ 0 & 0.1 & -20 \end{bmatrix}$$

To find the steady-state probabilities P_1, P_2, and P_3, we solve

$$\begin{bmatrix} 0 \\ 0 \\ 1 \end{bmatrix} = \begin{bmatrix} -0.2 & 10 & 0 \\ 0.2 & -10.1 & 20 \\ 1 & 1 & 1 \end{bmatrix} \begin{bmatrix} P_1 \\ P_2 \\ P_3 \end{bmatrix}$$

The result is $P_1 = 0.980296$, $P_2 = 0.019606$, and $P_3 = 0.000098$.   ■■

7.4 SOME USEFUL MODELS AND CONCEPTS

7.4.1 Binary Model for a Single Repairable Component

In the binary model for a single repairable component, the component is assumed to exist in one of two states—the *up* state S_0 or the *down* state S_1. The transition rates between the two states are assumed to be constant. The state-space or transition diagram will then be as shown in Figure 7.1. The transition rates λ and μ can be estimated from available data as follows:

$$\lambda = \frac{\text{number of failures in a given period of time}}{\text{total operating time in the same period}} \qquad (7\text{–}16)$$

$$\mu = \frac{\text{number of repairs in a given period of time}}{\text{total repair time in the same period}} \qquad (7\text{–}17)$$

It is obvious from these equations that $1/\lambda$ and $1/\mu$ are estimates of the mean time to failure (MTTF) and the mean time to repair (MTTR), respectively, and that λ and μ are estimates of the failure and repair rates, respectively. The larger the

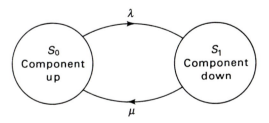

Figure 7.1   State-space diagram for a single repairable component.

amount of data used to find these estimates, the closer we approach their true values.

The ρ-matrix is

$$\begin{bmatrix} \rho_{00} & \rho_{01} \\ \rho_{10} & \rho_{11} \end{bmatrix} = \begin{bmatrix} 0 & \lambda \\ \mu & 0 \end{bmatrix}$$

and the Markov differential equations are

$$\begin{bmatrix} P_0'(t) \\ P_1'(t) \end{bmatrix} = \begin{bmatrix} -\lambda & \mu \\ \lambda & -\mu \end{bmatrix} \begin{bmatrix} P_0(t) \\ P_1(t) \end{bmatrix} \tag{7-18}$$

Let the component be in state S_0 (the operating state) at time $t = 0$. Then

$$\begin{bmatrix} P_0(0) \\ P_1(0) \end{bmatrix} = \begin{bmatrix} 1 \\ 0 \end{bmatrix} \tag{7-19}$$

Since

$$\mathbf{A} = \begin{bmatrix} -\lambda & \mu \\ \lambda & -\mu \end{bmatrix} \tag{7-20}$$

and

$$\mathbf{A}^2 = \mathbf{AA} = \begin{bmatrix} (\lambda^2 + \mu\lambda) & -(\lambda\mu + \mu^2) \\ -(\lambda^2 + \mu\lambda) & (\lambda\mu + \mu^2) \end{bmatrix} \tag{7-21}$$

application of the series method results in

$$e^{\mathbf{A}t}\mathbf{X}(0) = \begin{bmatrix} \left(1 - \lambda t + \lambda(\lambda + \mu)\dfrac{t^2}{2!} - \cdots\right)\left(\mu t - \mu(\lambda + \mu)\dfrac{t^2}{2!} + \cdots\right) \\ \\ \left(\lambda t - \lambda(\mu + \lambda)\dfrac{t^2}{2!} + \cdots\right)\left(1 - \mu t + \mu(\lambda + \mu)\dfrac{t^2}{2!} - \cdots\right) \end{bmatrix} \begin{bmatrix} 1 \\ 0 \end{bmatrix}$$

$$= \mathbf{X}(t) = \begin{bmatrix} P_0(t) \\ P_1(t) \end{bmatrix}$$

After completing the necessary matrix multiplications, we get

$$P_0(t) = 1 - \lambda t + \lambda(\lambda + \mu)\frac{t^2}{2!} - \cdots \tag{7-22a}$$

and

$$P_1(t) = \lambda t - \lambda(\mu + \lambda)\frac{t^2}{2!} + \cdots \tag{7-22b}$$

These are the same as

$$P_0(t) = \frac{\mu}{\lambda + \mu} + \frac{\lambda}{\lambda + \mu} e^{-(\lambda + \mu)t} \qquad (7\text{--}23a)$$

and

$$P_1(t) = \left(\frac{\lambda}{\lambda + \mu}\right) [1 - e^{-(\lambda + \mu)t}] \qquad (7\text{--}23b)$$

By letting $t \rightarrow \infty$, the steady-state or limiting state probabilities can easily be found. Denoting them simply as P_0 and P_1, we have

$$P_0 = \lim_{t \rightarrow \infty} P_0(t) = \frac{\mu}{\lambda + \mu} \qquad (7\text{--}24a)$$

$$P_1 = \lim_{t \rightarrow \infty} P_1(t) = \frac{\lambda}{\lambda + \mu} \qquad (7\text{--}24b)$$

If m and r are the MTTF and MTTR, respectively, then

$$m = \frac{1}{\lambda} , r = \frac{1}{\mu} , P_0 = \frac{m}{m + r} , \quad \text{and} \quad P_1 = \frac{r}{m + r}$$

These probabilities are referred to as the steady-state or limiting availability A and unavailability U, respectively. The time-dependent functions $A(t)$ and $U(t)$ are simply

$$A(t) = P_0(t) \qquad (7\text{--}25a)$$

and

$$U(t) = P_1(t) \qquad (7\text{--}25b)$$

For a component with a constant hazard (failure rate) of λ, we saw earlier that the time-dependent reliability function is

$$R(t) = e^{-\lambda t} \qquad (7\text{--}26)$$

if the component is good at time $t = 0$. However, with the possibility of repair included, the probability of finding the component in the *up* state at time $t = 0$ is

$$A(t) = P_0(t) = \frac{\mu}{\lambda + \mu} + \left(\frac{\lambda}{\lambda + \mu}\right) e^{-(\lambda + \mu)t} \qquad (7\text{--}27)$$

The differences between these two functions are illustrated in Figure 7.2, drawn for $\lambda = 0.1$ and for two different values of μ: $\mu = 1$ and $\mu = 10$. Although both $A(t)$ and $R(t)$ start at the same value (equal to 1) at $t = 0$, as time progresses, $R(t)$ is much more stringent than $A(t)$. Moreover, as μ increases (equivalent to decreasing repair times), $A(t)$ increases and approaches unity.

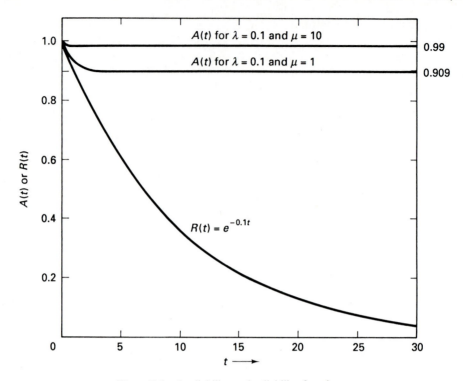

Figure 7.2   Availability and reliability functions.

7.4.2 Two Dissimilar Repairable Components

In the model for two dissimilar repairable components, we assume that each component can be in either the *up* state or the *down* state. Let λ_1, λ_2 and μ_1, μ_2 be the failure and repair rates, respectively, of the two components. Since each component can exist in one of two states, and since there are 2 such components, there are 2^2 or 4 possible states in which the system can exist. These are enumerated in the following table:

STATE	UNIT #1	UNIT #2
S_1	U	U
S_2	D	U
S_3	U	D
S_4	D	D

The corresponding state-space diagram is shown in Figure 7.3. Note that we are not allowing the possibility of transfer between S_1 and S_4 or between S_2 and S_3, because such transfers will require two simultaneous changes in the states of the components involved. The probabilities of such simultaneous occurrences are assumed to be negligibly small.

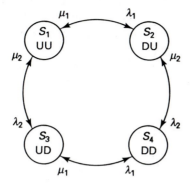

Figure 7.3 State-space diagram for two dissimilar repairable components.

The ρ-matrix is

$$
\begin{array}{c c}
 & \begin{array}{c c c c} S_1 & S_2 & S_3 & S_4 \end{array} \\
\begin{array}{c} S_1 \\ S_2 \\ S_3 \\ S_4 \end{array} &
\left[\begin{array}{c c c c}
0 & \lambda_1 & \lambda_2 & 0 \\
\mu_1 & 0 & 0 & \lambda_2 \\
\mu_2 & 0 & 0 & \lambda_1 \\
0 & \mu_2 & \mu_1 & 0
\end{array} \right]
\end{array}
$$

The stochastic transitional probability matrix (without the Δt's) is

$$
\mathbf{P} = \left[\begin{array}{c c c c}
1 - (\lambda_1 + \lambda_2) & \lambda_1 & \lambda_2 & 0 \\
\mu_1 & 1 - (\lambda_2 + \mu_1) & 0 & \lambda_2 \\
\mu_2 & 0 & 1 - (\lambda_1 + \mu_2) & \lambda_1 \\
0 & \mu_2 & \mu_1 & 1 - (\mu_1 + \mu_2)
\end{array} \right]
\tag{7--28}
$$

The Markov differential equations, in vector-matrix notation, are

$$
\begin{bmatrix} P_1'(t) \\ P_2'(t) \\ P_3'(t) \\ P_4'(t) \end{bmatrix} =
\begin{bmatrix}
-(\lambda_1 + \lambda_2) & \mu_1 & \mu_2 & 0 \\
\lambda_1 & -(\lambda_2 + \mu_1) & 0 & \mu_2 \\
\lambda_2 & 0 & -(\lambda_1 + \mu_2) & \mu_1 \\
0 & \lambda_2 & \lambda_1 & -(\mu_1 + \mu_2)
\end{bmatrix}
\begin{bmatrix} P_1(t) \\ P_2(t) \\ P_3(t) \\ P_4(t) \end{bmatrix}
\tag{7--29}
$$

The steady-state probabilities can be computed by the simultaneous solution of

$$
\alpha \mathbf{P} = \alpha
\tag{7--30}
$$

where

$$
\alpha = [P_1 \quad P_2 \quad P_3 \quad P_4]
\tag{7--31}
$$

and

$$
P_1 + P_2 + P_3 + P_4 = 1
\tag{7--32}
$$

Another approach to obtaining the steady-state probabilities is to solve the set of algebraic equations

$$
\begin{bmatrix} 0 \\ 0 \\ 0 \\ 1 \end{bmatrix} = \begin{bmatrix} -(\lambda_1 + \lambda_2) & \mu_1 & \mu_2 & 0 \\ \lambda_1 & -(\mu_1 + \lambda_2) & 0 & \mu_2 \\ \lambda_2 & 0 & -(\lambda_1 + \mu_2) & \mu_1 \\ 1 & 1 & 1 & 1 \end{bmatrix} \begin{bmatrix} P_1 \\ P_2 \\ P_3 \\ P_4 \end{bmatrix} \tag{7-33}
$$

The solution is

$$
P_1 = \frac{\mu_1 \mu_2}{(\lambda_1 + \mu_1)(\lambda_2 + \mu_2)}
$$

$$
P_2 = \frac{\lambda_1 \mu_2}{(\lambda_1 + \mu_1)(\lambda_2 + \mu_2)}
$$

$$
P_3 = \frac{\lambda_2 \mu_1}{(\lambda_1 + \mu_1)(\lambda_2 + \mu_2)} \tag{7-34}
$$

$$
P_4 = \frac{\lambda_1 \lambda_2}{(\lambda_1 + \mu_1)(\lambda_2 + \mu_2)}
$$

If the components are in series, then both must be good (or *up*) for system success, and the availability and unavailability are, respectively,

$$
A = P_1 \tag{7-35a}
$$

and

$$
U = P_2 + P_3 + P_4 \tag{7-35b}
$$

If the components are truly in parallel, then only one of the two needs to be good for the system to succeed. Thus, for this case,

$$
A = P_1 + P_2 + P_3 \tag{7-36a}
$$

and

$$
U = P_4 \tag{7-36b}
$$

■■ **Example 7–3**

In Example 7–2, we found the steady-state probabilities for a three-state process by using the coefficient matrix of Markov differential equations. An alternative method involves the use of the corresponding stochastic transitional probability matrix without the Δt's.

The coefficient matrix of Markov differential equations is

$$
\mathbf{A} = \begin{bmatrix} -0.2 & 10 & 0 \\ 0.2 & -10.1 & 20 \\ 0 & 0.1 & -20 \end{bmatrix}
$$

The corresponding stochastic transitional probability matrix without the Δt's is

$$\mathbf{P} = \begin{bmatrix} (1 - 0.2) & 0.2 & 0 \\ 10 & (1 - 10.1) & 0.1 \\ 0 & 20 & (1 - 20) \end{bmatrix}$$

Next, we let $\alpha = [P_1 \quad P_2 \quad P_3]$ and use $\alpha\mathbf{P} = \alpha$ in conjunction with $P_1 + P_2 + P_3 = 1$ to find the steady-state probabilities P_1, P_2, and P_3. The resulting equations are:

$$(1 - 0.2)P_1 + 10P_2 = P_1$$

$$0.2P_1 + (1 - 10.1)P_2 + 20P_3 = P_2$$

$$0.1P_2 + (1 - 20)P_3 = P_3$$

$$P_1 + P_2 + P_3 = 1$$

Simultaneous solution of these equations yields $P_1 = 10,000/10,201$, $P_2 = 200/10,201$, and $P_3 = 1/10,201$, which are the same as the values obtained by the other method.   ∎∎

7.4.3 Ternary Model for a Single Component

Markov processes can be used to determine component state probabilities, even if it is necessary to include the possibility of one or more forced derated states in the model. With just one derated state included, we get the ternary model shown in Figure 7.4.

There are several examples of components that may require a ternary model for better representation. A large hydrogen-cooled turbogenerator may be operated at a reduced output if the cooling system is not functioning at its full potential. A pump may deliver only a fraction of its rated output if certain parts are not functioning properly. A wind-electric conversion system may require sev-

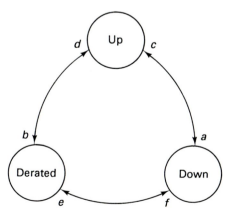

Figure 7.4 Ternary model for a repairable component.

eral derated states in the model to accommodate the variation of its output with varying wind speeds.

For the ternary model shown in Figure 7.4, the ρ-matrix is

$$
\begin{array}{c c}
 & \begin{array}{ccc} \text{Up} & \text{Down} & \text{Derated} \end{array} \\
\begin{array}{c} \text{Up} \\ \text{Down} \\ \text{Derated} \end{array} &
\begin{bmatrix}
0 & a & b \\
c & 0 & e \\
d & f & 0
\end{bmatrix}
\end{array}
$$

The Markov differential equations can now be written down by inspection:

$$
\begin{bmatrix}
P'_{up}(t) \\
P'_{down}(t) \\
P'_{derated}(t)
\end{bmatrix}
=
\begin{bmatrix}
-(a+b) & c & d \\
a & -(c+e) & f \\
b & e & -(d+f)
\end{bmatrix}
\begin{bmatrix}
P_{up}(t) \\
P_{down}(t) \\
P_{derated}(t)
\end{bmatrix}
\tag{7-37}
$$

The steady-state probabilities are obtained by solving the algebraic equations

$$
\begin{bmatrix}
0 \\
0 \\
1
\end{bmatrix}
=
\begin{bmatrix}
-(a+b) & c & d \\
a & -(c+e) & f \\
1 & 1 & 1
\end{bmatrix}
\begin{bmatrix}
P_{up} \\
P_{down} \\
P_{derated}
\end{bmatrix}
\tag{7-38}
$$

The solution is

$$
P_{up} = \frac{cd + cf + ed}{\mathcal{D}}
$$

$$
P_{down} = \frac{ad + af + bf}{\mathcal{D}}
\tag{7-39}
$$

$$
P_{derated} = \frac{bc + be + ae}{\mathcal{D}}
$$

where

$$
\mathcal{D} \equiv ad + ae + af + bc + be + bf + cf + dc + de
$$

7.4.4 Two Identical Repairable Components

Let the failure and repair rates of the components be λ and μ, respectively. The first step in the analysis is to identify the possible states of the system. They are:

State 1   Both components *up* (or good)
State 2   One component *up* and one component *down*
State 3   Both components *down*

Next, we construct the state-space diagram shown in Figure 7.5 and mark all the relevant values.

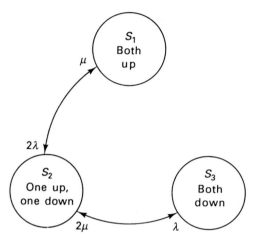

Figure 7.5 State-space diagram for two identical components.

Note that we allow only one change at a time. This stipulation is based on the assumption that the probability of the simultaneous occurrence of two events is so small that it can be neglected. The rates of departure ρ_{12} and ρ_{32} are 2λ and 2μ, respectively, because two components are available for failure or repair.

The ρ-matrix for this situation is

$$
\begin{array}{c c}
 & \begin{array}{c c c} 1 & 2 & 3 \end{array} \\
\begin{array}{c} 1 \\ 2 \\ 3 \end{array} &
\left[\begin{array}{c c c}
0 & 2\lambda & 0 \\
\mu & 0 & \lambda \\
0 & 2\mu & 0
\end{array}\right]
\end{array}
$$

The Markov differential equations can now be written by inspection as

$$
\begin{bmatrix} P_1'(t) \\ P_2'(t) \\ P_3'(t) \end{bmatrix} =
\begin{bmatrix}
-2\lambda & \mu & 0 \\
2\lambda & -(\lambda + \mu) & 2\mu \\
0 & \lambda & -2\mu
\end{bmatrix}
\begin{bmatrix} P_1(t) \\ P_2(t) \\ P_3(t) \end{bmatrix}
\tag{7--40}
$$

The stochastic transitional probability matrix $\mathbf{P}$, without the Δt's, is

$$
\mathbf{P} =
\begin{bmatrix}
1 - 2\lambda & 2\lambda & 0 \\
\mu & 1 - \lambda - \mu & \lambda \\
0 & 2\mu & 1 - 2\mu
\end{bmatrix}
\tag{7--41}
$$

Two approaches are available to find the limiting-state (or steady-state) probabilities. In the first, we use $\boldsymbol{\alpha}\mathbf{P} = \boldsymbol{\alpha}$, in which $\boldsymbol{\alpha} = [P_1 \ \ P_2 \ \ P_3]$. This equation leads to

$$
\left.
\begin{aligned}
(1 - 2\lambda)P_1 + \mu P_2 &= P_1 \\
2\lambda P_1 + (1 - \lambda - \mu)P_2 + 2\mu P_3 &= P_2 \\
\lambda P_2 + (1 - 2\mu)P_3 &= P_3
\end{aligned}
\right\}
\tag{7--42}
$$

Solving these three equations in conjunction with

$$P_1 + P_2 + P_3 = 1$$

yields

$$
\left.
\begin{aligned}
P_1 &= \left(\frac{\mu}{\lambda + \mu}\right)^2 \\[2mm]
P_2 &= \frac{2\lambda}{\mu}\left(\frac{\mu}{\lambda + \mu}\right)^2 \\[2mm]
P_3 &= \left(\frac{\lambda}{\lambda + \mu}\right)^2
\end{aligned}
\right\}
\tag{7–43}
$$

If both components are needed for system success (a series configuration), then the system availability and unavailability are, respectively,

$$A = P_1 = \left(\frac{\mu}{\lambda + \mu}\right)^2 \tag{7–44a}$$

and

$$U = P_2 + P_3 = \frac{2\lambda\mu + \lambda^2}{(\lambda + \mu)^2} \tag{7–44b}$$

If only one component is needed for system success (a parallel configuration), then the system availability and unavailability are, respectively,

$$A = P_1 + P_2 = \frac{\mu^2 + 2\lambda\mu}{(\lambda + \mu)^2} \tag{7–45a}$$

and

$$U = P_3 = \left(\frac{\lambda}{\lambda + \mu}\right)^2 \tag{7–45b}$$

In the second approach, we pick the first two of the three Markov differential equations and add to them $P_1 + P_2 + P_3 = 1$. Then we replace all the derivatives by zero to get

$$
\begin{bmatrix} 0 \\ 0 \\ 1 \end{bmatrix} =
\begin{bmatrix}
-2\lambda & \mu & 0 \\
2\lambda & -(\lambda + \mu) & 2\mu \\
1 & 1 & 1
\end{bmatrix}
\begin{bmatrix} P_1 \\ P_2 \\ P_3 \end{bmatrix}
\tag{7–46}
$$

Solution of Equation (7–46) yields the same results for P_1, P_2, and P_3 as before.

To find the MTTF, we designate S_3 as the absorbing state and rearrange **P** as follows:

$$
\begin{array}{c}
\\
3\\
1\\
2
\end{array}
\begin{bmatrix}
3 & 1 & 2\\
1 & 0 & 0\\
0 & (1 - 2\lambda) & 2\lambda\\
\lambda & \mu & (1 - \lambda - \mu)
\end{bmatrix}
$$

The truncated matrix is

$$
\mathbf{Q} = \begin{bmatrix} (1 - 2\lambda) & 2\lambda \\ \mu & (1 - \lambda - \mu) \end{bmatrix} \tag{7-47}
$$

from which it follows that

$$
[\mathbf{I} - \mathbf{Q}]^{-1} = \begin{bmatrix} 2\lambda & -2\lambda \\ -\mu & (\lambda + \mu) \end{bmatrix}^{-1}
$$

$$
= \left(\frac{1}{2\lambda^2}\right) \begin{bmatrix} (\lambda + \mu) & 2\lambda \\ \mu & 2\lambda \end{bmatrix} \tag{7-48}
$$

$$
= \text{fundamental matrix } \mathbf{N}
$$

Now we are in a position to find the answers to some practical questions. For example, if the process starts in state 1, then

$$
\text{MTTF} = \frac{(\lambda + \mu) + (2\lambda)}{2\lambda^2} = \frac{3\lambda + \mu}{2\lambda^2} \tag{7-49}
$$

If the process starts in state 2, then

$$
\text{MTTF} = \frac{\mu + 2\lambda}{2\lambda^2} \tag{7-50}
$$

Moreover,

$$
\mathbf{NR} = \begin{bmatrix} \left(\dfrac{\lambda + \mu}{2\lambda^2}\right) & \left(\dfrac{2\lambda}{2\lambda^2}\right) \\ \left(\dfrac{\mu}{2\lambda^2}\right) & \left(\dfrac{2\lambda}{2\lambda^2}\right) \end{bmatrix} \begin{bmatrix} 0 \\ \lambda \end{bmatrix} = \begin{bmatrix} 1 \\ 1 \end{bmatrix} \tag{7-51}
$$

Since we have only one absorbing state, whether the process starts in transient state 1 or 2, it will be absorbed by entering *the* absorbing state. This is the reason that all of the entries in **NR** are unity.

■■ **Example 7–4**

Suppose the two-component system just discussed is not repaired if both components are down. Also, let there be an additional failure mode that wipes out both units simultaneously with no possibility of repair. Then the state-space diagram should be modified as shown in Figure 7.6. Note that we now have two absorbing states and two transient states.

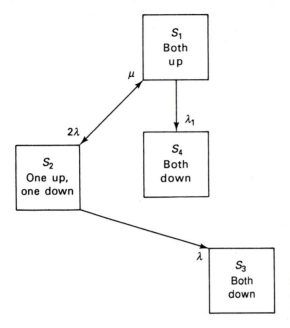

Figure 7.6 Modified state-space diagram for two repairable components with an additional failure mode.

The matrix of transition rates is

$$
\begin{array}{c}
\begin{array}{cccc} S_1 & S_2 & S_3 & S_4 \end{array} \\
\begin{array}{c} S_1 \\ S_2 \\ S_3 \\ S_4 \end{array}
\left[
\begin{array}{cccc}
0 & 2\lambda & 0 & \lambda_1 \\
\mu & 0 & \lambda & 0 \\
0 & 0 & 0 & 0 \\
0 & 0 & 0 & 0
\end{array}
\right]
\end{array}
$$

The corresponding Markov differential equations, in vector-matrix notation, are

$$
\begin{bmatrix}
P_1'(t) \\
P_2'(t) \\
P_3'(t) \\
P_4'(t)
\end{bmatrix}
=
\begin{bmatrix}
-(2\lambda + \lambda_1) & \mu & 0 & 0 \\
2\lambda & -(\lambda + \mu) & 0 & 0 \\
0 & \lambda & 0 & 0 \\
\lambda_1 & 0 & 0 & 0
\end{bmatrix}
\begin{bmatrix}
P_1(t) \\
P_2(t) \\
P_3(t) \\
P_4(t)
\end{bmatrix}
$$

The rearranged stochastic transitional probability matrix without the Δt's is

$$
\mathbf{P} =
\begin{array}{c}
\begin{array}{c} S_3 \\ S_4 \\ S_1 \\ S_2 \end{array}
\end{array}
\begin{array}{c}
\begin{array}{cccc} S_3 & S_4 & S_1 & S_2 \end{array} \\
\left[
\begin{array}{cc:cc}
1 & 0 & 0 & 0 \\
0 & 1 & 0 & 0 \\
\hdashline
0 & \lambda_1 & 1 - (2\lambda + \lambda_1) & 2\lambda \\
\lambda & 0 & \mu & 1 - (\lambda + \mu)
\end{array}
\right]
\end{array}
$$

The fundamental matrix is

$$\mathbf{N} = \begin{bmatrix} 2\lambda + \lambda_1 & -2\lambda \\ -\mu & (\mu + \lambda) \end{bmatrix}^{-1}$$

$$= \frac{1}{2\lambda^2 + \lambda_1(\mu + \lambda)} \begin{bmatrix} (\mu + \lambda) & 2\lambda \\ \mu & (2\lambda + \lambda_1) \end{bmatrix}$$

If the process starts in state S_1, then

$$\text{MTTF} = \frac{3\lambda + \mu}{2\lambda^2 + \lambda_1(\mu + \lambda)}$$

If the process starts in state S_2, then

$$\text{MTTF} = \frac{2\lambda + \lambda_1 + \mu}{2\lambda^2 + \lambda_1(\mu + \lambda)}$$

The probabilities of absorption through the two absorbing states for a given starting point (namely, state S_1 or state S_2) are given by the entries in the matrix **NR**. This matrix is calculated next:

$$\mathbf{NR} = \frac{1}{2\lambda^2 + \lambda_1(\mu + \lambda)} \begin{bmatrix} (\mu + \lambda) & 2\lambda \\ \mu & (2\lambda + \lambda_1) \end{bmatrix} \begin{bmatrix} 0 & \lambda_1 \\ \lambda & 0 \end{bmatrix}$$

$$= \frac{1}{2\lambda^2 + \lambda_1(\mu + \lambda)} \begin{bmatrix} 2\lambda^2 & \lambda_1(\mu + \lambda) \\ \lambda(2\lambda + \lambda_1) & \mu\lambda_1 \end{bmatrix}$$

If the process starts in state S_1, then the probabilities of absorption through the absorbing states S_3 and S_4 are

$$\frac{2\lambda^2}{2\lambda^2 + \lambda_1(\mu + \lambda)}$$

and

$$\frac{(\mu + \lambda)\lambda_1}{2\lambda^2 + \lambda_1(\mu + \lambda)}$$

respectively. As expected, these two probabilities add up to unity. If the process starts in state S_2, then the probabilities of absorption through S_3 and S_4 are

$$\frac{\lambda(2\lambda + \lambda_1)}{2\lambda^2 + \lambda_1(\mu + \lambda)}$$

and

$$\frac{\lambda_1\mu}{2\lambda^2 + \lambda_1(\mu + \lambda)}$$

respectively. These two probabilities also add up to unity.

Next, let us consider a numerical case. Suppose the two components under study are remote generators supplying electricity to an installation, and let the generator parameters be

$$\lambda = 0.001 \quad \text{failure/hr}$$

$$\mu = 0.01 \quad \text{repair/hr}$$

$$\lambda_1 = 0.0001 \text{ failure/hr}$$

If the process starts in S_1, then

$$\text{MTTF} = \frac{(3 \times 0.001) + (0.01)}{2(0.001)^2 + (0.0001)(0.01 + 0.001)}$$

$$= \frac{0.013}{[3.1 \times 10^{-6}]}$$

$$= 4,193.55 \text{ hours}$$

$$P(\text{absorption through } S_3) = \frac{2 \times (0.001)^2}{3.1 \times 10^{-6}}$$

$$= 0.64516$$

$$P(\text{absorption through } S_4) = \frac{0.011 \times 0.0001}{3.1 \times 10^{-6}}$$

$$= 0.35484$$

If the process starts in S_2, then

$$\text{MTTF} = \frac{(2 \times 0.001) + 0.0001 + 0.01}{3.1 \times 10^{-6}}$$

$$= 3,903.23 \text{ hours}$$

$$P(\text{absorption through } S_3) = \frac{(0.001)(0.002 + 0.0001)}{3.1 \times 10^{-6}} = 0.67742$$

$$P(\text{absorption through } S_4) = \frac{(0.0001)(0.01)}{3.1 \times 10^{-6}} = 0.32258 \qquad \blacksquare\blacksquare$$

7.5 INTRODUCTION TO FREQUENCY AND DURATION TECHNIQUES

In addition to finding the time-dependent and steady-state probabilities of the various system states, it will be highly useful to know how long the system resides in a state, how much time is required by the system to go from one state to another, and how much time it takes to complete an "in" and a "not-in" cycle for various states.

Let us consider an *n*-state Markov process and focus on two states, S_i and S_j. We introduce the following definitions:

- *Passage time* is the time taken to go from S_i to S_j; it is a continuous random variable, and its expected value is called the *expected passage time*.
- *Residence time* in a state is the passage time from this state to *any other* state; it is a continuous random variable, and its expected value is the *expected residence time*.
- *Cycle time* for a state is the time required to complete an "in" and a "not-in" cycle for that state. Cycle time is also a continuous random variable. Its expected value is equal to the sum of the expected residence time and the expected time between residences in the given state.

If the process has only two states, the passage time from S_1 to S_2 is also the residence time in S_1, and the residence time in S_2 is the time between residences in S_1. Therefore, the cycle times are the same for both states.

The reciprocal of the cycle time of a state is called the expected *frequency of occurrence* of that state.

Now, consider a repairable component with a binary model, as shown in Figure 7.7(a). To find the expected value of the passage time T_{12} from S_1 to S_2,

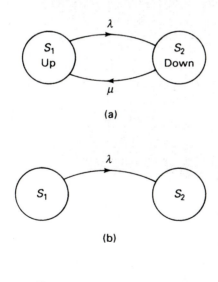

(a)

(b)

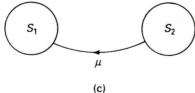

(c)

Figure 7.7 Models for one repairable component. (a) Binary model and state-space diagram. (b) Modified model, assuming state S_2 is absorbing. (c) Modified model, assuming state S_1 is absorbing.

suppose S_2 to be an absorbing state, and redraw the state-space diagram as shown in Figure 7.7(b). The corresponding Markov differential equations are

$$\begin{bmatrix} P_1'(t) \\ P_2'(t) \end{bmatrix} = \begin{bmatrix} -\lambda & 0 \\ \lambda & 0 \end{bmatrix} \begin{bmatrix} P_1(t) \\ P_2(t) \end{bmatrix} \qquad (7\text{--}52)$$

Let the system be in state S_1 at time $t = 0$. Then

$$\begin{bmatrix} P_1(0) \\ P_2(0) \end{bmatrix} = \begin{bmatrix} 1 \\ 0 \end{bmatrix} \qquad (7\text{--}53)$$

and the time-dependent probabilities are

$$P_1(t) = e^{-\lambda t}$$
$$P_2(t) = 1 - e^{-\lambda t} \qquad (7\text{--}54)$$

Note that $P_2(t)$ is the probability that the system will be in S_2 at time t. This is the same as saying that $P_2(t)$ is the probability that the passage time $T_{12} \leq t$. Therefore, by the definition of the distribution function of a continuous random variable, we conclude that $P_2(t)$ is the cumulative distribution function $F_{T_{12}}(t)$ of the random variable T_{12}. Thus,

$$F_{T_{12}}(t) = P_2(t) = 1 - e^{-\lambda t} \qquad (7\text{--}55)$$

The corresponding density function is obtained by differentiating the distribution function. Thus,

$$f_{T_{12}}(t) = \frac{d}{dt} F_{T_{12}}(t) = \lambda e^{-\lambda t} \qquad (7\text{--}56)$$

The expected value of T_{12} can now be computed:

$$E[T_{12}] = \int_0^\infty t f_{T_{12}}(t)\, dt = \int_0^\infty \lambda t e^{-\lambda t}\, dt$$
$$= \frac{1}{\lambda} \qquad (7\text{--}57)$$

Similarly, the expected passage time $E[T_{21}]$ from S_2 to S_1 can be obtained by considering S_1 as the absorbing state, as shown in Figure 7.7(c). With initial conditions $P_1(0) = 0$ and $P_2(0) = 1$, we get

$$P_1(t) = 1 - e^{-\mu t}$$
$$P_2(t) = e^{-\mu t} \qquad (7\text{--}58)$$

Proceeding as before, we can see that

$$E[T_{21}] = \frac{1}{\mu} \tag{7-59}$$

$$\text{Cycle time for } S_1 = \left\{ \begin{array}{c} \text{expected residence} \\ \text{time in } S_1 \end{array} \right\} + \left\{ \begin{array}{c} \text{expected time between} \\ \text{residences in } S_1 \end{array} \right\}$$

$$= E[T_{12}] + E[T_{21}]$$

$$= \frac{1}{\lambda} + \frac{1}{\mu}$$

$$= \frac{\lambda + \mu}{\lambda\mu} \tag{7-60}$$

Similarly,

$$\text{Cycle time for } S_2 = E[T_{21}] + E[T_{12}] \tag{7-61}$$

$$= \frac{\lambda + \mu}{\lambda\mu}$$

With only two states, if the system is not in S_1, it should be in S_2 and vice versa. Therefore, the cycle times are the same for both states.

The expected frequency of occurrence of state S_1 is

$$f_1 = \frac{1}{\text{cycle time for } S_1}$$

$$= \frac{\lambda\mu}{\lambda + \mu} = \left(\frac{\mu}{\lambda + \mu}\right)\lambda \tag{7-62}$$

The expected frequency of occurrence of state S_2 is

$$f_2 = \frac{1}{\text{cycle time for } S_2}$$

$$= \frac{\lambda\mu}{\lambda + \mu} = \left(\frac{\lambda}{\lambda + \mu}\right)\mu \tag{7-63}$$

Note that, in either case, the expected frequency of occurrence can be expressed as the product of the steady-state probability of being in that state and the rate of departure from it. It is also easy to see that the expected frequency of occurrence of a state is equal to the product of the steady-state probability of not being in that state and the rate of entry into it.

7.6 GENERAL APPROACH TO FINDING THE EXPECTED PASSAGE TIME

Suppose a system can reside in one of $(n - 1)$ transient states or an absorbing state. We would like to be able to compute the expected passage time from any one of the transient states to the absorbing state. In the context of system reliability, the absorbing state is the failed state, and all the other states correspond to the system being up. Then the expected passage time from any of the transient states to the absorbing state is also the expected value of the time between residences in the absorbing state, which is the same as the expected time the system is up.

Assuming the nth state to be the absorbing state, we have $\rho_{n1} = \rho_{n2} = \cdots = \rho_{nn} = 0$. Therefore, the Markov differential equations will have the form

$$
\begin{bmatrix} P_1'(t) \\ P_2'(t) \\ \vdots \\ P_{n-1}'(t) \\ P_n'(t) \end{bmatrix} = \begin{bmatrix} -\sum_{j=2}^{n} \rho_{1j} & \rho_{21} & \cdots & \rho_{n-1,1} & 0 \\ \rho_{12} & -\sum_{\substack{j=1 \\ j\neq 2}}^{n} \rho_{2j} & \cdots & \rho_{n-1,2} & 0 \\ \vdots & \vdots & \vdots & \vdots & \vdots \\ \rho_{1,n-1} & \rho_{2,n-1} & \cdots & -\sum_{\substack{j=1 \\ j\neq(n-1)}}^{n} \rho_{n-1,j} & 0 \\ \rho_{1n} & \rho_{2n} & \cdots & \rho_{n-1,n} & 0 \end{bmatrix} \begin{bmatrix} P_1(t) \\ P_2(t) \\ \vdots \\ P_{n-1}(t) \\ P_n(t) \end{bmatrix} \quad (7\text{--}64)
$$

We will also assume that at time $t = 0$, the system is not in the failed state and that $P_1(0), P_2(0), \ldots, P_{n-1}(0)$ are the probabilities of the system being in states $S_1, S_2, \ldots, S_{n-1}$, respectively. In vector notation, the initial conditions vector can be written as

$$
\mathbf{P}(0) = \begin{bmatrix} P_1(0) \\ P_2(0) \\ \vdots \\ P_{n-1}(0) \\ 0 \end{bmatrix}
$$

Let D be the determinant of the $(n-1)$th-order square matrix obtained by deleting the nth row and the nth column in the coefficient matrix of the Markov differential equations. That is,

$$D = \begin{vmatrix} -\sum_{j=2}^{n} p_{1j} & \cdots & p_{n-1,1} \\ \\ p_{1,n-1} & \cdots & -\sum_{\substack{j=1 \\ j\neq(n-1)}}^{n} p_{n-1,j} \end{vmatrix} \qquad (7\text{-}65)$$

We will define an nth-order determinant $\tilde{D}$ as follows:

$$\tilde{D} = \begin{vmatrix} -\sum_{j=2}^{n} p_{1j} & \cdots & p_{n-1,1} & P_1(0) \\ \\ p_{12} & \cdots & p_{n-1,2} & P_2(0) \\ \\ p_{1,n-1} & \cdots & -\sum_{\substack{j=1 \\ j\neq(n-1)}}^{n} p_{n-1,j} & P_{n-1}(0) \\ \\ 1 & \cdots & 1 & 0 \end{vmatrix} \qquad (7\text{-}66)$$

Then we have:

$$\left.\begin{array}{l}\text{Expected passage time from any one of}\\ \text{the transient states to the absorbing state}\end{array}\right\} = \frac{\tilde{D}}{D} \qquad (7\text{-}67)$$

As discussed previously, this quantity is also equal to the expected value of the time between residences in the absorbing (failed) state.

7.7 GENERAL EXPRESSION FOR THE FREQUENCY OF A STATE

For the ith state of an n-state Markov process, let

$E[T_i]$ = expected value of the cycle time

$E[T_i']$ = expected value of the time between residences

and

$E[T_i'']$ = expected value of the residence time

Then

$$E[T_i] = E[T_i'] + E[T_i''] \qquad (7\text{-}68)$$

and the frequency of encounter (or occurrence) for state i is

$$f_i = \frac{1}{E[T_i]} \qquad (7\text{-}69)$$

The expression for frequency can also be written as

$$f_i = \frac{E[T_i'']}{E[T_i]} \frac{1}{E[T_i'']} \tag{7-70}$$

Using what is known as renewal theory (see Shooman, sec. 6.10.3), we can show that the steady-state probability of being in state S_i is

$$P_i = \frac{E[T_i'']}{E[T_i]} \tag{7-71}$$

Alternatively, it is obvious that the steady-state probability of being in a state is equal to the ratio of the expected residence time to the expected cycle time. Thus,

$$f_i = \frac{P_i}{E[T_i'']} \tag{7-72}$$

To find the expected value of the residence time, $E[T_i'']$, we lump all the other states together into one state, as shown in Figure 7.8, and consider this state to be absorbing. Then the expected value of the passage time from S_i to the lumped absorbing state is the expected residence time in S_i.

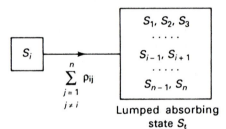

Lumped absorbing
state S_t

Figure 7.8 Procedure for finding the mean residence time, $E[T_i'']$.

Assuming that at time $t = 0$, the system is in S_i, the following procedure is adopted to find $E[T_i'']$. First, we obtain the ρ-matrix, which is

$$\begin{array}{cc} & \begin{array}{cc} S_i & S_\ell \end{array} \\ \begin{array}{c} S_i \\ S_\ell \end{array} & \begin{bmatrix} 0 & \sum \\ 0 & 0 \end{bmatrix} \end{array}$$

where $\sum$ is shorthand for the sum $\displaystyle\sum_{\substack{j=1 \\ j \neq i}}^{n} \rho_{ij}$.

The coefficient matrix of Markov equations is

$$\begin{bmatrix} -\sum & 0 \\ \sum & 0 \end{bmatrix}$$

with initial conditions

$$\begin{array}{c} S_i \\ S_\ell \end{array} \begin{bmatrix} 1 \\ 0 \end{bmatrix}$$

We then have

$$D = |-\Sigma| = -\Sigma$$

$$\tilde{D} = \begin{vmatrix} -\Sigma & 1 \\ 1 & 0 \end{vmatrix} = -1$$

$$E[T_i''] = \frac{\tilde{D}}{D} = \frac{1}{\Sigma}$$

and

$$f_i = (P_i)(\Sigma)$$

or

$$f_i = P_i \sum_{\substack{j=1 \\ j \neq i}}^{n} \rho_{ij} \qquad (7\text{-}73)$$

Once again, we see that the expected frequency of occurrence of a state is equal to the steady-state probability of being in that state times the total rate of departure from it.

■■ **Example 7–5**

Consider again the case of two identical repairable components discussed in Section 7.4.4 and shown in Figure 7.5. The frequencies of occurrence (also known as *frequencies of encounter*) of the various states are:

$$f_1 = P_1(2\lambda) = (2\lambda)\left(\frac{\mu}{\lambda + \mu}\right)^2$$

$$f_2 = P_2(\lambda + \mu) = (2\lambda)\left(\frac{\mu}{\lambda + \mu}\right)$$

$$f_3 = P_3(2\mu) = (2\mu)\left(\frac{\lambda}{\lambda + \mu}\right)^2$$

Assuming that at time $t = 0$, the system is in state S_1, we want to find the expected passage time from S_1 to S_3. We first designate S_3 as absorbing and modify the Markov equations to

$$\begin{bmatrix} P_1'(t) \\ P_2'(t) \\ P_3'(t) \end{bmatrix} = \begin{bmatrix} -2\lambda & \mu & 0 \\ 2\lambda & -(\lambda + \mu) & 0 \\ 0 & \lambda & 0 \end{bmatrix} \begin{bmatrix} P_1(t) \\ P_2(t) \\ P_3(t) \end{bmatrix}$$

The initial condition vector is

$$\begin{bmatrix} 1 \\ 0 \\ 0 \end{bmatrix}$$

Therefore,

$$D = \begin{vmatrix} -2\lambda & \mu \\ 2\lambda & -(\lambda + \mu) \end{vmatrix} = 2\lambda^2$$

and

$$\tilde{D} = \begin{vmatrix} -2\lambda & \mu & 1 \\ 2\lambda & -(\lambda + \mu) & 0 \\ 1 & 1 & 0 \end{vmatrix} = 3\lambda + \mu$$

The expected passage time from S_1 to S_3 is

$$\frac{\tilde{D}}{D} = \frac{3\lambda + \mu}{2\lambda^2}$$

This expression is the same as that for the MTTF obtained earlier in Equation (7–49), assuming that the process starts in state S_1.

If at time $t = 0$, the system is in state S_2, then the initial condition vector is

$$\begin{bmatrix} 0 \\ 1 \\ 0 \end{bmatrix}$$

and

$$\tilde{D} = \begin{vmatrix} -2\lambda & \mu & 0 \\ 2\lambda & -(\mu + \lambda) & 1 \\ 1 & 1 & 0 \end{vmatrix} = 2\lambda + \mu$$

Therefore, the expected passage time from S_2 to S_3 is $(2\lambda + \mu)/(2\lambda^2)$, which is the same as the MTTF obtained in Equation (7–50), assuming that the process starts in state S_2.

If we are not sure about where the system is at time $t = 0$, but only have estimated probabilities of finding the system in states S_1 and S_2, we let the initial condition vector be, for example,

$$\begin{bmatrix} 0.6 \\ 0.4 \\ 0 \end{bmatrix}$$

Now,

$$\tilde{D} = \begin{vmatrix} -2\lambda & \mu & 0.6 \\ 2\lambda & -(\lambda + \mu) & 0.4 \\ 1 & 1 & 0 \end{vmatrix} = 2.6\lambda + \mu$$

and the expected passage time to S_3 is $(2.6\lambda + \mu)/(2\lambda^2)$. It can easily be seen that this value is the weighted sum of the previous two passage times, the weighting factors being the probabilities of the system being in states S_1 and S_2 at time $t = 0$.

To find the expected passage time from state S_3 to S_1, we designate S_1 as absorbing and modify the Markov differential equations as follows:

$$\begin{bmatrix} P_2'(t) \\ P_3'(t) \\ P_1'(t) \end{bmatrix} = \begin{bmatrix} -(\lambda + \mu) & 2\mu & 0 \\ \lambda & -2\mu & 0 \\ \mu & 0 & 0 \end{bmatrix} \begin{bmatrix} P_2(t) \\ P_3(t) \\ P_1(t) \end{bmatrix}$$

Now,

$$D = \begin{vmatrix} -(\lambda + \mu) & 2\mu \\ \lambda & -2\mu \end{vmatrix} = 2\mu^2$$

Assuming that the system is in state S_3 at $t = 0$,

$$\tilde{D} = \begin{vmatrix} -(\lambda + \mu) & 2\mu & 0 \\ \lambda & -2\mu & 1 \\ 1 & 1 & 0 \end{vmatrix} = \lambda + 3\mu$$

Therefore, the expected passage time from S_3 to S_1 is $(\lambda + 3\mu)/(2\mu^2)$.

To find the expected passage time from S_3 to S_2, we designate S_2 as absorbing. The modified Markov equations are then

$$\begin{bmatrix} P_1'(t) \\ P_3'(t) \\ P_2'(t) \end{bmatrix} = \begin{bmatrix} -2\lambda & 0 & 0 \\ 0 & -2\mu & 0 \\ 2\lambda & 2\mu & 0 \end{bmatrix} \begin{bmatrix} P_1(t) \\ P_3(t) \\ P_2(t) \end{bmatrix}$$

Now,

$$D = 4\lambda\mu$$

$$\tilde{D} = \begin{vmatrix} -2\lambda & 0 & 0 \\ 0 & -2\mu & 1 \\ 1 & 1 & 0 \end{vmatrix} = 2\lambda$$

Therefore, the expected passage time from S_3 to S_2 is $2\lambda/4\lambda\mu = 1/(2\mu)$. This value should not come as a surprise, since we know that the expected residence time in a state is equal to the reciprocal of the total rate of departure from that state. For state S_3,

$$\text{expected residence time} = \frac{1}{2\mu}$$

Since the system has to enter S_3 from S_2 only,

$$\text{expected time between residences in } S_3 = \frac{2\lambda + \mu}{2\lambda^2}$$

The expected value of the cycle time for state S_3 is the sum of these two quantities:

$$\text{expected cycle time} = \frac{1}{2\mu} + \frac{2\lambda + \mu}{2\lambda^2}$$

$$= \frac{(\lambda + \mu)^2}{2\lambda^2\mu}$$

which is the reciprocal of f_3, found earlier.

By now, it should be obvious to us that by employing the general techniques discussed, we can answer a variety of questions regarding the behavior of many a system under study.   ■■

7.8 FLOW-TYPE SYSTEMS AND CUMULATED STATES

There are many practical engineering systems in which different states lead to different outputs (flows or throughputs), instead of a simple success or failure. Such systems are modeled using a set of possible outputs, ranging from zero to full "capacity." Examples of flow-type systems are pumping stations with several pumps, generating stations with several units, chemical and processing industries, and the like.

After identifying the various possible states in which the system can exist, we order the states and label them S_1 through S_n, with S_1 corresponding to the state with the highest (full) capacity or output and S_n being the state with the lowest capacity (typically, zero).

Consider the ith state, S_i, for $1 \leq i \leq n$. Let λ_{+i} be the total rate of departure from S_i to states of higher capacity, i.e., $S_{i-1}, S_{i-2}, \ldots , S_1$. Also, let λ_{-i} be the total rate of departure from S_i to states of lower capacity, i.e., $S_{i+1}, S_{i+2}, \ldots , S_n$. Then the frequency of encountering S_i is

$$f_i = P_i(\lambda_{+i} + \lambda_{-i}) \tag{7–74}$$

where P_i is the steady-state probability of the system existing in S_i.

A *cumulative state $S_{i'}$*, is defined as the state with a capacity corresponding to S_i *or lower*. The probability and frequency of a cumulative state are obtained by *merging* state S_i and all the states with lower capacities. When two mutually exclusive states are merged, the probability of the merged state is equal to the sum of the probabilities of the states merged. The frequency of encountering the merged state is equal to the sum of the frequencies of the states merged minus the frequency of mutual encounters because we are interested only in transitions in and out of the merged state.

Let us begin with the lowest capacity state S_n and start merging the states as we work our way up to the full-capacity state S_1. Let

$P_{i'}$ = probability of occurrence of all states i and below

= cumulative probability of the system existing in S_i or S_{i+1} or ... or S_n

$f_{i'}$ = frequency of encountering all states i and below

= cumulative frequency of encountering S_i or S_{i+1} or ... or S_n

Then

$$P_{n'} = P_n$$

$$P_{(n-1)'} = P_{n'} + P_{(n-1)}$$

$$P_{(n-2)'} = P_{(n-1)'} + P_{(n-2)}$$

or, in general,

$$P_{i'} = P_i + P_{(i+1)'} = P_i + \sum_{k=(i+1)}^{n} P_k \qquad (7\text{–}75)$$

Moreover,

$$f_{n'} = f_n$$

$$f_{(n-1)'} = f_{n-1} + f_n - [P_n\lambda_{+n} + P_{n-1}\lambda_{-(n-1)}]$$

$$f_{(n-2)'} = f_{n-2} + f_{(n-1)'} - [P_{n-1}\lambda_{+(n-1)} + P_{n-2}\lambda_{-(n-2)}]$$

$$= f_{(n-1)'} + P_{n-2}(\lambda_{+(n-2)} + \lambda_{-(n-2)}) - [P_{n-1}\lambda_{+(n-1)} + P_{n-2}\lambda_{-(n-2)}]$$

$$= f_{(n-1)'} + P_{n-2}\lambda_{+(n-2)} - P_{n-1}\lambda_{+(n-1)}$$

But, under steady-state conditions,

$$P_{n-1}\lambda_{+(n-1)} = P_{n-2}\lambda_{-(n-2)}$$

Therefore,

$$f_{(n-2)'} = f_{(n-1)'} + P_{n-2}[\lambda_{+(n-2)} - \lambda_{-(n-2)}]$$

Generalizing, we can write

$$f_{i'} = f_{(i+1)'} + P_i(\lambda_{+i} - \lambda_{-i}) \qquad (7\text{–}76)$$

Since the frequency of a cumulative state is equal to the product of the steady-state probability of the cumulative state and the rate of departure from the cumulative state, we can write

$$\left.\begin{array}{r}\text{rate of departure from a}\\ \text{cumulative state}\end{array}\right\} = \frac{\text{cumulative frequency}}{\text{cumulative probability}} \qquad (7\text{–}77)$$

The reciprocal of the rate of departure from a cumulative state is the average duration associated with that state.

7.8.1 States with Identical Capacity

In the enumeration of the various possible states of a system, it is possible for several of them to have identical outputs. However, if we allow only one change and neglect the simultaneous occurrence of more than one change, it can be seen that there will be no direct links between the states of identical capacity in the state-space diagram. In that case, the merging of such states can be accomplished as follows:

$$\left.\begin{array}{l}\text{steady-state probability}\\\text{of the merged state}\end{array}\right\} = \Sigma \left\{\begin{array}{l}\text{steady-state probabilities}\\\text{of the states merged}\end{array}\right.$$

$$\left.\begin{array}{l}\text{frequency of the}\\\text{merged state}\end{array}\right\} = \Sigma \left\{\begin{array}{l}\text{frequencies of the}\\\text{states merged}\end{array}\right.$$

$$\left.\begin{array}{l}\text{capacity (or output)}\\\text{of the merged state}\end{array}\right\} = \left\{\begin{array}{l}\text{capacity (or output) of any one}\\\text{of the states of identical capacity}\end{array}\right\}$$

If S_i and S_j are states of identical capacity that are merged to form state S_k, then $C_k = C_i = C_j$; $P_k = P_i + P_j$; $f_k = f_i + f_j$; and $\lambda_{\pm k} = (P_i\lambda_{\pm i} + P_j\lambda_{\pm j})/P_k$.

■■ Example 7–6

Let us consider two dissimilar repairable units with failure rates λ_1 and λ_2 and repair rates μ_1 and μ_2. Also, let the capacity (power output if the unit is a generator, pumping capacity if it is a pump, etc.) of unit 1 be smaller than the capacity of unit 2. Numbering the states from S_1 through S_4 in descending order of capacities, we

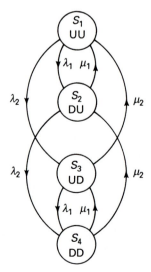

Figure 7.9 State-space diagram for two dissimilar repairable units.

see that S_1 corresponds to UU (meaning that both units are up), S_2 to DU (unit 1 down and unit 2 up), S_3 to UD and S_4 to DD. The corresponding state-space diagram is shown in Figure 7.9.

We will start the cumulation process with S_4 and work our way up to S_1. From Section 7.4.2, we have

$$P_{4'} = P_4 = \frac{\lambda_1 \lambda_2}{\mathscr{D}} \tag{7-78}$$

where

$$\mathscr{D} \equiv (\lambda_1 + \mu_1)(\lambda_2 + \mu_2) \tag{7-79}$$

Moreover,

$$f_{4'} = f_4 = \frac{\lambda_1 \lambda_2 (\mu_1 + \mu_2)}{\mathscr{D}} \tag{7-80}$$

$$P_{3'} = P_3 + P_{4'} = \frac{\lambda_2(\mu_1 + \lambda_1)}{\mathscr{D}} \tag{7-81}$$

$$f_{3'} = f_{4'} + P_3(\lambda_{+3} - \lambda_{-3})$$

$$= f_{4'} + \frac{\lambda_2 \mu_1}{\mathscr{D}}(\mu_2 - \lambda_1) \tag{7-82}$$

$$= \frac{\lambda_2 \mu_2}{\lambda_2 + \mu_2}$$

$$P_{2'} = P_2 + P_{3'} = \frac{\lambda_1 \mu_2 + \lambda_2 \mu_1 + \lambda_1 \lambda_2}{\mathscr{D}} \tag{7-83}$$

$$f_{2'} = f_{3'} + P_2(\lambda_{+2} - \lambda_{-2})$$

$$= f_{3'} + \frac{\lambda_1 \mu_2}{\mathscr{D}}(\mu_1 - \lambda_2) \tag{7-84}$$

$$= \frac{\mu_1 \mu_2 (\lambda_1 + \lambda_2)}{\mathscr{D}}$$

$$P_{1'} = P_1 + P_{2'} = 1 \tag{7-85}$$

and

$$f_{1'} = f_{2'} + P_1(\lambda_{+1} - \lambda_{-1})$$

$$= f_{2'} + \frac{\mu_1 \mu_2}{\mathscr{D}}[0 - (\lambda_1 + \lambda_2)] \tag{7-86}$$

$$= 0$$

Since S_1 is the highest capacity state, the probability of all states, S_1 and below, is the probability of all possibilities, which is obviously unity. Moreover, the system has to be in one of the four states at all times, and, as such, the frequency of the cumulated state $P_{1'}$ is zero.

The mean durations of the cumulated states are obtained by dividing the cumulative probability by the cumulative frequency:

$$m_{i'} = \frac{P_{i'}}{f_{i'}}, \quad i = 1, 2, 3, 4 \tag{7-87}$$

or

$$m_{4'} = \frac{1}{\mu_1 + \mu_2} \tag{7-88}$$

$$m_{3'} = \frac{1}{\mu_2} \tag{7-89}$$

$$m_{2'} = \frac{\lambda_1 \mu_2 + \lambda_2 \mu_1 + \lambda_1 \lambda_2}{\mu_1 \mu_2 (\lambda_1 + \lambda_2)} \tag{7-90}$$

and

$$m_{1'} = \infty \tag{7-91}$$

The system will always be in cumulative state $P_{1'}$ and therefore, the mean duration of this state is infinity.

■■

■■ **Example 7-7**

In Example 4-1, we assumed that we knew the probability of failure for each of three generators in a power plant. Let us reexamine this example in more detail. The failure and repair rates for each of the three units and their availabilities and unavailabilities are listed in the following table:

Unit	Capacity, MW	λ, day^{-1}	μ, day^{-1}	$A = \mu/(\lambda + \mu)$	$U = \lambda/(\lambda + \mu)$
1	100	0.00505	0.5	0.99	0.01
2	150	0.01020	0.5	0.98	0.02
3	200	0.01237	0.4	0.97	0.03

Our objective is to develop exact and cumulative models for this flow-type system. Such models are called *capacity outage probability* (COP) tables. Since each unit can be up or down, there are 2^3, or 8, possibilities. We will label and arrange the possible states from S_1 through S_8, in decreasing order of available capacities. The rates of departure to states of higher and lower capacity can be found easily by remembering that we allow only one change at any one transition. For example, starting with state GBG (the first and third units good, or up, and the second unit bad, or down), the only transition that will take us to a higher capacity is the repair of unit 2, and the transitions that will take us to states of lower capacity are the failures of unit 1 and unit 3. Therefore, the total rate of departure to higher capacity states is μ_2, and total rate of departure to lower capacity states is $(\lambda_1 + \lambda_2)$. Also, the frequency of encountering state S_i is equal to the steady-state probability of being in that state times the total rate of departure from the state; that is, $f_i = P_i[\lambda_{+i} + \lambda_{-i}]$. Using all this information, we can construct what we call the *exact COP table*:

Index i	Exact state	Status of units	Capacity in, MW	Capacity out, MW	Probability P_i	λ_{+i}, day^{-1}	λ_{-i}, day^{-1}	f_i, day^{-1}
1	S_1	GGG	450	0	0.941094	0	0.02762	0.025993
2	S_2	BGG	350	100	0.009506	0.5	0.02257	0.004968
3	S_3	GBG	300	150	0.019206	0.5	0.01742	0.009938
4	S_4	GGB	250	200	0.029106	0.4	0.01525	0.012086
5	S_5	BBG	200	250	0.000194	1.0	0.01237	0.000196
6	S_6	BGB	150	300	0.000294	0.9	0.01020	0.000268
7	S_7	GBB	100	350	0.000594	0.9	0.00505	0.000538
8	S_8	BBB	0	450	0.000006	1.4	0	0.000008

The cumulative model is a table that lists cumulative probabilities, cumulative frequencies, and the rate of departure from a cumulative state. We complete this table by starting from the bottom and working upwards, using the recursive equations (7–75) and (7–76). For simplicity, we use S_{ic} for $S_{i'}$. The completed table is as follows:

Index n	Cumulative state	Capacity in, MW	Capacity out, MW	Cumulative probability	Cumulative frequency, day^{-1}	Total rate of departure, day^{-1}	Mean duration, day
1	S_{1c}	450	0	1.000000	0	0	∞
2	S_{2c}	350	100	0.058906	0.0259986	0.44136	2.26572
3	S_{3c}	300	150	0.049400	0.0214602	0.43442	2.30192
4	S_{4c}	250	200	0.030194	0.0121917	0.40378	2.47660
5	S_{5c}	200	250	0.001088	0.0009932	0.91287	1.09545
6	S_{6c}	150	300	0.000894	0.0008016	0.89664	1.11527
7	S_{7c}	100	350	0.000600	0.0005400	0.90000	1.11111
8	S_{8c}	0	450	0.000006	0.0000084	1.40000	0.71429

The mean duration of a cumulative state is the reciprocal of the total rate of departure from that cumulative state.   ■■

7.9 MIXED PRODUCT APPROACH

We saw earlier that, for a single repairable component with failure and repair rates of λ and μ, respectively,

$$\text{availability} = A = \frac{\mu}{\lambda + \mu} \tag{7–92a}$$

$$\text{unavailability} = U = \frac{\lambda}{\lambda + \mu} \tag{7–92b}$$

and

$$\text{failure (or success) frequency} = \nu = A\lambda = U\mu \tag{7–93}$$

In the case of a two-out-of-three system, assuming that the components are dissimilar, the availability and unavailability can be expressed in the true polynomial forms

$$A = A_1A_2 + A_2A_3 + A_3A_1 - 2A_1A_2A_3 \tag{7–94a}$$

and

$$U = U_1U_2 + U_2U_3 + U_3U_1 - 2U_1U_2U_3 \tag{7–94b}$$

Note that the expressions for A and U are given strictly in terms of the component availabilities and unavailabilities, respectively. To calculate the failure frequency ν for this system, we can do one of two things:

1. Multiply every product term $A_iA_j \ldots$ by $(\lambda_i + \lambda_j + \ldots)$; or
2. Multiply every product term $U_iU_j \ldots$ by $(\mu_i + \mu_j + \ldots)$.

Thus,

$$\nu = A_1A_2(\lambda_1 + \lambda_2) + A_2A_3(\lambda_2 + \lambda_3) + A_3A_1(\lambda_3 + \lambda_1)$$
$$- 2A_1A_2A_3(\lambda_1 + \lambda_2 + \lambda_3) \tag{7–95a}$$

or

$$\nu = U_1U_2(\mu_1 + \mu_2) + U_2U_3(\mu_2 + \mu_3) + U_3U_1(\mu_3 + \mu_1)$$
$$- 2U_1U_2U_3(\mu_1 + \mu_2 + \mu_3) \tag{7–95b}$$

In many cases, system availability or unavailability is expressed as mixed polynomials of the component availabilities and unavailabilities. In such cases, the system failure frequency ν is calculated as follows (see Schneeweiss, 1981):

- Starting with a mixed product form of expression for A, multiply each mixed product term $A_iA_j \ldots U_mU_n \ldots$ by $(\lambda_i + \lambda_j + \ldots - \mu_m - \mu_n - \ldots)$; or
- Starting with a mixed product form of expression for U, multiply each mixed product term $U_iU_j \ldots A_mA_n \ldots$ by $(\mu_i + \mu_j + \ldots - \lambda_m - \lambda_n - \ldots)$.

The validity of the above procedure can easily be proved. For this, let

$$U' = U_i U_j \ldots \tag{7-96}$$

and

$$\mu' = \mu_i + \mu_j + \ldots \tag{7-97}$$

Then

$$\nu' = U' \mu' \tag{7-98}$$

Next, let us consider the mixed product $U'A_k$. We have

$$U'A_k = U'(1 - U_k)$$

$$= U' - U'U_k$$

The failure frequency ν associated with this mixed product is

$$\nu = U'\mu' - U'U_k(\mu' + \mu_k)$$

$$= U'(1 - U_k)\mu' - U'(U_k \mu_k)$$

$$= U'A_k\mu' - U'(A_k \lambda_k)$$

or

$$\nu = U'A_k(\mu' - \lambda_k) \tag{7-99}$$

Next, we can consider $(U'A_k)$ as the new U' and $(\mu' - \lambda_k)$ as the new μ' and proceed. Therefore, by induction, we arrive at the following result:

The contribution to system frequency from the term $U_i U_j \ldots A_m A_n \ldots$ in the expression for U is given by $(U_i U_j \ldots A_m A_n \ldots)(\mu_i + \mu_j + \ldots - \lambda_m - \lambda_n - \ldots)$.

By duality, we conclude that, corresponding to the mixed product term $A_i A_j \ldots U_m U_n \ldots$ in the expression for A, the contribution to the system frequency is

$$(A_i A_j \ldots U_m U_n \ldots)(\lambda_i + \lambda_j + \ldots - \mu_m - \mu_n - \ldots)$$

Referring back to the two-out-of-three system and expressing A as a mixed product, we get

$$A = A_1 A_2 A_3 + A_1 A_2 U_3 + A_1 U_2 A_3 + U_1 A_2 A_3 \tag{7-100a}$$

and the system failure frequency is

$$\nu = A_1 A_2 A_3 (\lambda_1 + \lambda_2 + \lambda_3) + A_1 A_2 U_3 (\lambda_1 + \lambda_2 - \mu_3) \\ + A_1 U_2 A_3 (\lambda_1 + \lambda_3 - \mu_2) + U_1 A_2 A_3 (\lambda_2 + \lambda_3 - \mu_1) \tag{7-100b}$$

which, after simplification, can be shown to be the same as Equation (7-95a).

Alternatively, we can start with

$$U = U_1 U_2 U_3 + U_1 U_2 A_3 + U_2 U_3 A_1 + U_3 U_1 A_2 \tag{7-101a}$$

and calculate

$$\nu = U_1 U_2 U_3(\mu_1 + \mu_2 + \mu_3) + U_1 U_2 A_3(\mu_1 + \mu_2 - \lambda_3) \qquad (7\text{--}101b)$$
$$+ U_2 U_3 A_1(\mu_2 + \mu_3 - \lambda_1) + U_3 U_1 A_2(\mu_1 + \mu_3 - \lambda_2)$$

which, after simplification, can be shown to be the same as Equation (7–95b).

If $(MTTF)_s$ and $(MTTR)_s$ are, respectively, the mean time to failure and mean time to repair for a system, then

$$\text{system availability} = A_s = \frac{(MTTF)_s}{(MTTF)_s + (MTTR)_s} \qquad (7\text{--}102)$$

$$\text{system unavailability} = U_s = \frac{(MTTR)_s}{(MTTF)_s + (MTTR)_s} \qquad (7\text{--}103)$$

and

$$\text{system failure frequency} = \nu_s = \frac{1}{(MTTF)_s + (MTTR)_s} \qquad (7\text{--}104)$$

From these, it follows that

$$(MTTF)_s = \frac{A_s}{\nu_s} \qquad (7\text{--}105)$$

and

$$(MTTR)_s = \frac{U_s}{\nu_s}$$
$$= \frac{(1 - A_s)}{\nu_s} \qquad (7\text{--}106)$$

Typically, A_s and ν_s are calculated first, and using these, $(MTTF)_s$ and $(MTTR)_s$ are obtained.

■■ **Example 7–8**

For a truly parallel configuration consisting of two components, each with availability A, failure rate λ, and repair rate μ, we have

$$\text{System availability} = A^2 + AU + UA$$
$$= A^2 + 2AU$$
$$= 2A - A^2 \equiv A_s$$

$$\text{System unavailability} = U^2 \equiv U_s$$

$$\text{System failure frequency} = \nu_s = 2A(\lambda) - A^2(2\lambda)$$

Alternatively,

$$\nu_s = U^2(2\mu)$$

Using $A = \mu/(\lambda + \mu)$ and $U = \lambda/(\lambda + \mu)$, it is easy to show that the two expressions for ν_s are identical.

With $\lambda = 0.01$ hr^{-1}, and $\mu = 0.49$ hr^{-1}, we have

$$A = 0.98$$

$$U = 0.02$$

$$U_s = 0.0004$$

$$A_s = 0.9996$$

$$\nu_s = 0.000392 \text{ hr}^{-1}$$

Therefore,

$$(\text{MTTF})_s = \frac{0.9996}{0.000392}$$

$$= 2{,}550 \text{ hr}$$

and

$$(\text{MTTR})_s = \frac{0.0004}{0.000392}$$

$$= 1.02041 \text{ hr}$$

The dramatic improvement realized in all reliability parameters by using two components in parallel is obvious.   ■■

■■ **Example 7–9**

For the case of three identical components in series, each with availability of A, unavailability U, failure rate λ, and repair rate μ,

$$\text{system availability} = A_s = A^3$$

$$\text{system failure frequency} = \nu_s = 3\lambda A^3$$

$$\text{system unavailability} = U_s = 1 - A^3 = 1 - (1 - U)^3 = U^3 - 3U^2 + 3U$$

$$\text{system failure frequency} = \nu_s = 3\mu U^3 - 6\mu U^2 + 3\mu U$$

With $\lambda = 0.01$ hr^{-1} and $\mu = 0.09$ hr^{-1}, $A = 0.9$ and $U = 0.1$. Moreover, $\nu_s = 3(0.01)(0.9)^3 = 0.02187$ hr^{-1}.

Also, $\nu_s = 3(0.09)(0.1)^3 - 6(0.09)(0.1)^2 + 3(0.09)(0.1) = 0.02187$ hr^{-1}.   ■■

7.10 FREQUENCY BALANCE APPROACH

Steady-state probabilities can easily be obtained from state-space diagrams without the standard matrix formulation of Markov equations by using the frequency balance approach. This approach is based on the fact that, for any state in an ergodic system, the expected frequency of entering the state must be equal to the expected frequency of leaving the state. Using this principle, frequency balance equations can be written down for each state. These equations, coupled with the equation stating that the probabilities of all states must add up to unity, give us the required equations to solve for steady-state probabilities. However, this method becomes cumbersome if there is a high degree of communication between the states. If the system is relatively small, with a large number of one-way transitions, then the method becomes convenient and fast. Example 7–10 illustrates the procedure for several different cases.

■■ **Example 7–10**

For a single component with a binary model (see Figure 7.10a), we have just one frequency balance equation, namely,

$$P_1 \lambda = P_2 \mu$$

Also,

$$P_1 + P_2 = 1$$

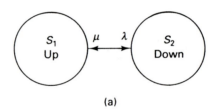

(a)

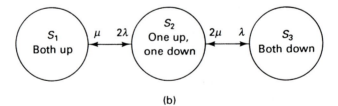

(b)

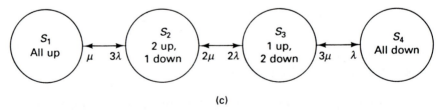

(c)

Figure 7.10 State-space diagrams for (a) a single binary component, (b) two identical repairable components, and (c) three identical repairable components.

Solving for P_1 and P_2, we have $P_1 = \mu/(\lambda + \mu)$ and $P_2 = \lambda/(\lambda + \mu)$.

For two identical components (see Figure 7.10b), the frequency balance equations are

$$P_1 2\lambda = P_2 \mu$$

$$P_2(\lambda + \mu) = P_1 2\lambda + P_3 2\mu$$

$$P_3 2\mu = P_2 \lambda$$

In addition,

$$P_1 + P_2 + P_3 = 1$$

Solving these equations simultaneously, we obtain

$$P_1 = \mu^2/\mathscr{D}; \quad P_2 = 2\lambda\mu/\mathscr{D}; \quad P_3 = \lambda^2/\mathscr{D}$$

where $\mathscr{D} \equiv (\lambda + \mu)^2$.

For three identical components (see Figure 7.10c), the frequency balance equations are

$$3\lambda P_1 = \mu P_2$$

$$P_2(\mu + 2\lambda) = 3\lambda P_1 + 2\mu P_3$$

$$P_3(2\mu + \lambda) = 2\lambda P_2 + 3\mu P_4$$

$$3\mu P_4 = \lambda P_3$$

Also,

$$P_1 + P_2 + P_3 + P_4 = 1$$

Simultaneous solution of these equations yields

$$P_1 = \mu^3/\mathscr{D}; \quad P_2 = 3\lambda\mu^2/\mathscr{D}; \quad P_3 = 3\lambda^2\mu/\mathscr{D}; \quad P_4 = \lambda^3/\mathscr{D}$$

where $\mathscr{D} \equiv (\lambda + \mu)^3$.

We could have arrived at the same results by the application of the binomial distribution as follows:

$$(P + Q)^3 = P^3 + 3P^2Q + 3PQ^2 + Q^3$$

where $P \equiv \mu/(\lambda + \mu)$ and $Q \equiv \lambda/(\lambda + \mu)$.

Therefore, $P_1 = P^3$, $P_2 = 3P^2Q$, $P_3 = 3PQ^2$, and $P_4 = Q^3$.   ∎∎

7.11 MORE APPLICATIONS OF MARKOV PROCESSES

The technique of identifying the different states, developing the state-space diagram, and setting up and solving Markov equations is so versatile that it has found application in many areas. A sampling of these is considered in this section.

7.11.1 Modeling Normal Repair and Preventive Maintenance

One approach to modeling a repairable component that also undergoes preventive maintenance is to use the state-space diagram of Figure 7.11. Several assumptions are operative in deriving this model:

1. No failure can occur during maintenance, and maintenance is not started during repair.
2. The time between consecutive maintenances is considered a continuous random variable.
3. All transition rates are assumed constant.

Even with these assumptions, this model provides a fair description of steady-state conditions. The frequency balance equations for the model are

$$P_N(\lambda + \lambda_M) = P_M\mu_M + P_R\mu$$

$$P_M\mu_M = P_N\lambda_M$$

$$P_R\mu = P_N\lambda$$

Also,

$$P_N + P_M + P_R = 1$$

Solution of these equations gives

$$P_N = \frac{\mu\mu_M}{\mathcal{D}} \tag{7-107}$$

$$P_M = \frac{\mu\lambda_M}{\mathcal{D}} \tag{7-108}$$

$$P_R = \frac{\lambda\mu_M}{\mathcal{D}} \tag{7-109}$$

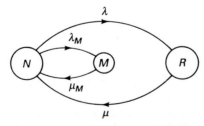

Normal   Under   Repair
operation   maintenance

Figure 7.11 State-space diagram for a repairable component, including preventive maintenance.

where

$$\mathscr{D} \equiv \mu\mu_M + \mu\lambda_M + \lambda\mu_M \qquad (7\text{-}110)$$

The probability of finding the component in the repair state is

$$P_R = \frac{\lambda\mu_M}{\mu\mu_M + \mu\lambda_M + \lambda\mu_M}$$

or

$$P_R = \frac{\lambda}{\mu + \lambda + \mu(\lambda_M/\mu_M)} \qquad (7\text{-}111)$$

For

$$\left(\frac{\lambda_M}{\mu_M}\right) \ll 1, \; P_R \simeq \frac{\lambda}{\lambda + \mu} \qquad (7\text{-}112)$$

A small λ_M means that the time between maintenances is very large, and a large μ_M means that the time taken for maintenance is very small. If both of these conditions obtain, then P_R is about the same as what we would get by completely neglecting maintenance.

7.11.2 Inclusion of Installation Time

Typically, after the repair of a component, it has to be reinstalled before it can start functioning again. This installation time is included in the state-space diagram shown in Figure 7.12.

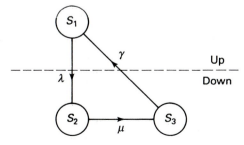

S_1: Unit up
S_2: Unit down
S_3: Unit repaired, but not yet installed

Figure 7.12 State-space diagram for a repairable component, including installation time.

Steady-state probabilities for this case can be found by the frequency balance approach or from the Markov equations. The values are:

$$P_1 = \frac{\gamma\mu}{\mathcal{D}} \qquad (7-113)$$

$$P_2 = \frac{\lambda\gamma}{\mathcal{D}} \qquad (7-114)$$

$$P_3 = \frac{\lambda\mu}{\mathcal{D}} \qquad (7-115)$$

where

$$\mathcal{D} \equiv \gamma\mu + \gamma\lambda + \mu\lambda \qquad (7-116)$$

Since state S_1 corresponds to *up* and S_2 and S_3 together constitute the *down* state, the availability $A = P_1$ and the unavailability $U = P_2 + P_3$. Also,

$$f_{\text{up}} = f_1 \qquad (7-117)$$

and

$$f_{\text{down}} = f_2 + f_3 - P_2\mu = f_{\text{up}} \qquad (7-118)$$

The mean residence times in states S_1, S_2, and S_3 are $1/\lambda$, $1/\mu$, and $1/\gamma$, respectively. The mean duration of up time is (A/f_{up}), which is equal to $1/\lambda$. The mean duration of down time is (U/f_{down}), and this is equal to $[(1/\mu) + (1/\gamma)]$, or the sum of the mean residence times in states S_2 and S_3.

7.11.3 Inclusion of a Spare

Outage times can be reduced by having a spare. In the event of a failure, the spare is installed while the failed component is undergoing repair. Since typical installation times are much shorter than typical repair times, the overall effect of the spare is to decrease the forced outage time and, consequently, increase the availability of the system.

Figure 7.13 shows the state-space diagram when a spare is included. Now there are five states. The transition rate from S_5 to S_2 depends on the nature of the repair facilities available. If resources are available to repair more than one failed component at a time, then this transition rate is 2μ, as shown. However, if only one failed component can be repaired at any particular time, then the transition rate is μ.

Analysis of this model can proceed in the usual manner, resulting in expres-

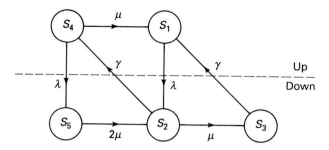

S_1: Unit up, one spare
S_2: Unit down, one spare
S_3: Unit down, two spares
S_4: Unit up, no spare
S_5: Unit down, no spare

Figure 7.13   State-space diagram for a repairable component with an identical spare.

sions for steady-state probabilities, frequencies, and mean residence times. Two of these results are as follows:

$$\text{Mean duration of up time} = \frac{1}{\lambda} \tag{7–119}$$

$$\text{Mean duration of down time} = \frac{1}{\gamma} + \frac{\lambda\gamma}{2\mu(\lambda + \mu)(\gamma + \mu)} \tag{7–120}$$

As the number of spares available increases, the number of states to be included in the state-space diagram increases also. With two spares backing a component, seven states are needed in the model. Moreover, the existing repair facilities must be scrutinized carefully, and appropriate transition rates must be used in the model.

As the number of spares increases, the system availability increases and its unavailability decreases. In the limit, as the number of spares tends to infinity, repair becomes meaningless, and whenever there is a failure, the component is simply replaced, taking an amount of time equal to the installation time. Therefore, in the limit,

$$\text{availability} = A \to A_\infty = \frac{\gamma}{\gamma + \lambda} \tag{7–121}$$

and

$$\text{unavailability} = U \to U_\infty = \frac{\lambda}{\lambda + \gamma} \tag{7–122}$$

The corresponding model and the nature of the variation of U are shown in Figure 7.14.

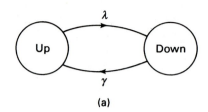

(a)

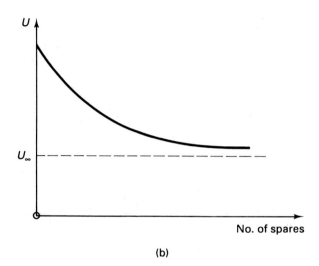

No. of spares

(b)

Figure 7.14 (a) Markov model with infinite spares. (b) Dependence of unavailability on the number of spares.

7.11.4 Modeling Standby Systems

Suppose we have a main unit C and a standby unit D. The standby unit is "cold" while not in service, and therefore, we can assume that it cannot fail during that period. The associated state-space diagram is shown in Figure 7.15, and only state S_4 corresponds to failure. An analysis of this model yields

$$P_4 = \left(\frac{\lambda_C \lambda_D}{\mu_C}\right) \frac{1 + (\lambda_C/\mu_D)}{\lambda_D + \left[1 + \dfrac{\lambda_C}{\mu_C} + \dfrac{\lambda_C \lambda_D}{\mu_C \mu_D}\right](\lambda_C + \mu_D + \mu_C)} \tag{7-123}$$

which, if we assume that

$$\lambda_C, \lambda_D \ll \mu_C, \mu_D \tag{7-124}$$

can be simplified to

$$P_4 \simeq \frac{\lambda_C \lambda_D}{\mu_C(\mu_C + \mu_D)} \tag{7-125}$$

The frequency of encountering the failed state is

$$f_4 = P_4(\mu_C + \mu_D) \simeq \frac{\lambda_C \lambda_D}{\mu_C} \tag{7-126}$$

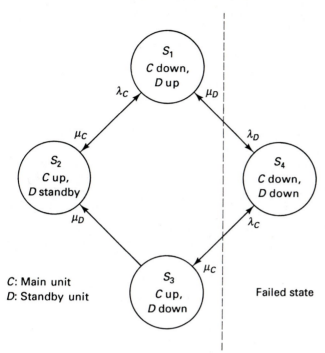

Figure 7.15 State-space diagram for a main unit and a standby unit.

If, in addition, $\mu_C = \mu_D \equiv \mu$, then

$$P_4 \simeq \frac{\lambda_C \lambda_D}{2\mu^2} \qquad (7\text{–}127)$$

and

$$f_4 \simeq \frac{\lambda_C \lambda_D}{\mu} \qquad (7\text{–}128)$$

We saw earlier that, for two dissimilar independent units, the probability of both units failing is

$$P(\text{units 1 and 2 failing}) = \frac{\lambda_1 \lambda_2}{(\lambda_1 + \mu_1)(\lambda_2 + \mu_2)} \qquad (7\text{–}129)$$

and the failure frequency is

$$f(\text{failure}) = \frac{(\mu_1 + \mu_2)\lambda_1 \lambda_2}{(\lambda_1 + \mu_1)(\lambda_2 + \mu_2)} \qquad (7\text{–}130)$$

Now suppose $\mu_1 = \mu_2 \equiv \mu$ and $\mu \gg \lambda_1, \lambda_2$. Then

$$P(\text{failure}) \simeq \frac{\lambda_1 \lambda_2}{\mu^2} \qquad (7\text{–}131)$$

and

$$f(\text{failure}) \simeq \frac{2\lambda_1\lambda_2}{\mu} \tag{7-132}$$

Comparing Equation (7–127) with Equation (7–131) and Equation (7–128) with Equation (7–132), we arrive at the conclusion that the probability and frequency of the double-failure state in the standby system are about one-half of the corresponding values for a system in which the same two components are operating independently and in parallel. This improvement is the result of the standby system operating only intermittently and not being subjected to the aging process while not in service.

The state-space diagram of Figure 7.15 can be modified to include the probability that the standby unit will fail to start. If, out of n attempts to start, the unit fails to start np times, then p is the probability of its failure to start, and this probability can be included in the model, as shown in Figure 7.16. Note that the total rate of departure from state S_2 is still equal to λ_C, except that it is now split two ways: $p\lambda_C$ and $(1 - p)\lambda_C$. Analysis of this state-space diagram with the assumption that

$$\lambda_C, \lambda_D \ll \mu_C, \mu_D$$

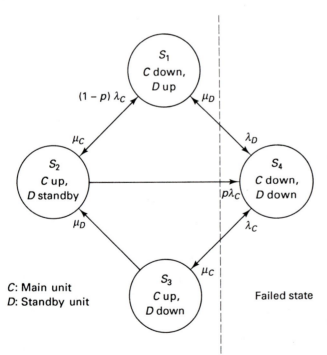

C: Main unit
D: Standby unit

Figure 7.16 State-space diagram for the system of Figure 7.15 with standby start-up failure included.

results, respectively, in the following expressions for the failure probability and failure frequency:

$$P_4 \simeq \frac{\lambda_C}{\mu_C + \mu_D} \left[\frac{\lambda_D}{\mu_C} + p \right] \qquad (7\text{--}133)$$

$$f_4 \simeq \lambda_C \left[\frac{\lambda_D}{\mu_C} + p \right] \qquad (7\text{--}134)$$

An alternative way of including the failure of the standby unit to start is shown in Figure 7.17. A new state, S_5, is introduced to represent hidden failures of the standby unit. Hidden failures will not be discovered until the unit is required to operate, and such failures are common with devices that are required to operate only infrequently. For this model, the following relationships apply:

$$p = \frac{\lambda_h}{\lambda_C + \lambda_h} \simeq \frac{\lambda_h}{\lambda_C} \qquad (7\text{--}135)$$

$$P_4 \simeq \left(\frac{1}{\mu_C + \mu_D} \right) \left(\frac{\lambda_C \lambda_D}{\mu_C} + \lambda_h \right) \qquad (7\text{--}136)$$

$$f_4 \simeq \frac{\lambda_C \lambda_D}{\mu_C} + \lambda_h \qquad (7\text{--}137)$$

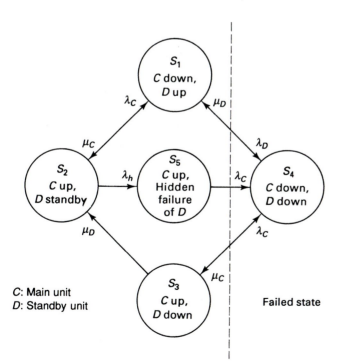

C: Main unit
D: Standby unit

Figure 7.17 State-space diagram for the system of Figure 7.15 with an alternative approach to including the start-up failure of the standby unit.

Periodic inspection and preventive maintenance improve the reliability of standby units. The state-space diagram with maintenance included in the model is shown in Figure 7.18, in which λ_m is the maintenance rate and $1/\lambda_m$ is the mean time between instances of maintenance. The mean time required for maintenance, T_m, is equal to $1/\mu_m$. States S_4 and S_7 are the failed states. They could be merged into one state S_F with a probability P_F and a frequency of encounter f_F. State S_2 is the "normal" state. If a hidden failure of standby unit D is discovered during inspection, the system is in S_5 at the outset of inspection. A transfer to S_3 is made,

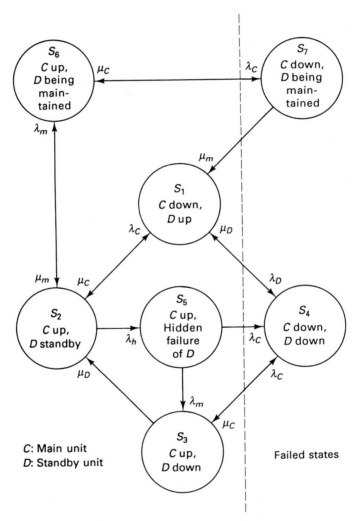

Figure 7.18   State-space diagram for the system of Figure 7.17 with maintenance included.

and a complete repair ensues, thus avoiding a transfer to S_4, one of the failure states. Assuming that

$$\lambda_h \ll \lambda_C; \lambda_m \ll \mu_C, \mu_D$$

the following expressions for P_F and f_F can be derived:

$$P_F \simeq \frac{\lambda_C}{\mu_C + \mu_D} \left[\frac{\lambda_D}{\mu_C} + \frac{\lambda_h}{\lambda_C + \lambda_m} \right] + \frac{\lambda_C \lambda_m}{\mu_m(\mu_C + \mu_m)} \qquad (7\text{--}138)$$

$$f_F \simeq \lambda_C \left[\frac{\lambda_D}{\mu_C} + \frac{\lambda_h}{\lambda_C + \lambda_m} + \frac{\lambda_m}{\mu_m} \right] \qquad (7\text{--}139)$$

Frequent inspections of short duration result, in general, in relatively few total failures of long duration and, eventually, in improved system reliability.

■■ Example 7–11

Let us consider a special case of a repairable standby system with two identical units and perfect switching. Suppose that repair is attempted only when one of the two units is working; thus, if both units are down, no repair is performed. Under these conditions, the modified state-space diagram is shown in Figure 7.19.

The associated Markov differential equations are

$$\begin{bmatrix} P_1'(t) \\ P_2'(t) \\ P_3'(t) \end{bmatrix} = \begin{bmatrix} -0.5 & 10 & 0 \\ 0.5 & -10.5 & 0 \\ 0 & 0.5 & 0 \end{bmatrix} \begin{bmatrix} P_1(t) \\ P_2(t) \\ P_3(t) \end{bmatrix}$$

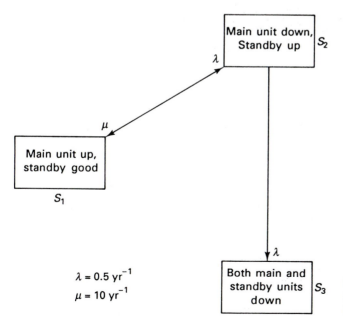

$\lambda = 0.5 \text{ yr}^{-1}$

$\mu = 10 \text{ yr}^{-1}$

Figure 7.19 Transition diagram for a repairable standby system with two identical units and perfect switching.

Considering only the first two equations, we find that the coefficient matrix of the Markov differential equations is

$$\mathbf{A} = \begin{bmatrix} -0.5 & 10 \\ 0.5 & -10.5 \end{bmatrix}$$

We will use the resolvent matrix method to solve the first two differential equations:

$$[s\mathbf{I} - \mathbf{A}]^{-1} = \begin{bmatrix} s + 0.5 & -10 \\ -0.5 & s + 10.5 \end{bmatrix}^{-1}$$

$$= \frac{1}{(s + 0.5)(s + 10.5) - 5} \begin{bmatrix} s + 10.5 & 10 \\ 0.5 & s + 0.5 \end{bmatrix}$$

Let

$$(s + 0.5)(s + 10.5) - 5 \equiv (s + a_1)(s + a_2)$$

where

$$a_1 = 0.0227744$$

and

$$a_2 = 10.9772256$$

By employing partial fraction techniques, we get

$$\frac{s + 10.5}{(s + a_1)(s + a_2)} = \frac{0.9564355}{s + a_1} + \frac{0.04356454}{s + a_2}$$

and

$$\frac{0.5}{(s + a_1)(s + a_2)} = (0.04564355)\left[\frac{1}{s + a_1} - \frac{1}{s + a_2}\right]$$

After taking the inverse Laplace transform of $[s\mathbf{I} - \mathbf{A}]^{-1}$, and with the initial condition

$$\begin{bmatrix} P_1(0) \\ P_2(0) \end{bmatrix} = \begin{bmatrix} 1 \\ 0 \end{bmatrix}$$

we obtain the solution,

$$P_1(t) = (0.9564355)\exp[-a_1 t] + (0.04356454)\exp[-a_2 t]$$

$$P_2(t) = (0.04564355)\{\exp[-a_1 t] - \exp[-a_2 t]\}$$

Therefore, $P_3'(t) = (0.5)P_2(t) = (0.022821775)\{\exp[-a_1 t] - \exp[-a_2 t]\}$. Integrating and using the initial condition $P_3(0) = 0$, we get

$$P_3(t) = 1 - 1.00208\exp[-a_1 t] + 0.00208\exp[-a_2 t]$$

The stochastic transitional probability matrix without the Δt's is

$$\mathbf{P} = \begin{bmatrix} 0.5 & 0.5 & 0 \\ 10 & -9.5 & 0.5 \\ 0 & 0 & 1 \end{bmatrix}$$

Rearranging, we get

$$
\mathbf{P} = \begin{array}{c} \\ S_3 \\ S_1 \\ S_2 \end{array}
\begin{array}{c}
\begin{array}{ccc} S_3 & S_1 & S_2 \end{array} \\
\left[\begin{array}{c|cc}
1 & 0 & 0 \\
\hline
0 & 0.5 & 0.5 \\
0.5 & 10 & -9.5
\end{array} \right]
\end{array}
$$

The associated fundamental matrix is

$$
\mathbf{N} = [\mathbf{I} - \mathbf{Q}]^{-1}
$$

$$
= \begin{bmatrix} 0.5 & -0.5 \\ -10 & 10.5 \end{bmatrix}^{-1}
$$

$$
= \begin{bmatrix} 42 & 2 \\ 40 & 2 \end{bmatrix}
$$

If the system starts in state S_1, the MTTF is equal to $(42 + 2)$, or 44, years, and if the system starts in state S_2, the MTTF is equal to $(40 + 2)$, or 42, years.   ■■

7.11.5 Modeling Multicomponent Systems with Only One Repair Facility

In our discussion of repairable components in this chapter thus far, we have not considered the limitations imposed on the operation of a system by the lack of multiple repair facilities. If more than one component was down at any one time, we tacitly assumed that facilities were available to repair all of them simultaneously. If this is not the case, we can still use the Markov method, but with a slight change in some of the transition rates.

Let us consider again the system shown in Figure 7.5. The assumed departure rate of 2μ from state S_3 to state S_2 implies that we can summon a multiple repair crew if necessary. If we have only one repair crew (facility), then the departure rate from S_3 to S_2 will be only μ instead of 2μ. The modified stochastic transitional probability matrix without the Δt's is then

$$
\mathbf{P} = \begin{bmatrix}
(1 - 2\lambda) & 2\lambda & 0 \\
\mu & (1 - \lambda - \mu) & \lambda \\
0 & \mu & (1 - \mu)
\end{bmatrix} \tag{7-140}
$$

The limiting-state probabilities are

$$
P_1 = \frac{\mu^2}{(\mu + \lambda)^2 + \lambda^2} \tag{7-141}
$$

$$
P_2 = \frac{2\lambda\mu}{(\mu + \lambda)^2 + \lambda^2} \tag{7-142}
$$

and

$$P_3 = \frac{2\lambda^2}{(\mu + \lambda)^2 + \lambda^2} \tag{7-143}$$

These expressions should be compared with Equations (7–43) to see the impact of having only one repair facility instead of two (see Example 7–12).

■■ **Example 7–12**

For $\lambda = 0.1$ and $\mu = 10$, steady-state probabilities with one and two repair facilities are calculated and listed in the following table:

	One repair facility	Two repair facilities
P_1	0.9802	0.980296
P_2	0.0196	0.019606
P_3	0.0002	0.000098

With only one repair crew, there is a significant increase in P_3, as compared to having two repair crews. The differences in the values of P_1 and P_2 are much smaller. Differences in the values of availability and unavailability depend on the system configuration, namely, whether the two components are in series or in parallel.   ■■

Extending this line of reasoning for a system with three identical components (see Figure 7.10c), if we have only one repair facility, then all the repair transitions will have a transition rate of μ only. The new values of limiting-state probabilities can be found in the usual manner (see Problem 7.30). The results are

$$P_1 = \frac{\mu^3}{\mathcal{D}} \tag{7-144}$$

$$P_2 = \frac{3\lambda\mu^2}{\mathcal{D}} \tag{7-145}$$

$$P_3 = \frac{6\lambda^2\mu}{\mathcal{D}} \tag{7-146}$$

$$P_4 = \frac{6\lambda^3}{\mathcal{D}} \tag{7-147}$$

where

$$\mathcal{D} \equiv (\lambda + \mu)^3 + 3\lambda^2\mu + 5\lambda^3 \tag{7-148}$$

Whenever we have only one repair facility, we tacitly assume that repair is done on a first-come, first-served basis.

7.12 MERGING OF STATES AND REDUCED MODELS

In the discussion of flow-type systems (see Section 7.8), we introduced the concept of merging a few of the states of a Markov model. This technique allows us to reduce a multistate model to a simple binary or ternary model. If two states that have no direct linkage in the state-space diagram are merged, then the frequency of the merged state is equal to the sum of the frequencies of the states merged, and the probability of existence of the merged state is equal to the sum of the probabilities of the states merged. If there is a direct linkage between the two states we want to merge, then there could be mutual encounters (transitions back and forth) between the two, and the frequency of the merged state will be equal to the sum of the frequencies of the states merged minus the frequency of mutual encounters. The procedure for obtaining the probability of the merged state does not change even if there are mutual encounters: It is always equal to the sum of the probabilities of the states being merged. We will illustrate these ideas using an example.

■■ **Example 7–13**

Let us consider once more the system shown in Figure 7.5. We will reduce this system to a binary model with the following state designations:

S_I: At least one component is up

S_{II}: Both components are down

While our new S_{II} is the same as our previous S_3, we have to merge the old S_1 and S_2 to obtain S_I.

For the merged state S_I,

$$P_I = P_1 + P_2 = \left(\frac{\mu}{\lambda + \mu}\right)^2 \left(1 + \frac{2\lambda}{\mu}\right)$$

$$f_I = f_1 + f_2 - \underbrace{[2\lambda P_1 + \mu P_2]}_{\text{mutual encounters}}$$

or

$$f_I = P_1(2\lambda) + P_2(\lambda + \mu) - (2\lambda P_1 + \mu P_2)$$

$$= P_2 \lambda$$

$$= \left(\frac{2\lambda^2}{\mu}\right)\left(\frac{\mu}{\lambda + \mu}\right)^2$$

In the binary model, the rate of departure λ_{12} from state S_I to state S_{II} is found as follows:

$$P_I \lambda_{12} = f_I$$

$$\lambda_{12} = \frac{f_I}{P_I}$$

$$= \frac{2\lambda^2}{2\lambda + \mu}$$

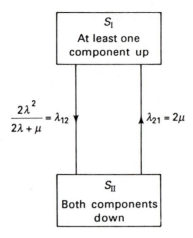

Figure 7.20   Binary model for two identical repairable components.

State S_{II} is the same as the old state S_3. Therefore,

$$P_{II} = P_3 = \left(\frac{\lambda}{\lambda + \mu}\right)^2$$

$$f_{II} = f_3 = 2\mu P_3$$

$$= 2\mu \left(\frac{\lambda}{\lambda + \mu}\right)^2$$

and

$$\lambda_{21} = \frac{f_{II}}{P_{II}}$$

$$= 2\mu$$

The resulting binary model is shown in Figure 7.20. In the case of a truly parallel configuration, S_I corresponds to *up* and S_{II} corresponds to *down* for the system. The mean durations of residence in the two binary states are easily found to be

$$m_I = \frac{1}{\lambda_{12}} = \frac{2\lambda + \mu}{2\lambda^2}$$

and

$$m_{II} = \frac{1}{\lambda_{21}} = \frac{1}{2\mu}$$

When a complex system has several subsystems with a large number of components, the techniques discussed here can be employed to derive simple binary or ternary models for the subsystems which, when put together, yield a manageable model for the entire system. ■■

7.13 MARKOV APPROACH TO NONREPAIRABLE SYSTEMS

In Chapters 3, 4, and 5, we discussed the procedures for evaluating the reliability of systems made up of nonrepairable components undergoing catastrophic failures. Although not generally used, state-space diagrams and the associated Markov models are valid even when we are dealing with nonrepairable systems. The only modification needed is to replace all the repair rates by zeros. We can go through the complete process of setting up the Markov differential equations and solving them to find the time-dependent probabilities, as before. The question of finding the steady-state probabilities becomes trivial since, eventually, the system will get into the failure state with no possibility of repair. We illustrate the procedure in some examples.

■■   **Example 7–14**

Consider again the case of two dissimilar repairable components with state-space diagram shown in Figure 7.3. With no repair, we replace all the μ's by zeros, and the Markov differential equations become (see Equation (7–29))

$$\begin{bmatrix} P_1'(t) \\ P_2'(t) \\ P_3'(t) \\ P_4'(t) \end{bmatrix} = \begin{bmatrix} -(\lambda_1 + \lambda_2) & 0 & 0 & 0 \\ \lambda_1 & -\lambda_2 & 0 & 0 \\ \lambda_2 & 0 & -\lambda_1 & 0 \\ 0 & \lambda_2 & \lambda_1 & 0 \end{bmatrix} \begin{bmatrix} P_1(t) \\ P_2(t) \\ P_3(t) \\ P_4(t) \end{bmatrix}$$

These equations can be solved once the initial conditions are known. With the initial conditions $P_1(0) = 1$, $P_2(0) = 0$, $P_3(0) = 0$, and $P_4(0) = 0$, the time-dependent state probabilities can easily be found to be

$$P_1(t) = e^{-(\lambda_1 + \lambda_2)t}$$

$$P_2(t) = e^{-\lambda_2 t} - e^{-(\lambda_1 + \lambda_2)t}$$

$$P_3(t) = e^{-\lambda_1 t} - e^{-(\lambda_1 + \lambda_2)t}$$

$$P_4(t) = 1 - e^{-\lambda_1 t} - e^{-\lambda_2 t} + e^{-(\lambda_1 + \lambda_2)t}$$

As time progresses, all of the probabilities except $P_4(t)$ tend towards zero and $P_4(t)$ tends towards unity, implying that the system will eventually undergo a transition to the failure state P_4.

If the two components are in series, then only state S_1 corresponds to success, and therefore, the system reliability is simply equal to $P_1(t)$. This value is just what we would have expected, since both components have constant hazards. If the two components are in parallel, then only state S_4 corresponds to system failure, and the system reliability is equal to the sum of $P_1(t)$, $P_2(t)$, and $P_3(t)$.   ■■

■■   **Example 7–15**

In the previous example, if the components are identical, then the state-space diagram of Figure 7.5 is applicable with $\mu = 0$, and the Markov equations become

$$\begin{bmatrix} P_1'(t) \\ P_2'(t) \\ P_3'(t) \end{bmatrix} = \begin{bmatrix} -2\lambda & 0 & 0 \\ 2\lambda & -\lambda & 0 \\ 0 & \lambda & 0 \end{bmatrix} \begin{bmatrix} P_1(t) \\ P_2(t) \\ P_3(t) \end{bmatrix}$$

The solution of these differential equations with initial conditions $P_1(0) = 1$, $P_2(0) = 0$, and $P_3(0) = 0$ is

$$P_1(t) = e^{-2\lambda t}$$

$$P_2(t) = 2(e^{-\lambda t} - e^{-2\lambda t})$$

$$P_3(t) = 1 - 2e^{-\lambda t} + e^{-2\lambda t}$$

For components in series,

$$R(t) = P_1(t) = e^{-2\lambda t}$$

For components in parallel,

$$R(t) = P_1(t) + P_2(t) = 2e^{-\lambda t} - e^{-2\lambda t}$$ ■■

7.14 STATE-SPACE TRUNCATION

When we have a large number of components in the system, each of which can be good or bad (up or down), we end up with a large number (2^N for N components) of states in the state space. We would like to omit those states with negligible probabilities. The aim is to find the level of truncation allowable for a required error in the system failure probability.

Let the system consist of N identical, independent components. Let the availability of a component be A and the unavailability be U. Then

$$\left.\begin{array}{l} \text{Probability of some state existing in which} \\ r \text{ out of } N \text{ components have failed} \end{array}\right\} = U^r A^{N-r} \qquad (7\text{–}149)$$

Since there are $\binom{N}{r}$ states representing an r-fold failure,

$$\text{Probability of an } r\text{-fold failure} = p_r = \binom{N}{r} U^r A^{N-r} \qquad (7\text{–}150)$$

Obviously, the number of failed components has a binomial distribution.

Assuming that the components have high reliability, say, $A \cong 1$, the probability of a *single* r-fold failure is much higher than the probability of a *single* $(r + 1)$-fold failure. Thus,

$$U^r A^{N-r} \gg U^{r+1} A^{N-r-1} \qquad (7\text{–}151)$$

However, the same does not *necessarily* apply for p_r and p_{r+1}, the probabilities for the *totality* of r-fold and $(r + 1)$-fold failure states. In addition, not every r-fold failure may result in *system failure*.

The error in system failure probability due to truncation is *not* measured by the probability of all the states omitted as against the probability of the total number of states. Rather, it is determined by the probability of system-failure states omitted vis-à-vis the probability of all the system failure states.

Let

$$\rho_r \equiv \begin{array}{l} \text{fraction of } r\text{-fold failures that} \\ \text{result in system failure} \end{array}$$

Then

$$\left.\begin{array}{l} \text{Probability of system failure} \\ \text{owing to exactly } r \text{ component} \\ \text{failures} \end{array}\right\} = p_{Fr} = \rho_r p_r \qquad (7\text{--}152)$$

and

$$\text{Total system failure probability} = P_F = \sum_{r=0}^{N} p_{Fr} \qquad (7\text{--}153)$$

Typically, $\rho_r = 0$ for $r \le r_1$, and $\rho_r = 1$ for $r \ge r_2$, and the variation of ρ_r with r will be as illustrated in Figure 7.21.

If we truncate the state space at $r = r_0$, then

$$\left.\begin{array}{l} \text{System failure probability calculated} \\ \text{from the truncated state space} \end{array}\right\} = \tilde{P}_F = \sum_{r=0}^{r_0} p_{Fr} \qquad (7\text{--}154)$$

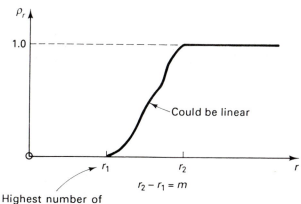

Highest number of
overlapping component
failures that still
cannot cause system
failure

Figure 7.21 Dependence of ρ_r on r.

Therefore,

$$\left.\begin{array}{l}\text{Percentage error introduced into the}\\\text{system failure probability}\\\text{due to truncation}\end{array}\right\} = 100 \left[1 - \frac{\tilde{P}_F}{P_F} \right] \equiv \epsilon \quad (7\text{--}155)$$

Using Equations (7–154) and (7–155), we can calculate the allowable r_0 for a given ϵ. The following general conclusions can be drawn from this discussion:

1. The value of r_0 depends on the size of the system and the interconnections among its components. It can be far in excess of the often assumed values of 1 or 2, especially if U is not small ($U \geq 0.1$).
2. The truncation level r_0 depends on r_1.
3. The value of r_0 is fairly insensitive to $m = r_2 - r_1$.

As a rule of thumb, for $\epsilon = 20\%$, $r_0 \cong NU + 1 + r_1$.

An accurate value of r_0 can be determined only by detailed calculations.

■■ **Example 7–16**

Suppose that we have a system with 10 identical components, each with an availability of 0.9. Suppose also that two overlapping failures can be tolerated without system failure. Then for a 20% error in the value of the system failure probability, we can truncate the state space after $r_0 = (10 \times 0.1) + 1 + 2 = 4$.   ■■

7.15 SUMMARY

State-space techniques and Markov models enable us to analyze and study a wide variety of systems from a probabilistic viewpoint. Most of the thought process goes into identifying the various states and finding the proper transition rates to use between different states. Once the state-space diagram is finalized, setting up the corresponding Markov differential equations, solving them, and finding the steady-state probabilities and the mean durations of residence in the different states become fairly routine, although special skills and the use of a computer may be required in the case of large systems. Frequency and duration techniques and the concept of a fundamental matrix help us to find residence times, passage times from one state to the other, and cycle times fairly easily. Flow-type systems and cumulated states require the use of merging of states. Such models are employed in evaluating the adequacy of power-generating plants, pumping stations, and the like. The frequency balance approach is a quick and easy way of finding steady-state probabilities if that is all we are interested in. The mixed product approach is a convenient way of finding the system failure frequency, starting with the expressions for system availability or unavailability. The versatility of the Markov approach has been demonstrated by its application to a large number of situations, such as repair and maintenance, the inclusion of installation time, the inclusion of a spare, standby system modeling, systems with restricted repair, and nonrepairable systems. Obviously, there are other applications the reader will be able to handle based on the knowledge and insight gained by studying these cases and examples. Finally, state-space truncation was considered briefly at the close of the chapter.

PROBLEMS

7.1 Compute $\exp[\mathbf{A}t]$ for

$$\mathbf{A} = \begin{bmatrix} 0 & -2 \\ 1 & -4 \end{bmatrix}$$

by (*i*) the resolvent matrix method, (*ii*) the eigenvalue method, and (*iii*) the series method.

7.2 A small electric generator fails, on average, once a year. From historical data, the mean repair time is estimated to be 5 days. Derive an expression for the availability of the generator, and calculate the reliability values for mission times of 400 hours and 800 hours. What is the unavailability in the long run?

7.3 Draw the complete state-space diagram for a system consisting of three nonidentical binary (*up* and *down*) components with failure and repair rates of λ_i and μ_i, respectively, for $i = 1, 2,$ and 3. Clearly mark all the transition rates in your diagram.

7.4 Draw the complete state-space diagram for a two-component system if each of the components has three states—*up, down,* and *derated*. Number the states, and use the notation λ_{ij} for the transition rate from state S_i to state S_j.

7.5 A system consists of two identical components in parallel, each with a failure rate of 1 per year. After the failure of one of the two, the other gets overloaded, and its failure rate increases to 1.5 per year. Draw a state-space diagram, and derive an expression for the reliability of the system.

7.6 Consider a six-diode double-way rectifier system, as shown in Figure P 7.6. The system is employed to obtain dc from a three-phase power supply. If just one diode

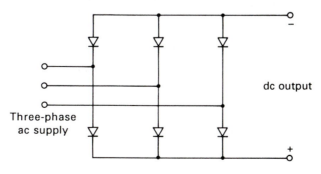

Figure P 7.6

fails, the system becomes inoperable. Every diode failure is accompanied by a repair process, followed by reinstallation of the repaired diode. When the bridge is down, all the diodes are de-energized, and in this mode, the probability of failure is negligibly small. Neglect the possibility of more than one diode failing simultaneously, and

   (*i*) Identify the states and draw a state-space diagram.

   (*ii*) Set up the Markov differential equations for the system.

   (*iii*) Find the steady-state probabilities of finding the system in each of the states.

For each of the six diodes, take the failure, repair, and installation rates to be λ, μ, and γ, respectively.

7.7 A system consists of two identical repairable components in series. If one of the two fails, the other cannot fail during the repair of the failed component. The failure and repair rates are 0.1 and 365 per year, respectively. Find (*i*) the probability of system failure, (*ii*) the frequency of system failure, and (*iii*) the average down time of the system.

7.8 One of the daily load models that is used in power system reliability evaluation assumes that the daily load consists of a peak load L_i for a fraction e of the day and a light load L_0 for the rest of the day. The peak load L_i can take on, at random, one of n different values, L_1, L_2, ... , and L_n. Assuming that the relative frequencies of occurrence of the n possible peak loads are β_i, $i = 1, 2, \ldots , n$, develop a Markov model for the load. Draw a neat state-space diagram, mark all the relevant values, and find the steady-state probabilities of all the load states. Also, find the mean duration of residence in each of the states.

7.9 A series system consists of n nonidentical components. If one of the components fails, the system is down, and no more failures can occur. The probability of more than one component failing simultaneously can be neglected. The failure and repair rates for the *i*th component are λ_i and μ_i, respectively.

   (*i*) Draw a state-space diagram, and include all the relevant information.

   (*ii*) Calculate the steady-state probabilities of the various states.

   (*iii*) What is the system availability?

   (*iv*) What is the system unavailability?

   (*v*) Find the MTTF, MTTR, and mean cycle time for the system.

   (*vi*) What is the system failure frequency?

7.10 Active safety systems are repaired only when they fail completely. Thus, they continue to operate, even if they are degraded due to environmental or other reasons. Eventually, they fail completely and are repaired. The possibility of a catastrophic failure completely shutting the system down cannot be ignored. A possible Markov model for such a system is shown in Figure P 7.10.

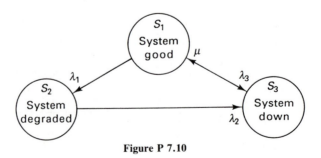

Figure P 7.10

Write down the Markov differential equations, and derive an expression for the steady-state availability of this system. Calculate the numerical values of availability and unavailability for $\lambda_1 = 0.001$ hr^{-1}, $\lambda_2 = 0.0015$ hr^{-1}, $\lambda_3 = 0.0001$ hr^{-1}, and $\mu = (1/300)$ hr^{-1}.

7.11 Develop a Markov model for the case of two identical components in a truly parallel arrangement. If one of the two fails, it is repaired and put back in operation. If both are down, repair is not done. The probability of simultaneous failure of both components is neglected. Use λ and μ for the failure and repair rates, respectively, of one component. Derive expressions for the time-dependent probabilities of the different states. At $t = 0$, both components are good and are operating.

7.12 For the system described in Problem 7.11, calculate the system MTTF by
    (*i*) integrating the system reliability function from 0 to ∞ and
    (*ii*) using the fundamental matrix.

7.13 For the system of Problem 7.11, if λ_1 represents the failure rate for both components failing simultaneously, (a) redraw the state-space diagram and set up the Markov differential equations and (b) derive an expression for the system MTTF.

7.14 Develop an exact and a cumulative probabilistic model for a power plant in which there are four generators operating in parallel with the following failure and repair data:

Unit	Capacity, MW	Failure Rate/Day	Repair Rate/Day
1	20	0.01	0.49
2	30	0.01	0.49
3	40	0.01	0.49
4	50	0.01	0.49

If there are any identical capacity states, merge them and develop reduced models.

7.15 A power plant has five units operating in parallel. Their capacities, failure rates, and repair rates are tabulated as follows:

Unit	Capacity, MW	Failure Rate/Day	Repair Rate/Day
1	20	0.01	0.49
2	20	0.01	0.49
3	30	0.01	0.49
4	30	0.01	0.49
5	40	0.01	0.49

    (*i*) Develop an exact COP table. Merge identical capacity states and derive the reduced model.
    (*ii*) Develop a cumulative model from the results of Part (*i*).

7.16 Consider a pumping station that has two identical pumps connected in parallel, each capable of pumping 1,500 gph. The failure and repair rates for each pump are 0.01 and 0.5 per hour, respectively.

(*i*) Construct the state-space diagram, and find the stochastic transitional probability matrix for the station.

(*ii*) Evaluate the mean hourly throughput of the pumping station.

(*iii*) Compare the result of Part (*ii*) with the throughput of a station having only one pump of capacity 3,000 gph and with the same failure and repair rates.

7.17 For the case of three identical components in parallel,

(*i*) Write a mixed-polynomial expression for the system availability A_s.

(*ii*) Write an expression for the system failure frequency ν_s using the result of Part (*i*).

(*iii*) Derive an expression for the system availability A_s in terms of the component availability A.

(*iv*) Write an expression for ν_s using the result of Part (*iii*).

(*v*) Write an expression for system unavailability U_s in terms of the component unavailability U.

(*vi*) Write an expression for ν_s using the result of Part (*v*).

Given that, for a component, $\lambda = 0.01 \text{ hr}^{-1}$ and $\mu = 0.49 \text{ hr}^{-1}$, calculate the numerical value of ν_s using the results of Parts (*ii*), (*iv*), and (*vi*). Also, calculate the values of $(\text{MTTF})_s$ and $(\text{MTTR})_s$.

7.18 A system has five identical components, of which three are required for system success.

(**a**) Write mixed-polynomial expressions for the system availability A_s and system unavailability U_s in terms of A and U, the availability and unavailability, respectively, for a single component.

(**b**) Derive expressions for the system frequency ν_s using A_s first and then U_s by employing Schneeweiss's method. The failure and repair rates for each component are λ and μ, respectively.

(**c**) If $\lambda = 1$ and $\mu = 99$, find the system frequency using the two expressions derived in Part (b), and show that the results are identical.

(**d**) Calculate the failure and repair rates for the system, using the values you obtained for A_s, U_s, and ν_s.

7.19 Systems that can fail in more than one manner can be modeled by a multistate Markov process. A typical example of such a system is an electrical system that can fail either in the open-circuit mode or in the short-circuit mode. Develop a three-state Markov model using the following parameters:

$$\lambda_o, \mu_o : \text{failure and repair rates for open-circuit failure}$$

$$\lambda_s, \mu_s : \text{failure and repair rates for short-circuit failure}$$

Set up the Markov differential equations in vector-matrix form. Find the steady-state probabilities using the frequency balance approach. Also, find the frequencies of encounter and mean durations for the three states.

7.20 In Problem 7.19, assuming that a short-circuit failure is not repairable, derive an expression for the system MTTF. Calculate its numerical value for $\lambda_o = 0.001 \text{ hr}^{-1}$, $\mu_o = 1 \text{ hr}^{-1}$, and $\lambda_s = 0.002 \text{ hr}^{-1}$.

7.21 In Problem 7.19, if no repair is possible no matter which way the system fails, set up the Markov equations and find the MTTF using the fundamental matrix. Also, find the probabilities of failing in the open-circuit mode and in the short-circuit mode.

7.22 Develop the state-space diagram for a system consisting of a main unit A and a standby unit B. The standby unit is assumed to be failproof in the standby mode. Switching from A to B can be considered to be perfect, and the possibility of repair is not to be included. If λ_a and λ_b are the failure rates of unit A and unit B, respectively, derive an expression for the reliability of the system. What is the MTTF?

7.23 Repeat Problem 7.22 for the case of $\lambda_a = \lambda_b \equiv \lambda$.

7.24 Consider a normally operating unit A backed up by a standby unit B. When A fails, B is automatically switched in by an ideal switch. The following failure rates are applicable:

> λ_a for unit A under normal operating conditions
>
> λ_b for unit B under normal operating conditions
>
> λ_{bs} for unit B in the standby mode

Neglect the possibility of repair and simultaneous failure of both units, and
  (*i*) Draw a state-space diagram.
  (*ii*) Derive expressions for the time-dependent probabilities of the various states.
  (*iii*) Find the reliability of the system.
  (*iv*) Find the system MTTF.

7.25 Repeat Problem 7.24 for the case of $\lambda_a = \lambda_b \equiv \lambda$.

7.26 Modify the state-space diagram of Figure 7.16 for the case of no repair. Derive expressions for the reliability and MTTF for this case. Comment on the special cases of $p = 0$ and $p = 1$.

7.27 Repeat Problem 7.26 for the special case of identical main and standby units.

7.28 Derive Equation (7–123) for the failure probability of a system with a main unit C backed up by a standby unit D. Employ the notation given in Figure 7.15.

7.29 For the case of three identical repairable components with only one repair facility, and if repair is not performed when all three components are down, derive an expression for the MTTF of the system. Assume that the system starts with all units good and working. Use λ and μ for the failure and repair rates, respectively, of one component.

7.30 Derive Equations (7–144) through (7–148) for the case of three identical repairable components with one repair crew. Compare these results with those derived for Example 7–10, and state your conclusions clearly.

7.31 An office has n personal computers, all of which are needed for its functioning. An identical unit is kept as a standby in case there is a failure of one of the units. Assuming that the repair and installation crew can handle only one computer (for either repair or installation) at a time, develop a Markov model and find (*i*) the steady-state probabilities, (*ii*) the frequencies of encounter, and (*iii*) the mean durations of the various states. For each computer, take $\lambda = $ failure rate, hr^{-1}; $\mu = $ repair rate, hr^{-1}; and $\alpha = $ installation rate, hr^{-1}.

   Calculate the numerical values in Parts (*i*) through (*iii*) for $n = 10$, $\lambda = 0.001$ hr^{-1}, $\alpha = 1 \ \mathrm{hr}^{-1}$, and $\mu = 0.1 \ \mathrm{hr}^{-1}$.

7.32 Reexamine the three-component system shown in Figure 7.10(c) for the case where two repair facilities are available. Find the steady-state probabilities using the frequency balance approach.

7.33 **(a)** A pumping station has four identical repairable pumps with unlimited repair facilities. Develop a state transition diagram, and indicate all the transition rates and what each state represents. For each pump, use λ and μ for the failure and repair rates, respectively.

    (b) Modify the diagram if the available repair facilities can accommodate only one failed pump and if repairs are done on a first-come, first-serve basis.

    (c) Repeat Part (b) if two failed pumps can be repaired simultaneously.

    (d) Repeat Part (a) if no repair is performed with all the pumps down. In each of Parts (a) through (d), write the Markov differential equations in vector-matrix notation.

7.34 Derive a binary model for a repairable component that also undergoes preventive maintenance (see Figure 7.11). Also, find the mean residence times in the up and down states of the binary model.

7.35 Derive a binary model for the system shown in Figure 7.12 in which the installation time after repair is included. Study the model as $\gamma \to \infty$, meaning that it takes zero time to install the component after repair.

7.36 For the case of a repairable component with an identical spare (see Figure 7.13), derive a binary model with states S_{up} and S_{down}. Find the associated transition rates. Show that the mean durations of up and down times are as given in Equations (7–119) and (7–120).

7.37 Derive a binary (two-state) model for a system consisting of four identical components, all of which are needed for system success. The probability of more than two components failing simultaneously can be neglected. The failure rate for any one component is λ_1, and the failure rate for a pair (there are two pairs) of components is λ_2. When a component or a pair of components fails, it is simply replaced. The installation rate for any one component is μ.

7.38 A system consists of two identical components in parallel, and the chance of both failing simultaneously is so small that it can be neglected. If one of the two components fails, the system is shut down completely until the faulty component is detected and isolated, after which the good one is returned to service. For each component, the failure, repair, and detection rates are λ, μ, and γ, respectively.

    (*i*) Draw a state-space diagram, identify the states, and mark all the important values clearly.

    (*ii*) Find the mean residence durations in each of the states.

    (*iii*) By merging the proper states, derive a binary model for this system.

    (*iv*) Find the system availability, unavailability, and frequency of failure.

7.39 Repeat Problem 7.38 if the repair facilities can handle only one failed component on a first-come, first-serve basis.

7.40 Draw the state-space diagram for a two-component system with a single repair crew. The crew can handle only one component at a time, and repairs are carried out on a first-come, first-serve basis. The two components are identical, are independent, and have failure and repair rates of λ and μ, respectively.

    (*i*) Set up the Markov differential equations for the system.

    (*ii*) Reduce the model to a ternary (three-state) model, with the states representing 0, 1, and 2 failures. Find all the necessary transition rates by suitable reasoning.

7.41 Derive expressions for the time-dependent probabilities of the various states of the system shown in Figure 7.10(c) if there are no repairs and the system is in state S_1 at time $t = 0$.

8

Approximate Methods

8.1 INTRODUCTION

The Markov approach to the evaluation of system reliability becomes involved and time consuming as the number of states required to model a system's behavior becomes large. However, it is possible to arrive at acceptable approximate equations for the series, parallel, and r-out-of-n configurations. These approximations can then be used in conjunction with the techniques developed to handle the combinatorial aspects of systems to evaluate the reliability of complex systems.

8.2 COMPONENTS IN SERIES

Consider the case of two components in series, and let λ_1, μ_1, and λ_2, μ_2 be the failure and repair rates of components 1 and 2, respectively. We would like to replace these components by an equivalent component with a failure rate of λ_s and a repair rate of μ_s.

We have seen earlier that, for two constant-hazard components in series, the system reliability is

$$R(t) = e^{-(\lambda_1 + \lambda_2)t} \tag{8-1}$$

321

The reliability of the equivalent component can be expressed as

$$R(t) = e^{-\lambda_s t} \tag{8-2}$$

Comparing the two equations, we conclude that

$$\lambda_s = \lambda_1 + \lambda_2 \tag{8-3}$$

From the example of two dissimilar repairable units, the steady-state probability of both components being good was found to be [see Equation (7–34)]

$$P_1 = \frac{\mu_1 \mu_2}{(\lambda_1 + \mu_1)(\lambda_2 + \mu_2)} \tag{8-4}$$

For the equivalent component, the steady-state probability of being in the good (or *up*) state is $\mu_s/(\lambda_s + \mu_s)$. Therefore, we require that

$$\frac{\mu_s}{\lambda_s + \mu_s} = \frac{\mu_1 \mu_2}{(\lambda_1 + \mu_1)(\lambda_2 + \mu_2)} \tag{8-5}$$

or

$$\frac{\lambda_s}{\mu_s} = \frac{(\lambda_1 + \mu_1)(\lambda_2 + \mu_2)}{\mu_1 \mu_2} - 1 \tag{8-6}$$

or

$$\mu_s = \frac{\lambda_s \mu_1 \mu_2}{\lambda_1 \lambda_2 + \lambda_1 \mu_2 + \lambda_2 \mu_1} \tag{8-7}$$

Expressing Equation (8–7) in terms of mean repair times r_1, r_2, and r_s, where

$$r_1 = \frac{1}{\mu_1}, \ r_2 = \frac{1}{\mu_2}, \text{ and } r_s = \frac{1}{\mu_s}$$

we get

$$r_s = \frac{\lambda_1 r_1 + \lambda_2 r_2 + \lambda_1 \lambda_2 r_1 r_2}{\lambda_1 + \lambda_2} \tag{8-8}$$

For component 1, the number of failures per unit time is λ_1, and every time the component is down, it takes, on average, r_1 time units to repair it. Therefore, $\lambda_1 r_1$ is the fraction of the time component 1 is down. For well-designed components, this number should be small. Similarly, $\lambda_2 r_2$ is also a small number, and we can make the approximation

$$\lambda_1 r_1 + \lambda_2 r_2 + \lambda_1 \lambda_2 r_1 r_2 \simeq \lambda_1 r_1 + \lambda_2 r_2 \tag{8-9}$$

so that

$$r_s \simeq \frac{\lambda_1 r_1 + \lambda_2 r_2}{\lambda_1 + \lambda_2} = \frac{\lambda_1 r_1 + \lambda_2 r_2}{\lambda_s} \tag{8-10}$$

Extending this line of thinking to n components in series, we get

$$\lambda_s = \sum_{i=1}^{n} \lambda_i \tag{8-11}$$

$$r_s \simeq \frac{1}{\lambda_s} \sum_{i=1}^{n} \lambda_i r_i \tag{8-12}$$

The system unavailability is

$$U_s \cong \lambda_s r_s = \sum_{i=1}^{n} \lambda_i r_i \tag{8-13}$$

and the availability is

$$A_s = 1 - U_s \tag{8-14}$$

■■ **Example 8-1**

Consider three components with the following parameters operating in series:

Component	1	2	3
Failure rate, year^{-1}	0.1	0.12	0.08
Repair time, hr	2	5	7

For the equivalent component,

$$\text{failure rate} = \lambda_s = 0.1 + 0.12 + 0.08 = 0.3 \text{ per year}$$

$$\text{repair time} \cong \left(\frac{1}{0.3}\right) [(0.1 \times 2) + (0.12 \times 5) + (0.08 \times 7)]$$

$$= 4.533 \text{ hr}$$

$$\text{Unavailability} \cong \frac{0.3 \times 4.533}{8,760} = 1.5525 \times 10^{-4}$$

$$\text{Availability} \cong 1 - 1.5525 \times 10^{-4} = 0.9998447 \qquad ■■$$

■■ **Example 8-2**

For two identical components with failure rate of one per year and repair time of 0.01 year, the equivalent repair time computed using the exact expression in Equation (8-8) is

$$r_s = \frac{2(1 \times 0.01) + (1 \times 0.01)^2}{2} = 0.01005 \text{ year}$$

and the approximate expression gives a value of 0.01 year. The error is only about 0.5%, which may be acceptable in many practical cases.

For four such components in series, a similar calculation yields an exact value of 0.010151 year, and the approximate expression once again gives a value of 0.01 year. Now the error is about 1.5%, which is still reasonable.

We see that the error introduced in the value of r_s by using the approximate expression increases with the number of components in series.

Now consider the unavailability values. For two components in series, the exact and approximate values of the unavailability U_s are 0.0201 and 0.02, respectively. The difference, again, is about 0.5%. For four components in series, the corresponding values are 0.040604 and 0.04, with the difference again being about 1.5%. So the percent errors in values of U_s are the same as the percent errors in values of r_s. The approximation procedure makes the series system look a little better than what it actually is, because we are neglecting multiple failures; the more components there are in series, the larger is the error in neglecting multiple failures. ■■

8.3 COMPONENTS IN PARALLEL

In the case of two components in parallel with full redundancy, the system will fail only if both components fail. From the example of two dissimilar repairable units, the steady-state probability of finding both components down is [see Equation (7–34)]

$$Q_p = \frac{\lambda_1\lambda_2}{(\lambda_1 + \mu_1)(\lambda_2 + \mu_2)} \tag{8-15}$$

For the equivalent component, $Q_p = \lambda_p/(\lambda_p + \mu_p)$, and therefore,

$$\frac{\lambda_p}{\lambda_p + \mu_p} = \frac{\lambda_1\lambda_2}{(\lambda_1 + \mu_1)(\lambda_2 + \mu_2)} \tag{8-16}$$

or

$$\frac{\mu_p}{\lambda_p} = \frac{(\lambda_1 + \mu_1)(\lambda_2 + \mu_2)}{\lambda_1\lambda_2} - 1 \tag{8-17}$$

or

$$\lambda_p = \frac{\mu_p\lambda_1\lambda_2}{\lambda_1\mu_2 + \lambda_2\mu_1 + \mu_1\mu_2} \tag{8-18}$$

Since repairing either component leads to system success, the equivalent repair rate is equal to the sum of the two individual repair rates. That is,

$$\mu_p = \mu_1 + \mu_2 \tag{8-19}$$

Using this equation in the expression for λ_p, we get

$$\lambda_p = \frac{(\mu_1 + \mu_2)(\lambda_1\lambda_2)}{\lambda_1\mu_2 + \lambda_2\mu_1 + \mu_1\mu_2} \tag{8-20}$$
$$= \frac{(r_1 + r_2)(\lambda_1\lambda_2)}{1 + \lambda_1r_1 + \lambda_2r_2}$$

As with series components, $\lambda_1 r_1 \ll 1$, $\lambda_2 r_2 \ll 1$, and therefore,

$$\lambda_p \simeq \lambda_1 \lambda_2 (r_1 + r_2) \tag{8-21}$$

The repair time for the equivalent component is

$$r_p = \frac{1}{\mu_p} = \frac{1}{\mu_1 + \mu_2} = \frac{r_1 r_2}{r_1 + r_2} \tag{8-22}$$

Also, $U_p = \lambda_p r_p$ and $A_p = (1 - U_p)$.

The equivalent failure rate can also be written as

$$\lambda_p \simeq \lambda_1 (\lambda_2 r_1) + \lambda_2 (\lambda_1 r_2) \tag{8-23}$$

A parallel system will fail if component 2 fails during the repair of component 1 or if component 1 fails during the repair of component 2. Every failure of component 1 will keep that component down for a period r_1, and the number of failures of component 2 during this time is given by $\lambda_2 r_1$. Since component 1 has a failure rate of λ_1 per unit time, the number of system failures due to component 2 failing during the repair of component 1 is $\lambda_1(\lambda_2 r_1)$ per unit time. Similarly, $\lambda_2(\lambda_1 r_2)$ is the number of system failures per unit time due to component 1 failing during the repair of 2. By adding these two contributions, we get an estimate for λ_p, the failure rate for the two-component parallel system.

We can extend the analysis to three components in parallel. Let the failure rates of the three components be λ_1, λ_2, and λ_3 and the repair times be r_1, r_2, and r_3, as shown in Figure 8.1(a). We then have two possible approaches to dealing with the problem. (1) Replace components 2 and 3 by an equivalent component

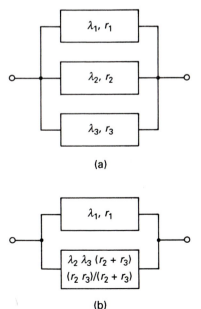

(a)

(b)

Figure 8.1   (a) Three components in parallel. (b) Two-component equivalent of (a). (c) Four components in parallel.

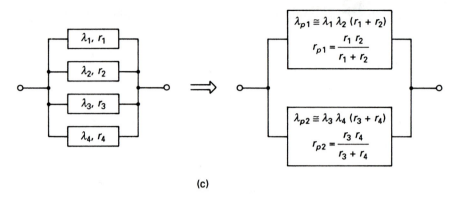

(c)

Figure 8.1 *(Continued)*

[see Figure 8.1 (b)]. Then replace the two components in parallel by an equivalent component. In that case,

$$\lambda_p = (\lambda_1)[\lambda_2\lambda_3(r_2 + r_3)]\left[r_1 + \frac{r_2r_3}{r_2 + r_3}\right]$$

$$= (\lambda_1\lambda_2\lambda_3)[r_1r_2 + r_2r_3 + r_3r_1] \tag{8-24}$$

and

$$r_p = \frac{r_1\left(\dfrac{r_2r_3}{r_2 + r_3}\right)}{r_1 + \dfrac{r_2r_3}{r_2 + r_3}} = \frac{r_1r_2r_3}{r_1r_2 + r_2r_3 + r_3r_1} \tag{8-25}$$

(2) There are six ways the system can fail. These are listed in the following table:

First Failure	Second Failure	During Repair of	Third Failure	During Overlapping Repair of
1	2	1	3	1 and 2
1	3	1	2	1 and 3
2	1	2	3	1 and 2
2	3	2	1	2 and 3
3	1	3	2	3 and 1
3	2	3	1	3 and 2

Writing approximate expressions for the failure rate due to each of the six possibilities and then summing them, we obtain

$$\lambda_p = \lambda_1(\lambda_2 r_1)\lambda_3 \frac{r_1 r_2}{r_1 + r_2}$$

$$+ \lambda_1(\lambda_3 r_1)\lambda_2 \frac{r_1 r_3}{r_1 + r_3}$$

$$+ \lambda_2(\lambda_1 r_2)\lambda_3 \frac{r_1 r_2}{r_1 + r_2}$$

$$+ \lambda_2(\lambda_3 r_2)\lambda_1 \frac{r_2 r_3}{r_2 + r_3} \qquad (8\text{--}26)$$

$$+ \lambda_3(\lambda_1 r_3)\lambda_2 \frac{r_3 r_1}{r_3 + r_1}$$

$$+ \lambda_3(\lambda_2 r_3)\lambda_1 \frac{r_3 r_2}{r_3 + r_2}$$

$$= \lambda_1 \lambda_2 \lambda_3 (r_1 r_2 + r_2 r_3 + r_3 r_1)$$

the same as Equation (8–24). Moreover,

$$\mu_p = \mu_1 + \mu_2 + \mu_3 \qquad (8\text{--}27)$$

or

$$\frac{1}{r_p} = \frac{1}{r_1} + \frac{1}{r_2} + \frac{1}{r_3} \qquad (8\text{--}28)$$

and

$$r_p = \frac{r_1 r_2 r_3}{r_1 r_2 + r_2 r_3 + r_3 r_1} \qquad (8\text{--}29)$$

the same as Equation (8–25). Also,

$$\text{System unavailability} = U_p = \lambda_p r_p = \lambda_1 \lambda_2 \lambda_3 r_1 r_2 r_3 \qquad (8\text{--}30)$$

and

$$\text{System availability} = A_p = 1 - U_p \qquad (8\text{--}31)$$

Extending the analysis to four components in parallel [see Figure 8.1(c)], we get

$$\lambda_{eq} \simeq \lambda_{p1}\lambda_{p2}(r_{p1} + r_{p2})$$

$$= \lambda_1\lambda_2\lambda_3\lambda_4(r_1 + r_2)(r_3 + r_4)\left[\frac{r_1r_2}{r_1 + r_2} + \frac{r_3r_4}{r_3 + r_4}\right]$$

$$= \lambda_1\lambda_2\lambda_3\lambda_4[(r_3 + r_4)(r_1r_2) + (r_1 + r_2)(r_3r_4)] \qquad (8\text{--}32)$$

$$= \lambda_1\lambda_2\lambda_3\lambda_4[r_1r_2r_3 + r_1r_2r_4 + r_3r_4r_1 + r_2r_3r_4]$$

Finally, for the general case of $(n - r + 1)$ identical components in parallel, we have

$$\lambda_{eq} \cong \lambda^{(n-r+1)}(n - r + 1)(\text{MTTR of one component})^{n-r} \qquad (8\text{--}33)$$

■■ **Example 8–3**

For two identical components in parallel, each with a failure and repair rate of λ and μ, respectively, the exact expression for the equivalent failure rate is $2\lambda^2/(2\lambda + \mu)$. The approximate value is $2\lambda^2/\mu$. Obviously, the approximation makes the system look a little worse than it actually is, because not every failure of one of the two components may result in the failure of the other component and the consequent system failure.

    The percent error (based on the exact value) in the equivalent failure rate can be calculated to be $200\lambda/\mu$. For $\lambda = 0.1$ failure per year and $\mu = 1,000$, the percent error is only 0.02. If $\lambda = 1$ per year and $\mu = 500$, then the percent error jumps to 0.4. For easy-to-repair, well-designed components (meaning low failure rates and small values for repair times), this approximation is quite acceptable.   ■■

■■ **Example 8–4**

If the components listed in Example 8–1 are operating in parallel with new failure rates of 1, 2, and 3 per year, then, for the equivalent component,

$$\lambda_p = \frac{1 \times 2 \times 3}{8,760^2}[(2 \times 5) + (5 \times 7) + (7 \times 2)]$$

$$= 4.613 \times 10^{-6} \text{ failure/year}$$

$$r_p = \frac{2 \times 5 \times 7}{(2 \times 5) + (5 \times 7) + (7 \times 2)} = 1.1864 \text{ hr}$$

$$\text{Unavailability} \cong \lambda_p r_p = 6.2476 \times 10^{-10}$$

$$\text{Availability} \cong 1 - 6.2476 \times 10^{-10} \qquad\qquad ■■$$

■■   **Example 8–5**

For five identical components in parallel, each with a failure rate of one per year and a repair time of 30 hours, the equivalent failure rate is

$$\lambda_{eq} = 1^5 \times 5 \times \left[\frac{30}{8,760} \right]^4$$

$$= 6.8776 \times 10^{-10} \text{ failure per year} \qquad ■■$$

8.4 AN *r*-OUT-OF-*n* STRUCTURE

We saw earlier that the system reliability for an *r*-out-of-*n* structure can be expressed as

$$R = \sum_{k=r}^{n} {}_nC_k p^k (1 - p)^{n-k} \qquad (8\text{–}34)$$

where *p* is the probability of success of a component. If each component has a constant hazard of λ, then

$$R(t) = \sum_{k=r}^{n} {}_nC_k e^{-k\lambda t}(1 - e^{-\lambda t})^{n-k} \qquad (8\text{–}35)$$

For the system to fail, $(n - r + 1)$ components must fail. Therefore, the equivalent repair rate must be the same as the equivalent repair rate of $(n - r + 1)$ identical components in parallel, and we conclude that

$$\mu_{eq} = \mu + \mu + \dots + \mu, (n - r + 1) \text{ times} \qquad (8\text{–}36)$$

$$= (n - r + 1)\mu$$

and

$$r_{eq} = \frac{1}{(n - r + 1)\mu} \qquad (8\text{–}37)$$

In other words,

$$\text{MTTR}_{eq} = \frac{\text{MTTR of one component}}{n - r + 1} \qquad (8\text{–}38)$$

To find the MTTF$_{eq}$, let us first get λ_{eq}. For the system to fail, $(n - r + 1)$ components should fail. There are ${}_nC_{(n-r+1)}$ ways we can select $(n - r + 1)$

components out of a collection of n components. Therefore, using Equation (8–33), we obtain

$$\lambda_{eq} = [_nC_{(n-r+1)}] \begin{Bmatrix} \text{equivalent failure rate for} \\ (n-r+1) \text{ components in parallel} \end{Bmatrix}$$

$$= \left[\frac{n!}{(n-r+1)!(r-1)!} \right] [\lambda^{n-r+1}(n-r+1)(\text{MTTR of one component})^{n-r}]$$

which, after simplification, yields

$$\lambda_{eq} = \frac{n!}{(n-r)!(r-1)!} \lambda^{n-r+1}(\text{MTTR of one component})^{n-r} \qquad (8\text{–}39)$$

Therefore,

$$MTTF_{eq} = \frac{1}{\lambda_{eq}}$$

$$= \begin{pmatrix} \text{MTTF of} \\ \text{one} \\ \text{component} \end{pmatrix}^{n-r+1} \frac{1}{\begin{pmatrix} \text{MTTR of} \\ \text{one} \\ \text{component} \end{pmatrix}^{n-r}} \left[\frac{(n-r)!(r-1)!}{n!} \right] \qquad (8\text{–}40)$$

$$= \begin{pmatrix} \text{MTTF} \\ \text{of one} \\ \text{component} \end{pmatrix} \left(\frac{\text{MTTF}}{\text{MTTR}} \right)^{n-r} \left\{ \frac{(n-r)!(r-1)!}{n!} \right\}$$

■■ **Example 8–6**

For the case of a truly parallel configuration, we can apply Equation (8–40) with $n = 2$ and $r = 1$ thus:

$$MTTF_{eq} = \frac{(MTTF)^2}{(MTTR)} \left[\frac{1!0!}{2!} \right]$$

Replacing MTTF by $1/\lambda$ and MTTR by $1/\mu$, we have

$$MTTF_{eq} = \frac{\mu}{2\lambda^2}$$

The equivalent failure rate is

$$\lambda_{eq} = \frac{1}{MTTF_{eq}} = \frac{2\lambda^2}{\mu}$$

which is the same as Equation (8–23) with $\lambda_1 = \lambda_2 = \lambda$ and $r_1 = r_2 = r = 1/\mu$.

■■

■■   **Example 8–7**
For a three-out-of-five configuration consisting of components with MTTF = 2,000 hours and MTTR = 16 hours, we have

$$MTTF_{eq} = (2,000) \left\{ \frac{2,000}{16} \right\}^{5-3} \left\{ \frac{(5-3)!(3-1)!}{5!} \right\}$$

$$= 1,041,666.667 \text{ hours}$$

and

$$MTTR_{eq} = \frac{16}{5-3+1}$$

$$= 5.333 \text{ hours}$$ ■■

■■   **Example 8–8**
A simplified block diagram of a computer system is shown in Figure 8.2. The numbers inside the blocks are the MTTF and MTTR, in hours. Calculate the six-month and one-year reliabilities of the system and also the availability of the overall system.

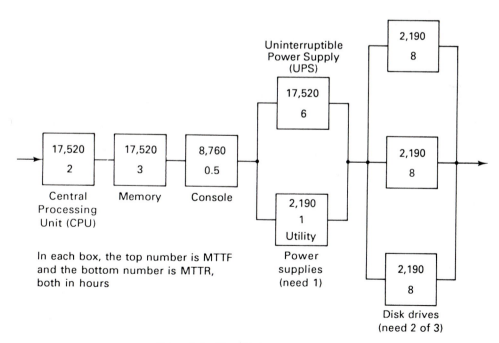

Figure 8.2 Simplified computer system.

Let us group the components into three subsystems:

I.   Central processing unit (CPU), memory, and console, in series
II. Uninterruptible power supply (UPS) and utility power supply, in parallel
III. Three disk drives, in parallel, of which only two are required for system operation

For subsystem I, for the equivalent component,

$$\frac{1}{(MTTF)_e} = \frac{1}{17,520} + \frac{1}{17,520} + \frac{1}{8,760} = \frac{2}{8,760} \text{ /hr}$$

$$(MTTF)_e = 4,380 \text{ hr}$$

$$(MTTR)_e = \left(\frac{2}{17,520} + \frac{3}{17,520} + \frac{0.5}{8,760}\right) \div \left(\frac{2}{8,760}\right)$$

$$= 1.5 \text{ hr}$$

$$R_I(t) = \exp\left[\frac{-t}{4,380}\right]$$

$$R_I(4,380) = e^{-1} = 0.367879$$

$$R_I(8,760) = e^{-2} = 0.135335$$

$$\text{Availability} = A_I = \frac{4,380}{4,381.5} = 0.999658$$

For subsystem II, for the equivalent component,

$$\frac{1}{(MTTF)_e} = \frac{6}{17,520 \times 2,190} + \frac{1}{2,190 \times 17,520}$$

$$= \frac{7}{17,520 \times 2,190} \text{ /hr}$$

$$(MTTF)_e = 5,481,257 \text{ hours}$$

$$\frac{1}{(MTTR)_e} = \frac{1}{6} + \frac{1}{1} = \frac{7}{6} \text{ /hr}$$

$$(MTTR)_e = \frac{6}{7} \text{ hour}$$

$$R_{II}(t) = \exp\left[\frac{-t}{5,481,257}\right]$$

$$R_{II}(4,380) = 0.999201$$

$$R_{II}(8,760) = 0.998407$$

$$\text{Availability} = A_{II} = \frac{5,481,257}{5,481,257.857} = 0.9999998$$

For subsystem III,

$$(\text{MTTF})_e = (2{,}190) \left[\frac{2{,}190}{8}\right]^{3-2} \left[(3-2)!\,\frac{(2-1)!}{3!}\right]$$

$$= 99{,}918.75 \text{ hours}$$

$$(\text{MTTR})_e = \frac{8}{(3-2)+1} = 4 \text{ hours}$$

$$R_{\text{III}}(t) = \exp\left[\frac{-t}{99{,}918.75}\right]$$

$$R_{\text{III}}(4{,}380) = 0.957111$$

$$R_{\text{III}}(8{,}760) = 0.916062$$

$$\text{Availability} = A_{\text{III}} = \frac{99{,}918.75}{99{,}922.75} = 0.9999599$$

The complete system consists of the three subsystems in series. Accordingly, the equivalent failure rate for the system is

$$\lambda_s = \frac{1}{4{,}380} + \frac{1}{5{,}481{,}257} + \frac{1}{99{,}918.75}$$

$$= 2.38501 \times 10^{-4}/\text{hr}$$

The system MTTF is equal to the reciprocal of λ_s. Therefore,

$$(\text{MTTF})_s = 4{,}192.85 \text{ hours}$$

The system MTTR is calculated as follows:

$$\lambda_s(\text{MTTR})_s = \frac{1.5}{4{,}380} + \frac{0.857}{5{,}481{,}257} + \frac{4}{99{,}918.75}$$

or

$$(\text{MTTR})_s = 1.6044 \text{ hours}$$

Since the subsystems are in series, the system reliabilities are obtained by multiplying the subsystem reliabilities together. This results in

$$R_s(4{,}380) = 0.35182$$

and

$$R_s(8{,}760) = 0.12378$$

Also,

$$A_s = A_\text{I}A_\text{II}A_\text{III} = 0.9996177$$

The weakest part of the system is the series combination of CPU, memory, and the console. ∎∎

8.5 INFLUENCE OF WEATHER

Components working outdoors are obviously subjected to the influence of weather conditions, and we can expect their failure rates to be higher during stormy weather as compared to normal weather conditions. In addition, it is possible to have several failures within a short period of time due to severe weather conditions. This is called the *bunching effect*, and it should be included in the reliability evaluation of engineering systems.

The Markov approach can be used to consider the influence of weather on failures of components. We will assume that the weather randomly transits between "stormy" and "normal." Let λ and λ' be the failure rates of a component during normal and stormy weather, respectively. The variation of λ and λ' with time is illustrated in Figure 8.3 in the form of a plot of failure rate versus time.

We can construct a state-space diagram for a single repairable component under two-state weather conditions by making the following assumptions:

1. The failure and repair times are exponentially distributed during both normal and stormy weather conditions. We let

   λ, μ = normal weather failure and repair rates, and

   λ', μ' = stormy weather failure and repair rates

2. The normal and stormy weather durations are also exponentially distributed. We let

   N, S = expected values of normal and stormy weather durations

   $n = 1/N$ = rate of departure from normal weather state to stormy weather state, and

   $m = 1/S$ = rate of departure from stormy weather state to normal weather state

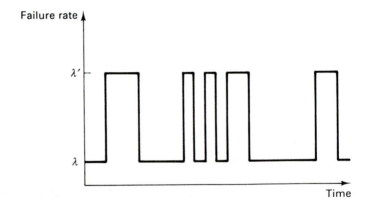

Figure 8.3   Variation of failure rate with time under a two-state weather condition.

Six parameters $(N, S, \lambda, \lambda', \mu, \mu')$ are needed to construct a Markov model for a repairable component that includes the influence of weather. Figure 8.4 shows the Markov model. The corresponding Markov differential equations in standard vector-matrix form are

$$
\begin{bmatrix} P_0'(t) \\ P_1'(t) \\ P_2'(t) \\ P_3'(t) \end{bmatrix} =
\begin{bmatrix}
-(n+\lambda) & m & \mu & 0 \\
n & -(m+\lambda') & 0 & \mu' \\
\lambda & 0 & -(\mu+n) & m \\
0 & \lambda' & n & -(m+\mu')
\end{bmatrix}
\begin{bmatrix} P_0(t) \\ P_1(t) \\ P_2(t) \\ P_3(t) \end{bmatrix}
$$

$$(8\text{-}41)$$

After deleting the Δt's, the stochastic transitional probability matrix is

$$
\mathbf{P} = \begin{array}{c} 0 \\ 1 \\ 2 \\ 3 \end{array}
\begin{array}{cccc}
 0 & 1 & 2 & 3 \\
\end{array}
\begin{bmatrix}
1-(n+\lambda) & n & \lambda & 0 \\
m & 1-(m+\lambda') & 0 & \lambda' \\
\mu & 0 & 1-(\mu+n) & n \\
0 & \mu' & m & 1-(m+\mu')
\end{bmatrix}
$$

$$(8\text{-}42)$$

To find the MTTF, we consider states S_2 and S_3 as absorbing and rearrange **P** as follows:

$$
\mathbf{P} = \begin{array}{c} 3 \\ 2 \\ 1 \\ 0 \end{array}
\begin{array}{cc}
3 \quad\quad 2 & \quad 1 \quad\quad\quad\quad\quad 0 \\
\end{array}
\left[
\begin{array}{cc:cc}
1 & 0 & 0 & 0 \\
0 & 1 & 0 & 0 \\
\hdashline
\lambda' & 0 & 1-(m+\lambda') & m \\
0 & \lambda & n & 1-(n+\lambda)
\end{array}
\right]
$$

$$(8\text{-}43)$$

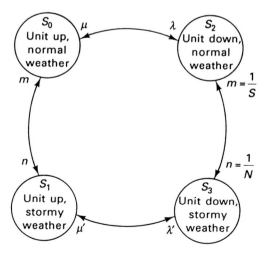

Figure 8.4   State-space diagram for a repairable component under two-state weather conditions.

The corresponding fundamental matrix is

$$\mathbf{N} = [\mathbf{I} - \mathbf{Q}]^{-1} \tag{8-44}$$

in which

$$\mathbf{Q} = \begin{bmatrix} 1 - (m + \lambda') & m \\ n & 1 - (n + \lambda) \end{bmatrix} \tag{8-45}$$

is the truncated matrix.

Taking the inverse, we get

$$\mathbf{N} = \frac{1}{(m\lambda + n\lambda' + \lambda\lambda')} \begin{matrix} 1 & 0 \\ \begin{bmatrix} (n + \lambda) & m \\ n & (m + \lambda') \end{bmatrix} & \begin{matrix} 1 \\ 0 \end{matrix} \end{matrix} \tag{8-46}$$

Assuming that at time $t = 0$, the unit is up and the weather is normal,

$$\text{MTTF} = M_{0,2-3} = \frac{n + m + \lambda'}{m\lambda + n\lambda' + \lambda\lambda'} \tag{8-47}$$

If at time $t = 0$, the unit is up and the weather is stormy, then

$$\text{MTTF} = M_{1,2-3} = \frac{n + \lambda + m}{m\lambda + n\lambda' + \lambda\lambda'} \tag{8-48}$$

Since $\lambda' > \lambda$, it is obvious that $M_{0,2-3} > M_{1,2-3}$. The average failure rate (forced outage rate) λ_f is equal to the reciprocal of $M_{0,2-3}$. Thus,

$$\lambda_f = (M_{0,2-3})^{-1} = \frac{m\lambda + n\lambda' + \lambda\lambda'}{n + m + \lambda'} \tag{8-49}$$

It is reasonable to assume that

$$\lambda\lambda' \ll m\lambda + n\lambda' \tag{8-50}$$

and

$$\lambda' \ll m + n \tag{8-51}$$

With these assumptions, a good approximation to the average failure rate is

$$\lambda_f \simeq \frac{m\lambda + n\lambda'}{m + n}$$

$$= \left(\frac{m}{m + n}\right)\lambda + \left(\frac{n}{m + n}\right)\lambda'$$

Or, in terms of $N = 1/n$ and $S = 1/m$, we have

$$\lambda_f = \left(\frac{N}{N + S}\right)\lambda + \left(\frac{S}{N + S}\right)\lambda' \tag{8-52}$$

For the special case of one-state (normal) weather, $N \rightarrow \infty$, $S \rightarrow 0$, $n \rightarrow 0$, $m \rightarrow \infty$, and $\lambda_f \rightarrow \lambda$.

■■ **Example 8–9**

A component fails, on average, once in five years, and 80% of the failures occur during stormy weather. If the mean durations of normal and stormy weather are 250 hours and 2 hours, respectively, we have

$$\frac{250}{252}\lambda + \frac{2}{252}\lambda' = 0.2$$

and

$$\left(\frac{2}{252}\right)\lambda' = 4\left(\frac{250}{252}\right)\lambda$$

Solving for λ and λ', we get

$$\lambda = 0.04032 \text{ failure per year of normal weather}$$

$$\lambda' = 20.16 \text{ failures per year of stormy weather} \qquad ■■$$

■■ **Example 8–10**

Let the normal and stormy weather failure rates of a component be 0.01 year^{-1} and 5 year^{-1}, respectively. If the normal and stormy weather durations are 200 hours and 2 hours, respectively, then the average failure rate is

$$\lambda_f = \left(\frac{200}{202}\right)(0.01) + \left(\frac{2}{202}\right)(5)$$

$$= 0.05446 \text{ failure per year}$$

Also,

$$n = \frac{1}{N} = 43.8 \text{ year}^{-1}$$

and

$$m = \frac{1}{S} = 4{,}380 \text{ year}^{-1}$$

If we do not employ any approximations and use Equation (8–49), then

$$\lambda_f = \frac{(4{,}380)(0.01) + (43.8)(5) + (5 \times 0.01)}{43.8 + 4{,}380 + 5}$$

$$= 0.05935 \text{ failure per year}$$

The difference between the two values of λ_f is 8.24%, which, though not very small, is acceptable for studies involving the influence of weather. ■■

8.5.1 Components in Series

In two-state weather, for k dissimilar independent components in series, neglecting the probability of more than one component failing simultaneously (equivalent to neglecting bunching effects), the overall forced outage rate can be expressed as

$$\lambda_{fe} = \sum_{i=1}^{k} \lambda_{fi} \qquad (8\text{–}53)$$

where

$$\lambda_{fi} = \left(\frac{N}{N+S}\right)\lambda_i + \left(\frac{S}{N+S}\right)\lambda_i' \qquad (8\text{–}54)$$

If, in addition, we can assume that the normal and stormy weather repair rates are nearly the same (a somewhat questionable, but often acceptable, assumption), then $\mu \simeq \mu'$, and the expected value of repair time (downtime) for the equivalent component can be expressed as

$$(\text{MTTR})_e = r_{fe} \simeq \lambda_{fe}^{-1} \sum_{i=1}^{k} \lambda_{fi} r_i \qquad (8\text{–}55)$$

■■ **Example 8–11**

For k identical components in series, the expression for the overall forced outage rate is

$$\lambda_{fe} = k\lambda_f$$

where

$$\lambda_f = \frac{N}{N+S}\lambda + \frac{S}{N+S}\lambda'$$

The expected downtime can be approximated as

$$r_{fe} = r = \text{mean repair time for one component,}$$
$$\text{assumed to be the same under both weather conditions}$$

Let $\lambda = 0.01$ year^{-1}, $\lambda' = 0.1$ year^{-1}, $N = 300$ hours, $S = 2$ hours, $r = 8.76$ hours, and $k = 6$. Then

$$\lambda_f = \left(\frac{300}{302}\right)(0.01) + \left(\frac{2}{302}\right)(0.1)$$

$$= 0.010596 \text{ failure per year}$$

$$\lambda_{fe} = 6\lambda_f$$

$$= 0.063576 \text{ failure per year}$$

$$\text{Unavailability} = \lambda_{fe} r$$

$$= \frac{(0.063576)(8.76)}{8,760}$$

$$= 6.3576 \times 10^{-5}$$

$$\text{Availability} = 1 - 6.3576 \times 10^{-5}$$

$$= 0.9999364$$

The availability for one component is $[1 - (0.010596)(0.001)] = 0.9999894$, and, as expected,

$$(0.9999894)^6 = 0.9999364$$

$$= \text{availability for the series system} \qquad \blacksquare\blacksquare$$

8.5.2 Two Components in Parallel

With two components (say, 1 and 2) in parallel and a two-state weather model, a Markov approach will require eight states to consider all possibilities. This number of states is cumbersome, so we will discuss an approximate method to compute the total forced outage rate for the equivalent component.

There are four ways in which a forced outage can occur:

(*i*) Component 1 or 2 fails during normal weather, followed by a failure of component 2 or 1 during normal weather.

(*ii*) Component 1 or 2 fails during normal weather, followed by a failure of component 2 or 1 during stormy weather.

(*iii*) Component 1 or 2 fails during stormy weather, followed by a failure of component 2 or 1 during the *same* storm.

(*iv*) Component 1 or 2 fails during stormy weather, followed by a failure of component 2 or 1 during normal weather.

The applicable failure rates for each of these cases can be approximated as follows. For case (*i*),

$$\text{Failure rate} = \left(\frac{N}{N+S}\right)\lambda_1(1 - P_{s1})P_{21} + \left(\frac{N}{N+S}\right)\lambda_2(1 - P_{s2})P_{12} \qquad (8\text{--}56)$$

in which

P_{s1} = probability of occurrence of a storm during the repair of component 1

P_{s2} = probability of occurrence of a storm during the repair of component 2

P_{21} = probability of component 2 failing during the repair of component 1

P_{12} = probability of component 1 failing during the repair of component 2

Since

$$P_{s1} = (1 - e^{-r_1/N}) \simeq \frac{r_1}{N} \tag{8-57}$$

$$P_{s2} = (1 - e^{-r_2/N}) \simeq \frac{r_2}{N} \tag{8-58}$$

$$P_{21} = 1 - e^{-\lambda_2 r_1} \simeq \lambda_2 r_1 \tag{8-59}$$

and

$$P_{12} = 1 - e^{-\lambda_1 r_2} \simeq \lambda_1 r_2 \tag{8-60}$$

using these equations in Equation (8–56), we obtain, for the two failures grouped under case (i),

$$\text{Failure rate} = \left(\frac{N}{N + S}\right)\left[\lambda_1\left(1 - \frac{r_1}{N}\right)(\lambda_2 r_1) + \lambda_2\left(1 - \frac{r_2}{N}\right)(\lambda_1 r_2)\right]$$

Since $r_1 \ll N$ and $r_2 \ll N$, we can neglect r_1/N and r_2/N. We then have the approximation

$$\text{Failure rate} \simeq \left(\frac{N}{N + S}\right)\lambda_1\lambda_2 (r_1 + r_2) \tag{8-61}$$

For case (ii),

$$\text{Failure rate} = \left(\frac{N}{N + S}\right)\lambda_1 P_{s1} P_{2fs} + \left(\frac{N}{N + S}\right)\lambda_2 P_{s2} P_{1fs} \tag{8-62}$$

where

P_{2fs} = probability of component 2 failing during a storm $\cong \lambda_2' S$

and

P_{1fs} = probability of component 1 failing during a storm $\cong \lambda_1' S$

Using these values in Equation (8–62), we obtain, for the two failures grouped under case (ii),

$$\begin{aligned}\text{Failure rate} &= \left(\frac{N}{N + S}\right)\left[\lambda_1 \frac{r_1}{N}\lambda_2' S + \lambda_2 \frac{r_2}{N} \lambda_1' S\right] \\ &= \left(\frac{N}{N + S}\right)\left(\frac{S}{N}\right)[\lambda_1\lambda_2' r_1 + \lambda_2\lambda_1' r_2]\end{aligned} \tag{8-63}$$

For case (iii),

$$\text{Failure rate} = \left(\frac{1}{N + S}\right)[P_{1fs}P_{2fs} + P_{2fs}P_{1fs}] \tag{8-64}$$

in which $1/(N + S)$ is the number of storms per unit time. For the two failures grouped under case (*iii*), then, we have

$$
\begin{aligned}
\text{Failure rate} &= \left(\frac{1}{N + S}\right)[2\lambda_1' S\lambda_2' S] \\
&= \left(\frac{N}{N + S}\right)\left(\frac{2S^2}{N}\right)\lambda_1'\lambda_2'
\end{aligned}
\tag{8-65}
$$

For case (*iv*),

$$
\begin{aligned}
\text{Failure rate} &= \left(\frac{1}{N + S}\right)[P_{1fs}(1 - P_{2fs})P_{21} + P_{2fs}(1 - P_{1fs})P_{12}] \\
&= \left(\frac{1}{N + S}\right)[\lambda_1'S(1 - \lambda_2'S)\lambda_2 r_1 + \lambda_2'S(1 - \lambda_1'S)\lambda_1 r_2]
\end{aligned}
\tag{8-66}
$$

Since $\lambda_1'S \ll 1$ and $\lambda_2'S \ll 1$, for the two failures grouped under case (*iv*), we have

$$
\text{Failure rate} \simeq \left(\frac{N}{N + S}\right)\left(\frac{S}{N}\right)[\lambda_1'\lambda_2 r_1 + \lambda_2'\lambda_1 r_2]
\tag{8-67}
$$

The nature of the weather during the second failure determines whether the system failure is a normal-weather or a stormy-weather failure. Thus, cases (*i*) and (*iv*) are normal-weather failures and cases (*ii*) and (*iii*) are stormy-weather failures. For the equivalent component,

$$
\text{Normal-weather failure rate} = \lambda_e = \lambda_1\lambda_2(r_1 + r_2) + \frac{S}{N}(\lambda_1'\lambda_2 r_1 + \lambda_2'\lambda_1 r_2)
\tag{8-68}
$$

$$
\text{Stormy-weather failure rate} = \lambda_e' = (\lambda_1\lambda_2' r_1 + \lambda_2\lambda_1' r_2) + 2S\lambda_1'\lambda_2'
\tag{8-69}
$$

and

$$
\begin{aligned}
\frac{\text{Total forced}}{\text{outage rate}} = \lambda_{fe} &= \left(\frac{N}{N + S}\right)\lambda e + \left(\frac{S}{N + S}\right)\lambda_e' \\
&= \sum \text{failure rates for cases } (i), (ii), (iii), \text{ and } (iv)
\end{aligned}
\tag{8-70}
$$

If the repair rate is the same under normal and stormy weather conditions, then the overall repair rate is the sum of the individual repair rates. Therefore,

$$
\frac{1}{r_{fe}} = \frac{1}{r_1} + \frac{1}{r_2}
\tag{8-71}
$$

and

$$
(\text{MTTR})_e = \frac{r_1 r_2}{r_1 + r_2}
\tag{8-72}
$$

8.6 SCHEDULED MAINTENANCE OUTAGES

Just as in the case of forced outages, when a component is removed from service for scheduled maintenance, the state of the system changes. We can always include additional states in the state-space diagram to consider scheduled maintenances and proceed as usual. However, such a procedure will be very unwieldy, even for a small number of components. Therefore, some approximate methods are discussed based on the following constraints:

1. A component is not taken out for maintenance if (*i*) such an action will overload or cause failure of the system, or (*ii*) maintenance cannot be completed before a storm strikes, or (*iii*) the weather is stormy or going to be stormy.
2. The time between maintenances and the maintenance times are exponentially distributed, with

$$\lambda_i'' = \text{maintenance outage rate for the } i\text{th component}$$

and

$$r_i'' = \text{average maintenance time (same as expected}$$
$$\text{value of downtime for maintenance)}$$

8.6.1 Components in Series

For components in series, it is impossible to remove a component for maintenance without the system going down. Also, we need not maintain one component at a time; we can schedule several components for simultaneous maintenance. Assuming that only one component undergoes maintenance at any one time, pessimistic values of reliability parameters may be calculated. For the equivalent element, we have

$$\text{Overall outage rate due to maintenance} = \lambda_e'' = \sum_{i=1}^{k} \lambda_i'' \qquad (8\text{--}73)$$

and

$$\left. \begin{array}{l} \text{Expected value of downtime} \\ \text{due to maintenance outages} \end{array} \right\} = r_e'' = \frac{\displaystyle\sum_{i=1}^{k} \lambda_i'' r_i''}{\lambda_e''}$$

$$= \frac{\displaystyle\sum_{i=1}^{k} \lambda_i'' r_i''}{\displaystyle\sum_{i=1}^{k} \lambda_i''} \qquad (8\text{--}74)$$

8.6.2 Two Components in Parallel

For the case of two components in parallel, the system fails if one of the components is out for maintenance and the other component fails during the maintenance of the first component. Following the approximate method discussed earlier for two components in parallel, we can say that

$$\lambda_e'' \simeq \underset{\substack{\text{forced outage of component 2} \\ \text{during maintenance} \\ \text{of component 1}}}{\lambda_1''(\lambda_2 r_1'')} \quad + \quad \underset{\substack{\text{forced outage of component 1} \\ \text{during maintenance} \\ \text{of component 2}}}{\lambda_2''(\lambda_1 r_2'')} \qquad (8\text{–}75)$$

Also,

$$\left. \begin{array}{l} \text{Expected value of system downtime due to forced outage} \\ \text{of component 2 during maintenance of component 1} \end{array} \right\} = \frac{r_1'' r_2}{r_1'' + r_2} \quad (8\text{–}76)$$

and

$$\left. \begin{array}{l} \text{Expected value of system downtime due to forced outage} \\ \text{of component 1 during maintenance of component 2} \end{array} \right\} = \frac{r_2'' r_1}{r_2'' + r_1} \quad (8\text{–}77)$$

Therefore, the expected value of the downtime for the equivalent component r_e'' is obtained as the weighted sum as

$$r_e'' = \frac{\lambda_1''(\lambda_2 r_1'')}{\lambda_1''(\lambda_2 r_1'') + \lambda_2''(\lambda_1 r_2'')} \left(\frac{r_1'' r_2}{r_1'' + r_2} \right) + \frac{\lambda_2''(\lambda_1 r_2'')}{\lambda_1''(\lambda_2 r_1'') + \lambda_2''(\lambda_1 r_2'')} \left(\frac{r_2'' r_1}{r_2'' + r_1} \right) \quad (8\text{–}78)$$

8.7 OVERLOAD OUTAGES

The question of overload outages arises only for components in parallel that are not fully redundant. The overload of a series system points to a serious design deficiency. For two components in parallel, we will first of all assume that component maintenance outages are scheduled in such a way that they do not, by themselves, result in overload outages.

Let P_1 (P_2) be the probability that component 1 (component 2) alone will not be able to perform the task (or carry the load) on hand. Assuming a two-state weather, for the equivalent component, we get

$$\left. \begin{array}{l} \text{Normal-weather overload} \\ \text{outage rate} \end{array} \right\} = \lambda_{oe} = \lambda_1 P_2 + \lambda_2 P_1 \quad (8\text{–}79)$$
$$\text{failures/year of normal weather}$$

$$\left. \begin{array}{l} \text{Stormy-weather overload} \\ \text{outage rate} \end{array} \right\} = \lambda_{oe}' = \lambda_1' P_2 + \lambda_2' P_1 \quad (8\text{–}80)$$
$$\text{failures/year of stormy weather}$$

and

$$\left. \begin{array}{l} \text{Overall overload} \\ \text{outage rate} \end{array} \right\} = \lambda_{ofe} = \left(\frac{N}{N + S} \right) \lambda_{oe} + \left(\frac{S}{N + S} \right) \lambda_{oe}'/\text{year} \quad (8\text{–}81)$$

sum of n exponentially distributed durations. This model can be simplified by assuming that $\rho_1 = \rho_2 = \ldots = \rho_{n-1} = \rho_n \equiv \rho$. With this assumption, all the n exponentially distributed durations will be identical.

With $n = 1$, we have the conventional binary model with one up state and one down state, both exponentially distributed. Then

$$\text{Mean downtime} = T_D = \frac{1}{\rho} \qquad (8\text{--}89)$$

$$\left.\begin{array}{c} \text{Density function of the random} \\ \text{variable downtime} \end{array}\right\} = f_{t_D} = \rho e^{-\rho t} \equiv f_1(t) \qquad (8\text{--}90)$$

With $n = 2$, the down state consists of two stages in series, and the density function of the random variable called downtime becomes

$$f_{t_D} = \int_0^t f_1(\tau) f_1(t - \tau)\, d\tau = \rho^2 t e^{-\rho t} \equiv f_2(t) \qquad (8\text{--}91)$$

Proceeding in a similar fashion, with n stages in series constituting the down state, we obtain the density function for the random variable downtime as

$$f_{t_D} = \rho^n \frac{t^{n-1}}{(n-1)!} e^{-\rho t} \qquad (8\text{--}92)$$

This equation is the same as the special Erlangian distribution (see Equation (4–75) with α replaced by $1/\rho$. The mean downtime, with density as given in Equation (8–92), is

$$T_D = \int_0^\infty t \rho^n \frac{t^{n-1}}{(n-1)!} e^{-\rho t}\, dt$$

$$= \frac{\rho^n}{(n-1)!} \int_0^\infty t^n e^{-\rho t}\, dt$$

or

$$T_D = \frac{\rho^n}{(n-1)!} \frac{\Gamma(n+1)}{\rho^{n+1}} \qquad (8\text{--}93)$$

With an integer value for n, Equation (8–93) becomes

$$T_D = \frac{n}{\rho} \qquad (8\text{--}94)$$

The second moment of the random variable downtime with a special Erlangian distribution can easily be shown to be $[n(n+1)]/\rho^2$. By judiciously choosing the parameters ρ and n, we can select a variety of shapes for the density function representing the downtime.

As an example, suppose we want to model a Weibull distribution by the

While the uptimes of most components are exponentially distributed (or close to it), the downtimes typically have Weibull or some other nonexponential distributions. Prudent design practices suggest that components that fail frequently require the least time to repair, and those that fail infrequently require the most time to repair. Thus, the shortest repair times are the most frequent and the longest repair times are the least frequent.

One approach to modeling nonexponential distributions is to employ the method of stages. In this approach, the nonexponential repair state is divided into a number of substates called stages, these stages are traversed in a given sequence, and all transfers between them are assumed to be made at constant rates. The sequence of stages can be in series or in parallel or a combination thereof. Depending on the sequence and the rates of departure between them, the total time spent in the stages can assume a variety of distributions. Three commonly occurring special cases are of interest:

(*i*) a Weibull distribution with $\beta \geq 1$; for this distribution, we use stages in series.

(*ii*) a Weibull distribution with $\beta < 1$; for this distribution, we use stages in parallel.

(*iii*) a lognormal distribution; for this distribution, we use stages in series, terminated with two stages in parallel.

8.9.1 Stages in Series

Let us consider a simple two-state (up and down) model for a component with nonexponential downtime. The down state is replaced by n stages in series, as shown in Figure 8.6. (The variable ρ represents the transition rates between different stages.) The total time spent in the down state will then be equal to the

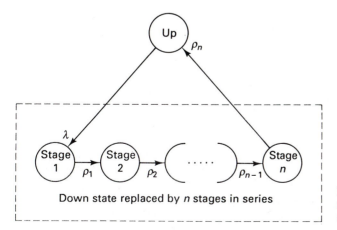

Down state replaced by *n* stages in series

Figure 8.6 Representation of a nonexponential downtime by stages in series.

The approximate expression given in Equation (8–81) slightly overestimates the number of outages because some component outages overlap, resulting in system failure.

The same technique can be extended to several components in parallel if the needed probability values can be deduced from system considerations. For example, for three components in parallel, we have

$$\lambda_{oe} = (\lambda_1 P_{2,3} + \lambda_2 P_{1,3} + \lambda_3 P_{1,2}) + (\lambda_{1,2} P_3 + \lambda_{2,3} P_1 + \lambda_{3,1} P_2) \qquad (8\text{–}82)$$

$$\lambda'_{oe} = (\lambda'_1 P_{2,3} + \lambda'_2 P_{1,3} + \lambda'_3 P_{1,2}) + (\lambda'_{1,2} P_3 + \lambda'_{2,3} P_1 + \lambda'_{3,1} P_2) \qquad (8\text{–}83)$$

In these equations,

$\lambda_{i,j}$ = normal-weather failure rate for the component equivalent to component i and component j in parallel

$\lambda'_{i,j}$ = stormy-weather failure rate for the component equivalent to component i and component j in parallel

$P_{i,j}$ = probability that components i and j in parallel will *not* be able to perform the task on hand when the third component is down

Finally, the overall overload outage rate can be expressed as

$$\lambda_{ofe} = \left(\frac{N}{N+S}\right)\lambda_{oe} + \left(\frac{S}{N+S}\right)\lambda'_{oe} \qquad (8\text{–}84)$$

8.8 TOTAL OUTAGE RATE

Now we can combine forced, maintenance, and overload outages to obtain expressions for the total outage rate using a two-state weather model.

For k components in series, we have

$$\lambda_{\text{total}} = \lambda_{fe} + \lambda''_e$$

$$= \sum_{i=1}^{k}\left[\left(\frac{N}{N+S}\right)\lambda_i + \left(\frac{S}{N+S}\right)\lambda'_i\right] + \sum_{i=1}^{k}\lambda''_i \qquad (8\text{–}85)$$

$$\text{Expected downtime} = r_{\text{total}} = \frac{\lambda_{fe} r_{fe} + \lambda''_e r''_e}{\lambda_{\text{total}}} \qquad (8\text{–}86)$$

For two components in parallel,

$$\lambda_{\text{total}} = \lambda_{fe} + \lambda''_e + \lambda_{ofe}$$

$$= \left[\left(\frac{N}{N+S}\right)\lambda_e + \left(\frac{S}{N+S}\right)\lambda'_e\right] + [\lambda''_1 \lambda_2 r''_1 + \lambda''_2 \lambda_1 r''_2] \qquad (8\text{–}87)$$

$$+ \left[\left(\frac{N}{N+S}\right)(\lambda_1 P_2 + \lambda_2 P_1) + \left(\frac{S}{N+S}\right)(\lambda'_1 P_2 + \lambda'_2 P_1)\right]$$

8.6.2 Two Components in Parallel

For the case of two components in parallel, the system fails if one of the components is out for maintenance and the other component fails during the maintenance of the first component. Following the approximate method discussed earlier for two components in parallel, we can say that

$$\lambda_e'' \simeq \underbrace{\lambda_1''(\lambda_2 r_1'')}_{\substack{\text{forced outage of component 2} \\ \text{during maintenance} \\ \text{of component 1}}} + \underbrace{\lambda_2''(\lambda_1 r_2'')}_{\substack{\text{forced outage of component 1} \\ \text{during maintenance} \\ \text{of component 2}}} \qquad (8\text{--}75)$$

Also,

$$\left.\begin{array}{r}\text{Expected value of system downtime due to forced outage} \\ \text{of component 2 during maintenance of component 1}\end{array}\right\} = \frac{r_1'' r_2}{r_1'' + r_2} \qquad (8\text{--}76)$$

and

$$\left.\begin{array}{r}\text{Expected value of system downtime due to forced outage} \\ \text{of component 1 during maintenance of component 2}\end{array}\right\} = \frac{r_2'' r_1}{r_2'' + r_1} \qquad (8\text{--}77)$$

Therefore, the expected value of the downtime for the equivalent component r_e'' is obtained as the weighted sum as

$$r_e'' = \frac{\lambda_1''(\lambda_2 r_1'')}{\lambda_1''(\lambda_2 r_1'') + \lambda_2''(\lambda_1 r_2'')}\left(\frac{r_1'' r_2}{r_1'' + r_2}\right) + \frac{\lambda_2''(\lambda_1 r_2'')}{\lambda_1''(\lambda_2 r_1'') + \lambda_2''(\lambda_1 r_2'')}\left(\frac{r_2'' r_1}{r_2'' + r_1}\right) \qquad (8\text{--}78)$$

8.7 OVERLOAD OUTAGES

The question of overload outages arises only for components in parallel that are not fully redundant. The overload of a series system points to a serious design deficiency. For two components in parallel, we will first of all assume that component maintenance outages are scheduled in such a way that they do not, by themselves, result in overload outages.

Let P_1 (P_2) be the probability that component 1 (component 2) alone will not be able to perform the task (or carry the load) on hand. Assuming a two-state weather, for the equivalent component, we get

$$\left.\begin{array}{r}\text{Normal-weather overload} \\ \text{outage rate}\end{array}\right\} = \lambda_{oe} = \lambda_1 P_2 + \lambda_2 P_1 \qquad (8\text{--}79)$$
$$\text{failures/year of normal weather}$$

$$\left.\begin{array}{r}\text{Stormy-weather overload} \\ \text{outage rate}\end{array}\right\} = \lambda_{oe}' = \lambda_1' P_2 + \lambda_2' P_1 \qquad (8\text{--}80)$$
$$\text{failures/year of stormy weather}$$

and

$$\left.\begin{array}{r}\text{Overall overload} \\ \text{outage rate}\end{array}\right\} = \lambda_{ofe} = \left(\frac{N}{N + S}\right)\lambda_{oe} + \left(\frac{S}{N + S}\right)\lambda_{oe}'/\text{year} \qquad (8\text{--}81)$$

$$r_{\text{total}} = \frac{1}{\lambda_{\text{total}}} \left\{ \lambda_{fe} r_{fe} + \lambda_e'' r_e'' \right.$$

$$+ r_1 \left[\left(\frac{N}{N+S}\right) \lambda_1 P_2 + \left(\frac{S}{N+S}\right) \lambda_1' P_2 \right]$$

$$\left. + r_2 \left[\left(\frac{N}{N+S}\right) \lambda_2 P_1 + \left(\frac{S}{N+S}\right) \lambda_2' P_1 \right] \right\} \tag{8–88}$$

■■ **Example 8–12**

An ideal electrical source is supplying three loads, as shown in Figure 8.5. For simplicity, we will assume that all the five components (the boxes represent sub-transmission systems) are identical with the following parameters:

Normal-weather failure rate = 0.3 failure/year of normal weather

Stormy-weather failure rate = 12.0 failures/year of stormy weather

Maintenance outages = 1 per year

$\left.\begin{array}{l}\text{Mean repair time under}\\\text{both weather conditions}\end{array}\right\}$ = 0.001 year

Mean maintenance time = 0.002 year

Assuming all the parallel configurations to be fully redundant, we will calculate the total number of outages per year and the mean downtime at load points A, B, and C using the approximate techniques discussed in this chapter. The mean durations of normal and stormy weather are 300 hours and 1.2 hours, respectively.

We will proceed in steps, as illustrated in Figure 8.5. The geometry of the configuration allows us to supply each load in more than one way, and this should be taken into account in our calculations.

Step 1

Replace components 1 and 2 in parallel by an equivalent component, designated as component 6. For this component, we have:

Normal-weather failure rate = λ_6

$$= (0.3)^2(0.002) + \left(\frac{1.2}{300}\right)(2 \times 12 \times 0.3 \times 0.001)$$

$$= 2.088 \times 10^{-4} \text{ failure/year of normal weather}$$

Stormy-weather failure rate = λ_6'

$$= 2(0.3 \times 12 \times 0.001) + 2\left(\frac{1.2}{8,760}\right) 12^2$$

$$= 0.046652 \text{ failure/year of stormy weather}$$

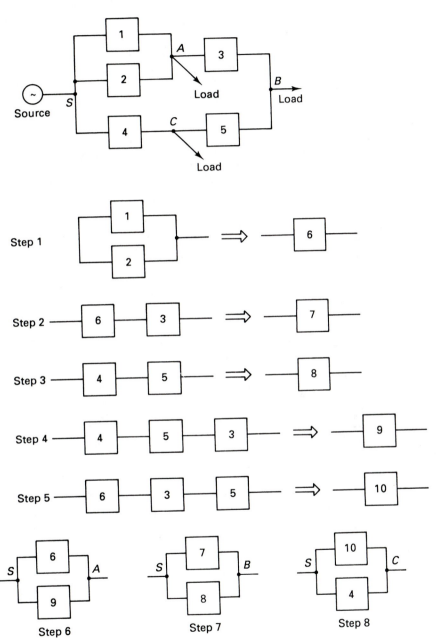

Figure 8.5 Simple transmission system.

Total forced outage rate $= \lambda_{f6}$

$$= \left(\frac{300}{301.2}\right)(2.088 \times 10^{-4}) + \left(\frac{1.2}{301.2}\right)(0.046652)$$

$$= 3.93833 \times 10^{-4} \text{ failure/year}$$

Maintenance outage rate $= \lambda_6''$

$$= 2(1)(0.3 \times 0.002)$$

$$= 12 \times 10^{-4} \text{ outage/year}$$

Overload outage rate $= \lambda_{0f6}$

$$= 0 \text{ because of full redundancy}$$

Total outage rate $= \lambda_{\text{total } 6}$

$$= 15.93833 \times 10^{-4} \text{ outage/year}$$

$\left.\begin{array}{r}\text{Expected value of repair}\\ \text{time after forced outages}\end{array}\right\} = r_{f6}$

$$= \frac{(0.001)(0.001)}{2(0.001)}$$

$$= 0.0005 \text{ year}$$

$\left.\begin{array}{r}\text{Expected value of downtime}\\ \text{after maintenance outages}\end{array}\right\} = r_6''$

$$= 2\left[\frac{1(0.3 \times 0.002)}{1(0.3 \times 0.002) + 1(0.3 \times 0.002)}\right.$$

$$\left. \times \frac{(0.002)(0.001)}{(0.002 + 0.001)}\right]$$

$$= 2\left\{\frac{1}{2}(6.667 \times 10^{-4})\right\} = 6.667 \times 10^{-4} \text{ year}$$

$\left.\begin{array}{r}\text{Mean value of total outage}\\ \text{duration}\end{array}\right\} = r_{\text{total } 6}$

$$= \left(\frac{10^4}{15.93833}\right)[3.93833 \times 10^{-4} \times 0.0005$$

$$+ 12 \times 10^{-4} \times 6.667 \times 10^{-4}]$$

$$= 6.2551 \times 10^{-4} \text{ year}$$

Step 2

Replace components 6 and 3 in series by an equivalent component, designated as component 7. For this component, we have:

$$\text{Normal-weather failure rate} = \lambda_7$$
$$= 0.3 + 2.088 \times 10^{-4}$$
$$= 0.3002088 \text{ failure/year of normal weather}$$

$$\text{Stormy-weather failure rate} = \lambda_7'$$
$$= 12 + 0.046652$$
$$= 12.046552 \text{ failures/year of stormy weather}$$

$$\left.\begin{array}{r}\text{Total forced outage rate}\\ \text{for component 7}\end{array}\right\} = \lambda_{f7}$$
$$= \left(\frac{300}{301.2}\right)(0.3002088) + \left(\frac{1.2}{301.2}\right)(12.046652)$$
$$= 0.347 \text{ failure/year}$$

$$\left.\begin{array}{r}\text{Total forced outage rate for}\\ \text{component 3}\end{array}\right\} = \left(\frac{300}{301.2}\right)0.3 + \left(\frac{1.2}{301.2}\right)(12)$$
$$= 0.3466 \text{ failure/year}$$

$$\text{Maintenance outage rate} = \lambda_7''$$
$$= 1 + 12 \times 10^{-4}$$
$$= 1.0012 \text{ outages/year}$$

$$\text{Total outage rate} = \lambda_{\text{total } 7}$$
$$= 1.3482 \text{ outages/year}$$

$$\left.\begin{array}{r}\text{Expected value of repair}\\ \text{time after forced outages}\end{array}\right\} = \left(\frac{1}{0.347}\right)[(3.93833 \times 10^{-4})(0.0005)$$
$$+ (0.3466)(0.001)]$$
$$= 9.994 \times 10^{-4} \text{ year}$$

$$\left.\begin{array}{r}\text{Expected value of downtime}\\ \text{after maintenance outages}\end{array}\right\} = \left(\frac{1}{1.0012}\right)[12 \times 10^{-4} \times 6.667$$
$$\times 10^{-4} + 1 \times 0.002]$$
$$= 3.862 \times 10^{-3} \text{ year}$$

$$\left.\begin{array}{r}\text{Total expected downtime}\\ \text{for component 7}\end{array}\right\} = r_{\text{total } 7}$$
$$= \frac{(0.347 \times 9.994 \times 10^{-4}) + (1.0012 \times 3.862 \times 10^{-3})}{1.3482}$$
$$= 3.1252 \times 10^{-3} \text{ year}$$

Step 3

Replace components 4 and 5 in series by an equivalent component, designated as component 8. For this component, we have:

Normal-weather failure rate $= \lambda_8$

$$= 0.6 \text{ failure/year of normal weather}$$

Stormy-weather failure rate $= \lambda_8'$

$$= 24 \text{ failures/year of stormy weather}$$

Total forced outage rate $= \lambda_{f8}$

$$= \left(\frac{300}{301.2}\right)(0.6) + \left(\frac{1.2}{301.2}\right)(24)$$

$$= 0.69323 \text{ failure/year}$$

Maintenance outage rate $= \lambda_8''$

$$= 2 \text{ outages/year}$$

Total outage rate $= \lambda_{\text{total } 8}$

$$= 2.69323 \text{ outages/year}$$

Mean outage duration $= r_{\text{total } 8}$

$$= \left(\frac{1}{2.69323}\right)[0.69323 \times 0.001 + 2 \times 0.002]$$

$$= 1.7426 \times 10^{-3} \text{ year}$$

Step 4

Replace components 4, 5, and 3 in series by an equivalent component, designated as component 9. For this component, we have:

Normal-weather failure rate $= \lambda_9$

$$= 0.9 \text{ failure/year of normal weather}$$

Stormy-weather failure rate $= \lambda_9'$

$$= 36 \text{ failures/year of stormy weather}$$

Total forced outage rate $= \lambda_{f9}$

$$= \left(\frac{300}{301.2}\right)(0.9) + \left(\frac{1.2}{301.2}\right)(36)$$

$$= 1.03984 \text{ failures/year}$$

Maintenance outage rate $= \lambda_9''$

$$= 3 \text{ outages/year}$$

Total outage rate $= \lambda_{\text{total } 9}$

$\qquad = 4.03984$ outages/year

Mean outage duration $= r_{\text{total } 9}$

$$= \left(\frac{1}{4.03984}\right)[1.03984 \times 0.001 + 3 \times 0.002]$$

$$= 1.7426 \times 10^{-3} \text{ year}$$

Step 5

Replace components 6, 3, and 5 in series by an equivalent component, designated as component 10. For this component, we have:

Normal-weather failure rate $= \lambda_{10}$

$\qquad = 0.6 + 2.088 \times 10^{-4}$

$\qquad = 0.60021$ failure/year of normal weather

Stormy-weather failure rate $= \lambda_{10}'$

$\qquad = 24.046652$ failures/year of stormy weather

Total forced outage rate $= \lambda_{f10}$

$$= \left(\frac{300}{301.2}\right)(0.60021) + \left(\frac{1.2}{301.2}\right)(24.046652)$$

$\qquad = 0.69362$ failure/year

Maintenance outage rate $= \lambda_{10}''$

$\qquad = 2.0012$ outages/year

Total outage rate $= \lambda_{\text{total } 10}$

$\qquad = 2.69482$ outages/year

Mean outage duration $= r_{\text{total } 10}$

$$= \left(\frac{1}{2.69482}\right)\{3.93833 \times 10^{-4} \times 0.0005 + 0.3466$$

$$\times 0.001 + 0.3466 \times 0.001 + 12 \times 10^{-4}$$

$$\times 6.667 \times 10^{-4} + 2 \times 0.002\}$$

$$= 1.74193 \times 10^{-3} \text{ year}$$

Step 6

Now we consider load point A. We have, in effect, components 6 and 9 in parallel between the source and the load. Therefore, the total number of outages per year

and the mean downtime at point A can be calculated using the approximate formulas derived for two components in parallel, designated with the subscript eA. We have:

Normal-weather failure rate $= \lambda_{eA}$

$$= (2.088 \times 10^{-4})(0.9)(0.0005 + 0.001)$$

$$+ \left(\frac{1.2}{300}\right)[(0.046652 \times 0.9 \times 0.0005) + (36$$

$$\times\ 2.088 \times 10^{-4} \times 0.001)]$$

$$= 3.9592 \times 10^{-7} \text{ failure/year of normal weather}$$

Stormy-weather failure rate $= \lambda'_{eA}$

$$= 2.088 \times 10^{-4} \times 36 \times 0.0005 + 0.9 \times 0.046652$$

$$\times\ 0.001 + 2\left(\frac{1.2}{8,760}\right)(0.046652)(36)$$

$$= 5.05875 \times 10^{-4} \text{ failure/year of stormy weather}$$

Total forced outage rate $= \lambda_{feA}$

$$= \left(\frac{300}{301.2}\right)(3.9592 \times 10^{-7})$$

$$+ \left(\frac{1.2}{301.2}\right)(5.05875 \times 10^{-4})$$

$$= 2.40978 \times 10^{-6} \text{ failure/year}$$

Maintenance outage rate $= \lambda''_{eA}$

$$= (12 \times 10^{-4})(0.9 \times 6.667 \times 10^{-4}) + (3)(2.088 \times$$

$$10^{-4} \times 0.002)$$

$$= 1.9728 \times 10^{-6} \text{ outage/year}$$

Overload outage rate $= \lambda_{0eA} = 0$

Total outage rate $= \lambda_{\text{total } eA}$

$$= (2.40978 + 1.9728)10^{-6} \text{ outage/year}$$

$$= 4.38258 \times 10^{-6} \text{ outage/year}$$

$$\left.\begin{array}{l}\text{Expected value of repair}\\ \text{time after forced outage}\end{array}\right\} = r_{feA}$$

$$= \frac{0.0005 \times 0.001}{0.0005 + 0.001}$$

$$= 3.333 \times 10^{-4} \text{ year}$$

$$\left.\begin{array}{l} \text{Expected value of downtime} \\ \text{after maintenance outage} \end{array}\right\} = r''_{eA}$$

$$r''_{eA} = \frac{12 \times 10^{-4}(0.9 \times 6.667 \times 10^{-4})}{\left\{\begin{array}{l} 12 \times 10^{-4}(0.9 \times 6.667 \times 10^{-4}) \\ + 3(2.088 \times 10^{-4} \times 0.002) \end{array}\right\}} \times \frac{6.667 \times 10^{-4} \times 0.001}{6.667 \times 10^{-4} + 0.001}$$

$$+ \frac{3(2.088 \times 10^{-4} \times 0.002)}{\left\{\begin{array}{l} 12 \times 10^{-4}(0.9 \times 6.667 \times 10^{-4}) \\ + 3(2.088 \times 10^{-4} \times 0.002) \end{array}\right\}} \times \frac{0.002 \times 0.0005}{0.002 + 0.0005}$$

$$= (0.364975)(4.00012 \times 10^{-4}) + (0.635025)(4 \times 10^{-4})$$

$$= 4.000044 \times 10^{-4} \text{ year}$$

$$\left.\begin{array}{l} \text{Expected value of total} \\ \text{outage duration} \end{array}\right\} = r_{\text{total } eA}$$

$$= \left(\frac{10^6}{4.38258}\right)[2.40978 \times 10^{-6} \times 3.333 \times 10^{-4}$$
$$+ 1.9728 \times 10^{-6} \times 4.000044 \times 10^{-4}]$$

$$= 3.63327 \times 10^{-4} \text{ year}$$

Step 7

Next, we consider load point B. We have components 7 and 8 in parallel between the source and the load. The total number of outages per year and the mean downtime at point B can be calculated as in step 6, using the approximate formulas derived for two components in parallel. These calculations are left as an exercise for the reader (see Problem 8.21).

Step 8

Finally, for a consideration of load point C, the procedure is similar to Step 6, with components 10 and 4 in parallel. This, too, is left as an exercise for the reader (see Problem 8.22).   ■■

8.9 MARKOV REPRESENTATION OF NONEXPONENTIAL DISTRIBUTIONS

State-space diagrams and their associated Markov equations are valid only if the residence times in various states have exponential distributions. If the distribution is not exponential, then the process becomes non-Markovian, and the expressions obtained for time-dependent probabilities are not valid. However, the steady-state or limiting probabilities obtained continue to be valid and applicable as long as the components of the system are statistically independent.

sum of n exponentially distributed durations. This model can be simplified by assuming that $\rho_1 = \rho_2 = \ldots = \rho_{n-1} = \rho_n \equiv \rho$. With this assumption, all the n exponentially distributed durations will be identical.

With $n = 1$, we have the conventional binary model with one up state and one down state, both exponentially distributed. Then

$$\text{Mean downtime} = T_D = \frac{1}{\rho} \tag{8-89}$$

$$\left.\begin{array}{c}\text{Density function of the random}\\ \text{variable downtime}\end{array}\right\} = f_{t_D} = \rho e^{-\rho t} \equiv f_1(t) \tag{8-90}$$

With $n = 2$, the down state consists of two stages in series, and the density function of the random variable called downtime becomes

$$f_{t_D} = \int_0^t f_1(\tau) f_1(t - \tau) \, d\tau = \rho^2 t e^{-\rho t} \equiv f_2(t) \tag{8-91}$$

Proceeding in a similar fashion, with n stages in series constituting the down state, we obtain the density function for the random variable downtime as

$$f_{t_D} = \rho^n \frac{t^{n-1}}{(n-1)!} e^{-\rho t} \tag{8-92}$$

This equation is the same as the special Erlangian distribution (see Equation (4–75) with α replaced by $1/\rho$. The mean downtime, with density as given in Equation (8–92), is

$$T_D = \int_0^\infty t \rho^n \frac{t^{n-1}}{(n-1)!} e^{-\rho t} \, dt$$

$$= \frac{\rho^n}{(n-1)!} \int_0^\infty t^n e^{-\rho t} \, dt$$

or

$$T_D = \frac{\rho^n}{(n-1)!} \frac{\Gamma(n+1)}{\rho^{n+1}} \tag{8-93}$$

With an integer value for n, Equation (8–93) becomes

$$T_D = \frac{n}{\rho} \tag{8-94}$$

The second moment of the random variable downtime with a special Erlangian distribution can easily be shown to be $[n(n+1)]/\rho^2$. By judiciously choosing the parameters ρ and n, we can select a variety of shapes for the density function representing the downtime.

As an example, suppose we want to model a Weibull distribution by the

method of stages in series. If M_1 and M_2 are the first and second moments of the Weibull distribution, we then equate

$$M_1 = \frac{n}{\rho} \qquad (8\text{–}95)$$

$$M_2 = \frac{n(n+1)}{\rho^2} \qquad (8\text{–}96)$$

and solve for n and ρ. The result is

$$n = \frac{M_1^2}{M_2 - M_1^2} \qquad (8\text{–}97)$$

$$\rho = \frac{M_1}{M_2 - M_1^2} \qquad (8\text{–}98)$$

The value calculated for n is rounded to the nearest integer.

Calculation of steady-state probabilities.   With $n = 1$, if λ and μ are the failure and repair rates, the steady-state probabilities and state frequencies are as follows:

$$P_{\text{up}} = \frac{\mu}{\lambda + \mu} \qquad (8\text{–}99)$$

$$P_{\text{down}} = \frac{\lambda}{\lambda + \mu} \qquad (8\text{–}100)$$

$$f_{\text{up}} = f_{\text{down}} = \frac{\lambda\mu}{\lambda + \mu} \qquad (8\text{–}101)$$

Note that with $n = 1$, Equations (8–97) and (8–98) require that $\rho = 1/M_1 \equiv \mu$.

Next, let us consider the same system, but with $n = 2$. Going back to Equations (8–97) and (8–98), we require that

$$2 = \frac{M_1^2}{M_2 - M_1^2}$$

or

$$M_2 = \frac{3}{2} M_1^2$$

and

$$\rho = \frac{M_1}{\frac{3}{2}M_1^2 - M_1^2} = \frac{2}{M_1} = 2\mu \qquad (8\text{–}102)$$

since $M_1 = 1/\mu$. The mean duration in each of the two down states is $1/(2\mu)$, and therefore, the mean duration in the two stages combined is once again $1/\mu$. The steady-state probabilities of the three states are easily found to be

$$P_{up} = \frac{\mu}{\lambda + \mu}$$

and

$$P_1 = P_2 = \frac{\lambda}{2(\lambda + \mu)}$$

Therefore,

$$P_{down} = P_1 + P_2 = \frac{\lambda}{\lambda + \mu}$$

and

$$f_{up} = f_{down} = \frac{\lambda\mu}{\lambda + \mu}$$

The conclusion that can be drawn from this discussion is that, as long as we maintain the same value of M_1, whether we use one or two or more stages to

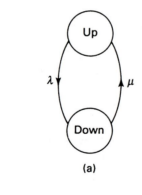

(a)

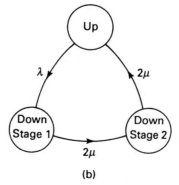

(b)

Figure 8.7 State-space diagrams for stages in series. (a) $n = 1$. (b) $n = 2$.

represent the down state, the limiting state probabilities (steady-state probabilities) and the state frequencies remain the same. Figure 8.7 shows the state-space diagrams for the cases of $n = 1$ and $n = 2$.

8.9.2 Stages in Parallel

Just as with stages in series, for stages in parallel, the down state is represented in terms of n stages, but after every stay in a stage, the component returns to the up state. Figure 8.8 illustrates the associated state-space diagram. We can think of the different stages as representing different types of failure, and every stage of the down state is approachable from the up state in accordance with a predetermined set of relative frequencies. The time spent in each of the n stages has an exponential distribution.

For the stage i, the time spent is t_i, and the density function of the random variable t_i is

$$f_{t_i} = \rho_i e^{-\rho_i t_i}, \quad i = 1, 2, \ldots, n \tag{8-103}$$

The density function of the downtime t_D, obtained as the weighted sum of the n density functions of the n stages, is

$$f_{t_D} = \sum_{i=1}^{n} \omega_i \rho_i e^{-\rho_i t_i} \tag{8-104}$$

Also, the mean downtime is

$$T_D = \int_0^{\infty} t f_{t_D} \, dt$$
$$= \sum_{i=1}^{n} \frac{\omega_i}{\rho_i} \tag{8-105}$$

By selecting suitable values for ω_i and ρ_i, we can adjust the shape of the density function of the downtime. For a model with two stages, there are three parameters

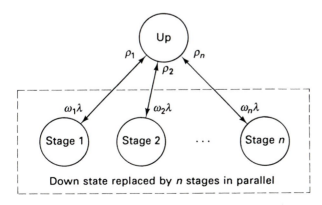

Down state replaced by n stages in parallel

Figure 8.8 Representation of a nonexponential downtime by stages in parallel.

to be found, namely, ω_1, ρ_1, and ρ_2, since $\omega_2 = 1 - \omega_1$. By equating the first three moments M_1, M_2, and M_3 of the downtime to the first three moments computed using the density function f_{tD}, we set up the necessary equations to find the parameter values. For $n = 2$,

$$f_{tD} = \omega_1 \rho_1 e^{-\rho_1 t_1} + \omega_2 \rho_2 e^{-\rho_2 t_2} \tag{8-106}$$

and

$$\omega_2 = 1 - \omega_1 \tag{8-107}$$

The rth moment is

$$m_r = \omega_1 \rho_1 \int_0^\infty t^r e^{-\rho_1 t}\, dt + \omega_2 \rho_2 \int_0^\infty t^r e^{-\rho_2 t}\, dt$$

$$= \omega_1 \rho_1 \frac{\Gamma(r + 1)}{\rho_1^{r+1}} + \omega_2 \rho_2 \frac{\Gamma(r + 1)}{\rho_2^{r+1}}$$

Since r is an integer, $\Gamma(r + 1) = r!$, and it follows that

$$m_r = \frac{\omega_1 r!}{\rho_1^r} + \frac{\omega_2 r!}{\rho_2^r} \tag{8-108}$$

Calculation of steady-state probabilities.   Figure 8.9 shows the state-space diagrams for stages in parallel with $n = 1$ and $n = 2$. For $n = 1$, the steady-state probabilities are

$$P_{\text{up}} = \frac{\mu}{\lambda + \mu} \tag{8-109}$$

and

$$P_{\text{down}} = \frac{\lambda}{\lambda + \mu} \tag{8-110}$$

Then

$$f_{\text{up}} = f_{\text{down}} = \frac{\lambda \mu}{\lambda + \mu} \tag{8-111}$$

as before.

For $n = 2$, assuming that $\rho_1 = \rho_2 \equiv \rho$, we can easily show that the steady-state probabilities and frequencies are the same as for $n = 1$. If only the steady-state (limiting) probabilities are required, we need not be concerned with the particular distribution of the downtime and can calculate these probabilities using the simple binary model.

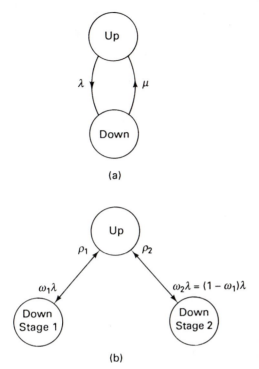

Figure 8.9 State-space diagrams for stages in parallel. (a) $n = 1$. (b) $n = 2$.

8.9.3 Stages in Series Terminated with Two Stages in Parallel

For stages in series terminated with two stages in parallel, the arrangement of the states and the associated parameters are illustrated in Figure 8.10. Five independent parameters—ρ, ρ_1, ρ_2, ω_1, and n—are required to complete this model. The necessary equations are derived by equating the first five moments of the nonexponential distribution with the first five moments corresponding to the density function

$$f_{tD} = \omega_1 \rho_1 \left[\frac{\rho}{\rho - \rho_1} \right]^n \left[e^{-\rho_1 t} - e^{-\rho t} \sum_{i=1}^{n} \frac{[(\rho - \rho_1)t]^{i-1}}{(i-1)!} \right]$$

$$+ \omega_2 \rho_2 \left[\frac{\rho}{\rho - \rho_2} \right]^n \left[e^{-\rho_2 t} - e^{-\rho t} \sum_{i=1}^{n} \frac{[(\rho - \rho_2)t]^{i-1}}{(i-1)!} \right]$$

(8–112)

An easier approach to finding the parameters is first to calculate the rth moments of the two paths (i) stages 1 through n and stage I and (ii) stages 1 through n and stage II and then take a weighted sum of these two moments using ω_1 and ω_2 as the weighting factors.

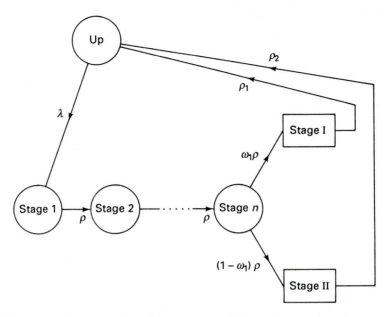

Figure 8.10 Representation of a nonexponential downtime by stages in series terminated with two stages in parallel.

■■   **Example 8–13**

The downtimes of an electromechanical component have a mean value of 8 hours and a standard deviation of 4 hours. Assuming that they have a special Erlangian distribution, we will develop a Markov model by employing the method of stages in series. The average failure rate is known to be one per five years. We therefore have

$$M_1 = \mu = 8 \text{ hr}$$

$$M_2 = \mu^2 + \sigma^2 = 80 \text{ hr}^2$$

$$\rho = \frac{8}{(80 - 64)} = 0.5 \text{ repair/hr, or } 4{,}380 \text{ repairs/year}$$

$$n = \frac{64}{(80 - 64)} = 4$$

The model is shown in Figure 8.11.

The steady-state probabilities and frequencies of the up and down states are calculated as follows: They are:

$$\text{Steady-state probability of the up state} = \frac{4{,}380}{4{,}380.2} = 0.999954$$

$$\text{Steady-state probability of the down state} = \frac{0.2}{4{,}380.2} = 4.566 \times 10^{-5}$$

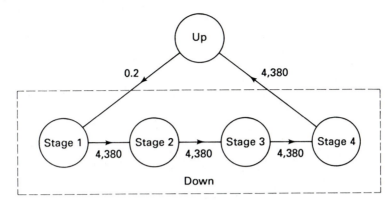

Figure 8.11   Markov model for Example 8–13.

Frequency of the up (or down) state $= 0.2 \times 0.999954$

$$= 0.199991 \text{ year}^{-1}$$ ■■

8.10 COMMON-MODE FAILURES

In general, the introduction of redundancy in design decreases system failure probabilities as long as the failures of the components are independent. However, there are many practical situations in which certain external events affect several components simultaneously and result in their malfunction or failure. Such events can nullify most of the benefit gained by the introduction of redundancy. Some examples of such common external causes resulting in multiple failures are as follows:

(*i*) A tornado sweeping through an outdoor substation, destroying all the equipment, such as transformers, circuit breakers, and switchgears.

(*ii*) A single fire in the control room of a power plant, causing the failure of a number of components.

(*iii*) Several identical components having the same built-in design error.

(*iv*) Errors made by a maintenance crew, causing the malfunction of a number of devices during otherwise normal operating conditions.

(*v*) Failure in the main power supply, resulting in the shutdown of a number of units.

(*vi*) Breakage of a dam, causing the failure of several generation units in a hydroelectric power plant.

(*vii*) The sudden introduction of undesirable environmental factors (dust, humidity, etc.), causing the failure of a number of devices.

A *common-mode* (or *common-cause*) failure is an event originated by a single external cause with several failure effects that are not consequences of each other. The possibility of common-mode failures can increase the probability of overall system failures by several orders of magnitude. All multiple events should occur within a specified time interval for them to be classified as common-mode failures. The length of this time interval depends strongly on the system under consideration and can range from a few seconds to a few hours.

Some of the common-mode failures can be identified and their probability of occurrence minimized by incorporating into the design features such as functional diversity, diversity of equipment, physical diversity, or even administrative diversity. Other such failures, due primarily to natural causes or to space or other limitations, have to be accepted. In that case, no effort must be spared to minimize their impact.

Common-mode failures can be included in Markov models by suitable modifications and/or the addition of states in the state-space diagram. Figures 8.12 and 8.13 illustrate two possibilities for the simple case of two components in parallel. In each, λ_{12} and μ_{12} denote the common-mode failure and repair rates, respectively. If the failed components can sometimes be repaired and returned to service simultaneously, then in Figure 8.12, μ_{12} is not zero. However, if the components cannot be returned to service simultaneously, then μ_{12} is set equal to zero. In the model shown in Figure 8.13, it is assumed that components failing under common-mode failures are repaired and returned to service simultaneously, but components failing individually are repaired and returned to service one at a time as soon as the repair is completed.

Once the appropriate assumptions are made, the corresponding state-space diagram is selected, and the required transition rates are found or estimated, then the rest of the analysis proceeds as before in a systematic manner. Probabilities and frequencies of failures are evaluated using standard procedures discussed in Chapters 6 and 7.

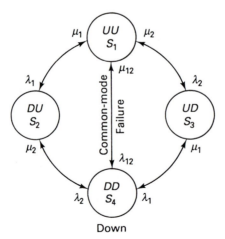

Figure 8.12 Inclusion of common-mode failures for the case of two components in parallel.

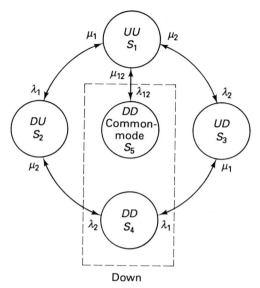

Figure 8.13 Alternative approach to including common-mode failures for the case of two components in parallel.

■■ **Example 8–14**

For the state-space diagram shown in Figure 8.13, the Markov differential equations are

$$
\begin{bmatrix} P_1'(t) \\ P_2'(t) \\ P_3'(t) \\ P_4'(t) \\ P_5'(t) \end{bmatrix} =
\begin{bmatrix}
-(\lambda_1 + \lambda_2 + \lambda_{12}) & \mu_1 & \mu_2 & 0 & \mu_{12} \\
\lambda_1 & -(\mu_1 + \lambda_2) & 0 & \mu_2 & 0 \\
\lambda_2 & 0 & -(\mu_2 + \lambda_1) & \mu_1 & 0 \\
0 & \lambda_2 & \lambda_1 & -(\mu_1 + \mu_2) & 0 \\
\lambda_{12} & 0 & 0 & 0 & -\mu_{12}
\end{bmatrix}
\begin{bmatrix} P_1(t) \\ P_2(t) \\ P_3(t) \\ P_4(t) \\ P_5(t) \end{bmatrix}
$$

Finding algebraic expressions for even the steady-state probabilities becomes rather involved. Therefore, to obtain approximate expressions for the system failure rate and unavailability, we recognize that we have simply two nonidentical components in parallel, with an additional failure mode. Based on Equation (8–21), when we include common-mode failures, we can see that the equivalent failure rate λ for the system can be approximated as

$$ \lambda \cong \lambda_1 \lambda_2 (r_1 + r_2) + \lambda_{12} $$

For two components in parallel, the unavailability is

$$ U_p \cong \lambda_p r_p = \lambda_1 \lambda_2 r_1 r_2 $$

To this expression, we add a term $\lambda_{12} r_{12}$ to include common-mode failures. Therefore, the system unavailability U can be expressed as

$$ U \cong \lambda_1 \lambda_2 r_1 r_2 + \lambda_{12} r_{12} $$

Finally, the system downtime is

$$r = \frac{U}{\lambda} = \frac{\lambda_1\lambda_2 r_1 r_2 + \lambda_{12} r_{12}}{\lambda_1\lambda_2(r_1 + r_2) + \lambda_{12}}$$ ■■

■■ **Example 8–15**

The state-space diagram for a system consisting of two identical components that are subjected to common-mode failures is shown in Figure 8.14. The corresponding Markov differential equations are

$$\begin{bmatrix} P_1'(t) \\ P_2'(t) \\ P_3'(t) \\ P_4'(t) \end{bmatrix} = \begin{bmatrix} -(2\lambda + \lambda_c) & \mu & 0 & \mu_c \\ 2\lambda & -(\lambda + \mu) & 2\mu & 0 \\ 0 & \lambda & -2\mu & 0 \\ \lambda_c & 0 & 0 & -\mu_c \end{bmatrix} \begin{bmatrix} P_1(t) \\ P_2(t) \\ P_3(t) \\ P_4(t) \end{bmatrix}$$

The steady-state probabilities of the states can be found by using the frequency balance method. The results are

$$P_1 = \frac{\mu^2\mu_c}{\mathcal{D}}$$

$$P_2 = \frac{2\lambda\mu_c}{\mathcal{D}}$$

$$P_3 = \frac{\lambda^2\mu_c}{\mathcal{D}}$$

and

$$P_4 = \frac{\lambda_c\mu^2}{\mathcal{D}}$$

where

$$\mathcal{D} \equiv \lambda^2\mu_c + \mu^2\lambda_c + \mu\mu_c(\mu + 2\lambda)$$

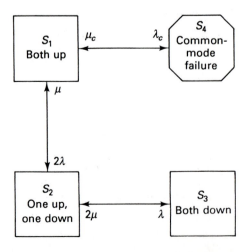

Figure 8.14 State-space diagram for two identical components, including common-mode failures.

If we assume no repair with both components down, then this example becomes the same as Example 7–4 with λ_1 replaced by λ_c. If the repair facilities can handle only one failed component at any one time, then the transition rate from S_3 to S_2 will be only μ instead of 2μ. In this case, we may also have to replace μ_c by μ, unless special repair procedures are employed subsequent to a common-mode failure.   ■■

Not every situation that requires the inclusion of common-mode failures warrants the Markov approach with the development of the associated state-space diagrams. For example, if we have a pumping station with four identical pumps, two of which are required for system success, and if all the pumps are supplied by a single power source, then we have a common-mode failure situation since, if the power supply fails, all the pumps will stop pumping. The system reliability for this case can be expressed as

$$R = R_{ps}\left[\sum_{i=2}^{4} {}_4C_i p^i (1 - p)^{4-i} \right] \tag{8–113}$$

where

$$p = \text{probability of success of one pump}$$

and

$$R_{ps} = \text{reliability of the power supply}$$

If all the pumps are needed for system success, then

$$R = R_{ps}p^4 \tag{8–114}$$

If only one pump is needed for success, then

$$R = R_{ps}[1 - (1 - p)^4] \tag{8–115}$$

If the power supply and the pumps can be modeled using constant-hazards λ_{ps} and λ, respectively, then the time-dependent reliability expression corresponding to Equation (8–113) is

$$R(t) = [\exp(-\lambda_{ps}t)]\left[\sum_{i=2}^{4} {}_4C_i e^{-i\lambda t}(1 - e^{-\lambda t})^{4-i} \right] \tag{8–116}$$

If the source of common-mode failures is within the components themselves, then the total failure rate of a component can be divided into two parts: an individual or regular failure rate λ_R and a common-mode failure rate λ_C. Then

$$\lambda = \lambda_R + \lambda_C \tag{8–117}$$

Such situations arise if the source of the common-mode failure is a change in the environment or in the maintenance procedures or if it is some other cause that affects all the components similarly. For two such identical components in parallel (a fully redundant system), the system reliability can be expressed as

$$R_p(t) = [1 - (1 - e^{-\lambda_R t})^2]e^{-\lambda_C t} \qquad (8\text{--}118)$$

or

$$R_p(t) = [2e^{-\lambda_R t} - e^{-2\lambda_R t}]e^{-\lambda_C t} \qquad (8\text{--}119)$$

In terms of β, the fraction of the total failures that is of common-mode origin, we have

$$\beta \equiv \frac{\lambda_C}{\lambda} = 1 - \frac{\lambda_R}{\lambda} \qquad (8\text{--}120)$$

and we can write Equation (8–119) as

$$R_p(t) = e^{-\lambda t}[2 - e^{-(1-\beta)\lambda t}] \qquad (8\text{--}121)$$

The reliability of a fully redundant system can decrease significantly even for small values of β, as compared with not considering common-mode failures ($\beta = 0$). Also, if we have more than two components in parallel, we have to consider several levels of common-mode failures, namely, all of them failing, two of them failing, three of them failing, and so on. The analysis gets considerably more complicated, and a Markov approach may become easier to use.

If the two identical components are in series (a nonredundant system), then the system reliability is

$$R_s(t) = [e^{-\lambda_R t}]^2 e^{-\lambda_C t} \qquad (8\text{--}122)$$

Alternatively,

$$R_s(t) = e^{-\lambda_{eq} t} \qquad (8\text{--}123)$$

where

$$\lambda_{eq} = 2\lambda_R + \lambda_C = (2 - \beta)\lambda \qquad (8\text{--}124)$$

We note that common-mode failures ($\beta \neq 0$) decrease the equivalent failure rate of a nonredundant system. If β is small, neglecting the common-mode failures is justified, especially since it underestimates the system reliability, thus leading to conservative designs.

■■ **Example 8–16**
A component with a failure rate of 0.0015 per hour will have, for a mission time of 100 hours, a reliability of exp[−0.0015 × 100], or 0.8607. Without the presence of common-mode failures, using two such components in parallel will lead to a system reliability of [1 − (1 − 0.8607)²], or 0.9806. However, if $\beta = 0.2$, the parallel system reliability (using Equation (8–121) will be reduced to

$$e^{-0.15}[2 - e^{-(1-0.2)(0.15)}] = 0.958$$

For each component, since $\beta = 0.2$, we have $\lambda_C = 0.0003$ hr^{-1} and $\lambda_R = (0.0015 - 0.0003) = 0.0012$ hr^{-1}. Using one more component in parallel will make the system reliability equal to

$$[1 - (1 - e^{-0.12})^3]e^{-0.03} = 0.969$$

This is less than the reliability of $[1 - (1 - 0.8607)^3] = 0.9973$ obtained with three components in parallel and no common-mode failures. Therefore, to achieve a system reliability of 0.97, instead of using three components in parallel, we should try somehow to reduce the number of common-mode failures.

Working backwards, we will calculate the value of β required for achieving a system reliability of 0.97 with only two components in parallel. Since we want

$$0.97 = e^{-0.15}[2 - e^{-(1-\beta)(0.15)}]$$

we conclude that $\beta = 0.09469$.

Our conclusion is that if we can reduce the number of common-mode failures such that $\beta \leq 0.09469$ instead of 0.2, then we can achieve a system reliability of not less than 0.97 with just two components in parallel. Note that, without this approach, simply using three components in parallel does not lead us to our goal of achieving a system reliability of 0.97.   ■■

8.11 RARE-EVENT APPROXIMATIONS

In the case of well-designed high-reliability components, the failure rates are expected to be very low, and therefore, the value of λt will be much less than 1 during their design life. Under these circumstances, failure events will be rare, and the rare-event approximations discussed in this section are commonly used.

For $\lambda t \ll 1$,

$$e^{-\lambda t} = 1 - (\lambda t) + \frac{1}{2!}(\lambda t)^2 - \frac{1}{3!}(\lambda t)^3 + \dots \qquad (8\text{--}125a)$$

$$\cong (1 - \lambda t) \qquad (8\text{--}125b)$$

The reliability function and the unreliability function are approximated respectively as

$$R(t) = e^{-\lambda t} \cong (1 - \lambda t) \qquad (8\text{--}126)$$

and

$$Q(t) = 1 - R(t) \cong \lambda t \qquad (8\text{--}127)$$

For n dissimilar components in series, the rare-event approximation can be used if

$$\sum_{i=1}^{n} \lambda_i t \ll 1 \qquad (8\text{--}128)$$

and the expressions for the system reliability and unreliability become

$$R(t) = \exp\left[-\sum_{i=1}^{n} \lambda_i t\right] \tag{8-129a}$$

$$\cong 1 - \sum_{i=1}^{n} \lambda_i t \tag{8-129b}$$

and

$$Q(t) \cong \sum_{i=1}^{n} \lambda_i t \tag{8-130}$$

For the case of n identical components in series, the rare-event approximations are

$$R(t) \cong 1 - n\lambda t \tag{8-131}$$

and

$$Q(t) \cong n\lambda t \tag{8-132}$$

For the case of n identical components in parallel,

$$R(t) = 1 - (1 - e^{-\lambda t})^n \tag{8-133a}$$

$$\cong 1 - (\lambda t)^n \tag{8-133b}$$

and

$$Q(t) \cong (\lambda t)^n \tag{8-134}$$

In the case of an r-out-of-n configuration, the system reliability can be expressed as

$$R(t) = \sum_{i=r}^{n} {}_nC_i[p(t)]^i[1 - p(t)]^{n-i} \tag{8-135}$$

in which $p(t)$ is the reliability function for one component. Letting $k = n - i$, we have $i = n - k$, and Equation (8–135) becomes

$$R(t) = \sum_{k=(n-r)}^{0} {}_nC_{(n-k)}[p(t)]^{(n-k)}[1 - p(t)]^k \tag{8-136}$$

Since

$$ {}_nC_{n-k} = {}_nC_k \tag{8-137}$$

and the order of summation can be reversed without affecting the outcome, we get

$$R(t) = \sum_{k=0}^{(n-r)} {_nC_k}[p(t)]^{(n-k)}[1 - p(t)]^k \tag{8-138}$$

Next, we note that the summation of the expression in Equation (8–138) from $k = 0$ to $k = n$ is unity. Therefore, we can express $R(t)$ as

$$R(t) = 1 - \sum_{k=(n-r+1)}^{n} {_nC_k}[p(t)]^{n-k}[1 - p(t)]^k \tag{8-139}$$

Now we will incorporate the rare-event approximation

$$p(t) \cong (1 - \lambda t) \tag{8-140}$$

into the analysis and obtain

$$Q(t) = 1 - R(t) \cong \sum_{k=(n-r+1)}^{n} {_nC_k}(1 - \lambda t)^{n-k}(\lambda t)^k \tag{8-141}$$

Further, since $(1 - \lambda t) \cong 1$, the rare-event approximation to $Q(t)$ becomes

$$Q(t) \cong \sum_{k=(n-r+1)}^{n} {_nC_k}(\lambda t)^k \tag{8-142}$$

Further simplification is obtained by dropping all the terms except the first one on the right-hand side of Equation (8–142). This leads to

$$Q(t) \cong {_nC_{(n-r+1)}}(\lambda t)^{n-r+1} \tag{8-143}$$

For the case of a truly parallel configuration, $r = 1$, and Equation (8–143) reduces to Equation (8–134).

The approximations discussed in Sections 8.2, 8.3, and 8.4 are for steady-state values, whereas the rare-event approximations considered in this section are applicable during the design life of high-reliability components.

■■ **Example 8–17**

Let T_d be the design life of a component. The error in using Equation (8–126) for computing $R(T_d)$ is

$$\left[\frac{(1 - \lambda T_d) - e^{-\lambda T_d}}{e^{-\lambda T_d}} \right] 100 = \left[\left(\frac{1 - \lambda T_d}{e^{-\lambda T_d}} \right) - 1 \right] 100\%$$

Since $\lambda = 1/(\text{MTTF})$, letting $[T_d/(\text{MTTF})] = \lambda T_d \equiv x$, we obtain the error in $R(T_d)$ as $[e^x(1 - x) - 1]100\%$.

The magnitude of the error is plotted against x in Figure 8.15. For the case of

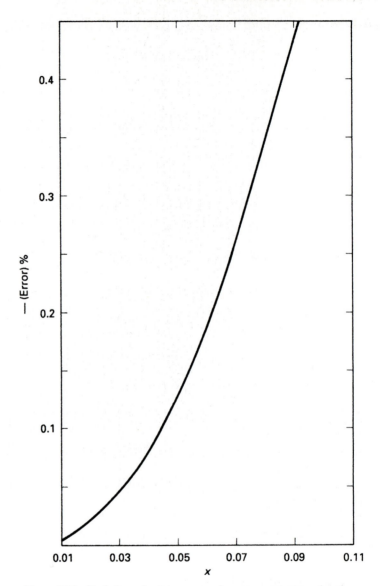

Figure 8.15 Variation of error versus x for rare-event approximation.

MTTF $= 20T_d$, the error is only -0.129%. This procedure consistently underestimates the reliability of the component. If $T_d =$ MTTF, the error becomes -100% ! ■■

■■ **Example 8–18**
Using Equation (8–133b), we can develop an expression for the number of components needed in parallel to achieve a required level of system design-life reliability

R_{sr}. We require that

$$R_{sr} \cong 1 - [Q(T_d)]^n$$

and therefore,

$$n \cong \frac{\log[1 - R_{sr}]}{\log Q(T_d)}$$

For $R_{sr} = 0.9998$ and $Q(T_d) = 0.04$, $n = 2.64$; in other words, we need three components in parallel.   ■■

■■   **Example 8–19**

A high-reliability unit designed for a mission duration of 6 months has a failure rate of 1 in 10 years. Estimate the reliability of a system employing two such units in parallel using the rare-event approximations.

We are given that

$$\lambda = 0.1 \text{ failure/year}$$

and

$$T_d = 0.5 \text{ year}$$

Therefore,

$$\lambda T_d = 0.05$$

Using Equation (8–133b) with $n = 2$, we obtain

$$\text{system reliability} \cong 1 - (0.05)^2$$

$$= 0.9975$$

If three such units are operated in parallel, then

$$\text{system reliability} \cong 1 - (0.05)^3$$

$$= 0.999875$$

If we use a two-out-of-three configuration, then, applying Equation (8–143), we get

$$Q(T_d) \cong {}_3C_{(3-2+1)}(0.05)^{3-2+1}$$

$$= 0.0075$$

and

$$R(T_d) \cong 0.9925$$ ■■

■■   **Example 8–20**

Use the rare-event approximations to evaluate the reliability of the system shown in Figure 8.16, assuming that all the units are identical and independent with a constant hazard of λ.

For the two-out-of-three branch I, the unreliability, from Equation (8–143), is ${}_3C_2(\lambda t)^2$. For the two units in parallel, from Equation (8–133b), the reliability is [1

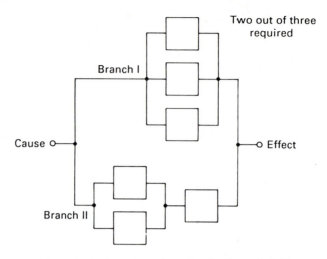

Figure 8.16   System configuration for Example 8–20.

$- (\lambda t)^2$], and therefore, the reliability of branch II is $[1 - (\lambda t)^2](1 - \lambda t)$. For the complete system, we have

$$\text{Unreliability} = [_3C_2(\lambda t)^2][1 - \{1 - (\lambda t)^2\}(1 - \lambda t)]$$

$$= 3[(\lambda t)^3 + (\lambda t)^4 - (\lambda t)^5]$$

For rare-event approximation, we take only the significant term and obtain

$$\left.\begin{array}{r}\text{System} \\ \text{unreliability}\end{array}\right\} \cong 3(\lambda t)^3$$

or

$$\left.\begin{array}{r}\text{System} \\ \text{reliability}\end{array}\right\} \cong 1 - 3(\lambda t)^3$$

If we want the system reliability to be at least 0.999, then we require that

$$3(\lambda t)^3 \le 0.001$$

or

$$(\lambda t) \le 0.06934$$

Thus, for a mission time of 10 units, the individual unit failure rate should be no greater than 6.93×10^{-3} per unit of time. ■■

8.12 SUMMARY

High-reliability components, with their associated low failure rates, allow us to use a collection of simple equations to evaluate the reliability of complex systems involving series, parallel, series-parallel, and r-out-of-n configurations. In the case of components operating outdoors, the influence of weather must be considered, and techniques to do this have been discussed. A subtransmission example worked out in detail has been employed to introduce procedures that include the influence of weather, along with scheduled maintenance and overload outages. Markov techniques to represent non-Markovian downtimes by the methods of stages in series and stages in parallel, as well as other possibilities, were discussed. Common-mode failures can greatly increase the overall failure rate in the case of redundant systems. Methods to analyze and study these failures were introduced with the aid of examples and problems. The chapter concluded with a series of rare-event approximations applicable to systems employing components with very high reliabilities.

PROBLEMS

8.1 For two components in series, what is the percent error introduced in the value of the equivalent repair time r_s by using the approximate method? Calculate the percent error for the following set of parameters:

(*i*) $\lambda_1 = \lambda_2 = 0.1$ failure/year; $r_1 = r_2 = 0.001$ year
(*ii*) $\lambda_1 = \lambda_2 = 1$ failure/year; $r_1 = r_2 = 0.01$ year

8.2 Two identical components, each having a failure rate of one per year, are in series. How long could the mean repair time for each component be if the magnitude of the error calculated using the approximate expression for the equivalent repair time r_s is not to exceed 1%?

8.3 For three dissimilar components in series, derive an expression for the equivalent repair time r_s. Do not employ any approximations.

8.4 If all seven identical components constituting a system must work for the system to succeed, and if the mean repair time for each component is 2 hours, how high can the failure rate for a component be to achieve an availability of 95% or better?

8.5 For two identical components in parallel with repair times of 0.01 year each, how large can the individual failure rates be if the magnitude of the error in the equivalent failure rate calculated using the approximate expression is not to exceed 1%?

8.6 Consider two identical components in parallel with full redundancy. The failure rate and mean repair time for each component are 0.1 failure/year and 0.001 year, respectively. What is the error in the equivalent failure rate λ_p if the approximate expression is used in the calculations? Repeat the problem for 1 failure/year and 0.01 year mean repair time for each component.

8.7 A load is supplied by an old transmission line with a failure rate of 2 per year and an expected repair time of 17.52 hours. In order to increase the reliability of the power supply, a new transmission line with a failure rate of 1 per year and an expected repair time of 8.76 hours is installed in parallel with the old line. Calculate the new values for the number of interruptions per year and the expected repair time.

8.8 An important load is supplied by three lines in parallel, as shown in Figure P 8.8. Each line has a failure rate of 1 per year and an expected repair time of 8.76 hours. Calculate (*i*) the decrease in the number of interruptions per year experienced by

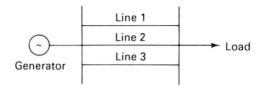

Figure P 8.8

the load, as compared to having only one line, and (*ii*) the decrease in the expected restoration time.

8.9 Two series strings consisting of four identical components in series in each are operating in parallel in a fully redundant configuration. For each component, the failure rate is 0.015 failure/year and the mean repair time is 21.9 hours. For the complete system, find (*i*) the failure rate, (*ii*) the mean downtime, and (*iii*) the unavailability. What is the percent error in the failure rate if the approximate expression is used?

8.10 A system has four identical components with failure and repair rates of 1 and 99 per year, respectively. Three out of the four components are required for system success. Calculate the equivalent failure and repair rates, the MTTF, and the MTTR using suitable approximate formulas.

8.11 For the case of a three-out-of-five system consisting of components with $\lambda = 1 \text{ hr}^{-1}$ and $\mu = 99 \text{ hr}^{-1}$, calculate the equivalent failure and repair rates using the approximate formulas derived in Chapter 8, and compare them with the exact values found in Problem 7.18.

8.12 Five identical components are in parallel. For each component, the failure rate is 1 per year and the expected value of the repair time is 0.001 year. Calculate the equivalent failure rate and repair time for the following cases:
  (*i*) Only one component out of five is required.
  (*ii*) Two components out of five are required.
  (*iii*) Three components out of five are required.
  (*iv*) Four components out of five are required.
  (*v*) All five components are required.

8.13 Consider a system consisting of two subsystems in a truly parallel configuration with the following parameters:

Subsystem A: 2 out of 3 identical components in parallel are required for success; $\lambda = 0.1$ per year and $r = 0.005$ year for each component.

Subsystem B: One out of two identical components in parallel is required for success; $\lambda = 0.05$ per year and $r = 0.01$ year for each component.

Evaluate the reliability indices (failure rate, mean downtime, and unavailability) for this system.

8.14 In modeling the outdoor performance of human beings, such as during driving and repairing outdoor equipment, the influence of weather cannot be ignored. The human error rate (number of errors committed per unit time) will increase under stormy weather conditions. The model illustrated in Figure P 8.14 has been suggested to study this aspect of human performance.

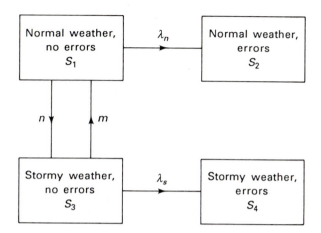

Figure P 8.14

(a) List all the assumptions implied in deriving this model.
(b) Find the time-dependent probabilities of the various states. Assume that $m = 1$, $n = 0.01$, $\lambda_s = 0.01$, and $\lambda_n = 0.001$, all in hours^{-1}. Also, assume that at time $t = 0$, the system is in state S_1.
(c) In the long run, how many times more likely is it for the person to err during normal weather conditions as compared to erring during stormy weather conditions?

8.15 For Problem 8.14, using the fundamental matrix $\mathbf{N}$ and the matrix $\mathbf{B} = \mathbf{NR}$, find the following:
  (*i*) Mean time to err, starting with state S_1.
  (*ii*) Mean time to err, starting with state S_3.
  (*iii*) Starting with state S_1, the number of times, in the long run, that a person is more likely to err during normal weather as compared to erring during stormy weather.
  (*iv*) Starting with state S_3, the number of times, in the long run, that a person is more likely to err during normal weather as compared to erring during stormy weather.

8.16 A component operating outdoors has f failures per year, and a fraction x of them occur during stormy weather. Derive expressions for the normal- and stormy-weather failure rates, λ and λ', in terms of f, x, and α, where α is the fraction of the year that normal weather prevails.

8.17 Calculate and plot the ratio of the stormy-weather failure rate to the normal-weather failure rate of a component as a function of the fraction of the failures occurring during stormy weather if normal weather prevails 90% of the time. Repeat the problem for normal weather that prevails 99.5% of the time.

8.18 Figure P 8.18 shows a load being supplied from a source through two identical transmission lines in series. Each line has an average failure rate of 1 per year, and it has been established from outage data that 90% of the failures occur during stormy weather. The expected durations of normal and stormy weather are 200 hours and 2 hours, respectively. The expected repair time for each line is 8.76 hours for both weather conditions. At the load point, calculate:

Figure P 8.18

(*i*) The number of failures per year of normal weather.
(*ii*) The number of failures per year of stormy weather.
(*iii*) The total number of failures per year.
(*iv*) The expected repair time.
(*v*) The average total outage time per year.

8.19 A system consists of three subsystems in series, with failure rates and repair times as shown in the following table:

Component	Failure Rate, No./Year	Repair Time, Hours	Weather Influence
1	0.2988	8.76	yes
2	0.2988	8.76	yes
3	0.5	17.52	no

For components 1 and 2, two-thirds of the failures occur during stormy weather. Component 3 is indoors, and weather has no influence on its failure rate. The expected durations of normal and stormy weather are 300 and 1.2 hours, respectively. The expected repair times can be assumed to be independent of the weather conditions. Calculate:

(*i*) The number of failures per year.
(*ii*) The number of failures per year of normal weather.
(*iii*) The number of failures per year of stormy weather.
(*iv*) The expected repair time.
(*v*) The expected outage time per year.

8.20 In the simple power system shown in Figure P 8.20, each line has an average failure rate of 1 per year, and 90% of the failures occur during stormy weather. The expected durations of normal and stormy weather are 200 and 2 hours, respectively. The expected repair time is 8.76 hours for both weather conditions for all three lines. Calculate:

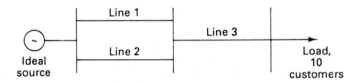

Figure P 8.20

(*i*) The number of interruptions of service per customer per year.

(*ii*) The average customer restoration time.

(*iii*) The average total interruption time per customer per year.

8.21 In Example 8–12, calculate the total number of outages per year and the mean downtime at load point *B*.

8.22 In Example 8–12, calculate the total number of outages per year and the mean downtime at load point *C*.

8.23 Evaluate the new values for the system reliability indices if each of the strings in the system described in Problem 8.9 is maintained twice a year with a mean maintenance downtime of four hours.

8.24 The failure rate for a component is constant and equal to 0.1 failure per year. However, the repair time for the component has a special Erlangian distribution with a mean value of 16 hours and a standard deviation of 6 hours. Set up a state-space diagram using the method of stages in series. Compute all the relevant values, and incorporate them into the state-space diagram.

8.25 Show that a downtime with the density function

$$\frac{\alpha(\alpha^2 + \beta^2)}{\beta^2} e^{-\alpha t}(1 - \cos \beta t)$$

can be represented by a cascade of three stages, as shown in Figure P 8.25. (*Hint:* Consider three random variables with the following density functions:

$$\alpha e^{-\alpha t}$$

$$(\alpha + j\beta)e^{-(\alpha + j\beta)t}$$

$$(\alpha - j\beta)e^{-(\alpha - j\beta)t}$$

Then, using Laplace transforms, show that the density function of the sum of these three random variables is the same as the given density function. Remember that convolution in the time domain corresponds to multiplication in the *s*-domain.)

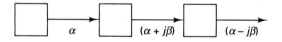

Figure P 8.25

8.26 Draw a state-space diagram for a component with a constant failure rate λ and a downtime as described in Problem 8.25. Find the steady-state probabilities for up and down states, and derive a binary model for the component.

8.27 Calculate the steady-state probabilities of up and down states for the case of two stages in parallel (see Figure 8.9) with $\rho_1 = \rho_2 \equiv \mu$.

8.28 A common-mode failure is an event having a single external cause with multiple failure effects that are not consequences of each other. Consider a system consisting of two components, 1 and 2, of which at least one must operate for system success. The pertinent data are as follows:

Individual failures and repairs: $\lambda_1 = 0.01$; $\lambda_2 = 0.02$

$$\mu_1 = 0.05; \mu_2 = 1$$

Common-mode failure and repair: $\lambda_{12} = 0.001$; $\mu_{12} = 0.01$

All the values are in numbers per year. Draw a state-space diagram with all the appropriate values marked. Employ the frequency balance approach to find the frequency of encounter and residence time for each state. Use the model shown in Figure 8.12.

8.29 Show that, for the case of two components in parallel, if common-mode failures are included as shown in Figure 8.12, the system failure rate λ and downtime r can be approximated as

$$\lambda \cong \lambda_1\lambda_2(r_1 + r_2) + \lambda_{12}$$

and

$$r \cong \frac{r_1 r_2 r_{12}}{r_1 r_2 + r_2 r_{12} + r_{12} r_1}$$

where $r_1 = 1/\mu_1$, $r_2 = 1/\mu_2$, and $r_{12} = 1/\mu_{12}$.

8.30 Develop a state transition diagram for two three-state devices in series. The devices can fail either in the open mode or in the short mode. Include common-cause failures. For each device, λ_0 is the open failure rate and λ_s is the short failure rate. Use λ_c for the common-cause failure rate. The failed components are not repaired. Common-cause failures and other failures can be assumed to be statistically independent. Starting with both components being good, show that, for the system,

$$\text{MTTF} = \frac{\lambda_o + 3\lambda_s}{(\lambda_o + \lambda_s)(\lambda_c + 2\lambda_o + 2\lambda_s)}$$

8.31 Consider the state-space diagram of Figure 8.12 for the case of two components in parallel. Derive an expression for the system MTTF by designating the DD state as absorbing and by employing the fundamental matrix technique. Calculate the MTTF value for $\lambda_1 = 0.01$, $\lambda_2 = 0.015$, $\lambda_{12} = 0.001$, and $\mu_1 = \mu_2 = 1$, all in hours^{-1}.

8.32 Repeat Problem 8.31 with all repairs ignored.

8.33 For the two-component system model shown in Figure 8.13, neglecting repair, find the system MTTF, starting with both components being good. Compare your result with the one obtained for Problem 8.32, and comment on your findings.

8.34 A system consists of three identical components, two of them in a truly parallel configuration and one in standby. The sensing and changeover device can be assumed perfect. Whenever we have two components in operation, there is a possibility of a common-mode failure, after which the system is not repaired. Such common-mode failures have not been observed when there was only one unit in operation. Also, the system is not repaired after a complete shutdown, even without a common-mode failure.

(*i*) Draw a state-space diagram with only one repair facility.

(*ii*) Find the MTTF, starting with all the components in good condition. Use λ and μ for the failure rate and repair rate, respectively, of one component, and use λ_c for the common-mode failure rate.

8.35 Repeat Problem 8.34 with no restrictions on the number of repair facilities available. Starting with all the components in good condition, show that

$$\text{MTTF} = \frac{8\lambda^2 + \lambda_c(\lambda + \mu) + \lambda\mu + \mu^2}{4\lambda^3 + \lambda_c^2(\lambda + \mu) + \lambda_c(4\lambda^2 + \lambda\mu + \mu^2) - 4\lambda^2\mu}$$

8.36 Consider components with a constant failure rate λ and a common-mode failure characterized by the factor β. For what value of β will a system consisting of two such components have the same reliability, whether they are in series or in parallel? Is this value realistic in practice?

8.37 The design-life reliability of a class of components is 0.91. What is the maximum allowable value for the β-factor associated with common-mode failures of these components if the design-life reliability for two such components in parallel is to be at least 0.98?

8.38 Starting with Equation (8–121), show that the MTTF for two components in parallel, each with a failure rate of λ and a common-mode failure characterized by the factor β, is

$$\text{MTTF} = \frac{3 - 2\beta}{\lambda(2 - \beta)}$$

8.39 Express the exponential terms in Equation (8–121) in terms of their series expansions, consider only the significant terms, and show that the reliability expression can be approximated as

$$R(t) \cong 1 - \beta(\lambda t) - \left\{ 1 - 2\beta + \frac{\beta^2}{2} \right\}(\lambda t)^2$$

Repeat Problem 8.37 using this approximation.

8.40 Consider a system consisting of two subsystems in series. The first subsystem has three units in parallel, and the second subsystem consists of two units in parallel. Assuming all the units to be identical and independent with $\lambda t = 0.01$, compute the system reliability, first by approximating $\exp[-\lambda t]$ by $(1 - \lambda t)$ and then by using the exact expressions. Repeat the problem for $\lambda t = 0.1$.

8.41 A military mission requires a minimum of three helicopters. A group of five choppers are sent to complete the mission. The total duration of the mission is 10 hours, and the failure rate for the class of choppers involved is 1 in 1,000 hours of operation.
  (*i*) Using the rare-event approximation, calculate the probability of the mission failing.
  (*ii*) Repeat part (*i*) using the exact expressions. Compare the two results and comment on them.

8.42 Suggest a rare-event approximation for computing the reliability of components with linearly increasing hazards. Consider two components with the same design life and a design-life reliability of 0.95, one having a constant hazard and the other having a linearly increasing hazard. With the aid of plots, compare the reliabilities computed using the corresponding rare-event approximations for $0 \leq t \leq T_d$, where T_d is the design life of the components.

8.43 (a) Derive rare-event approximations for the reliabilities of n identical, independent components in series and in parallel if the components have linearly increasing hazards (that is, $\lambda = kt$). (b) Repeat part (a) for an r-out-of-n configuration.

9

Reliability and Economics

9.1 INTRODUCTION

Interaction between reliability and economics occurs in all aspects of engineering. Modeling and analysis of this interaction are quite complex because of the need to quantify the cost of failure, which should include the monetary equivalent of nonmonetary items, and the need to assess the cost of improving reliability. In addition, expenditures incurred in improving reliability are certain and near term, whereas the cost of failure is uncertain and further away chronologically.

Let us focus on one narrow aspect of this vast topic, namely, the optimization of the performance measure of a system with respect to cost. We can immediately think of several questions that warrant answers: With a given expenditure of money, how much of an increase in performance can be achieved? What is the minimum cost to enhance the performance to a specified level? Where should we put our money—in other words, is it possible to economize at one point in the system and improve in another to meet the specified budget? On what basis do we compare the performance of competing designs? At some stage in the development process, answers to all these questions must be found, estimated, or assumed. Similar statements can be made with regard to system reliability, but here the questions become even more difficult to formulate and find answers to.

Equipment failure is associated with undesirable consequences that may be translated into economic terms. For example, the cost of a single day's outage of a large (>500 MWe) fossil-fuel-fired power plant can run into several hundred thousand dollars. Outages of large (1,000 MWe) nuclear power plants could cost as much as one million dollars per day, even if the reactor suffers no damage and there is no release of radioactivity. In the case of major accidents, such as the ones that occurred in Chernobyl in the former U.S.S.R. and in the chemical plant in Bhopal, India, the total cost is essentially incalculable, and their effects last a very long time. Similar examples abound in space technology, commercial and military aviation, and computer systems involved in sensitive defense installations or in international banking, to name a few cases.

Improving the reliability of a system costs money and is economically justified if the decrease in the expected cost of failure is at least equal to the expected cost of the improvement in reliability. In the past, only high-technology areas in aerospace and defense were the recipients of efforts at improving reliability. However, as we approach the 21st century, high technology is permeating almost all aspects of life. The need to maintain high levels of productivity and competitiveness in the international marketplace indicates that efforts at improving reliability can have a broad, positive impact in manufacturing and service industries as well. Once an improvement in reliability is shown to be justified economically, if the data inputs are realistic, then the question of whether it is too expensive becomes moot. In all our discussions, we should not overlook the ever-increasing cost of liability.

The options available for improving reliability obviously depend on the problem at hand. In most cases, the options can be grouped into three basic categories: (*i*) providing additional redundancy, (*ii*) designing-in environmental capability, and (*iii*) performing environmental testing. In practice, a combination of the three is preferred. The nature of the mix will depend on the ratio of incremental benefit to incremental cost as a function of cost for each of the three options, the role played by risk, and budgetary considerations. The final management decisions on design choices among several alternatives will almost invariably be based on an integrated cost-benefit trade-off. Such an approach forces the engineer to adjust to a less precise and more subjective environment of synergism.

It should be quite clear by now that the topic of reliability and economics is vast, multifaceted, and extremely important. Accordingly, it is impossible to do justice to this complex area in a single chapter, and it is not our intent to do so either. This chapter is by no means exhaustive; it simply attempts to introduce some of the analyses that have been employed in reliability-cost trade-offs and in optimization studies. The objective is to expose the reader to the important interrelationships between economics and the various aspects of reliability, using a few of the more commonly occurring situations.

9.2 THE ECONOMICS OF REDUNDANCY

We saw earlier that by introducing even a simple redundancy such as components in parallel, the reliability of a system can be improved to any desired level using components that are only average or even poor from a reliability point of view. We will examine the redundancy issue in detail here from an economic point of view.

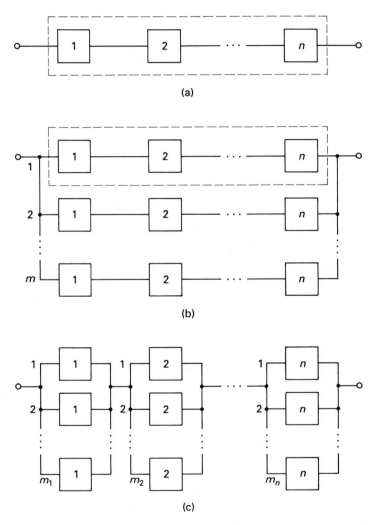

Figure 9.1 Redundancy configurations. (a) Basic (original) system; p_i = reliability of unit i, c_i = cost of unit i, $i = 1, 2, \ldots, n$. (b) System redundancy; parallel-series configuration, m identical systems in parallel. (c) Unit redundancy; series-parallel configuration, m_i = number of units in parallel replacing the original ith unit.

Redundant configurations can be broadly classified into two types: (*i*) those with system redundancy (parallel-series configuration) and (*ii*) those with unit redundancy (series-parallel configuration). An original or basic system [see Figure 9.1(a)] is assumed to consist of a number of units, all of which must succeed for the system to succeed. In essence, these units are logically in series. If the complete system is replicated one or more times and the replications are operated in parallel with it, as shown in Figure 9.1(b), then we have system redundancy. On the other hand, if the individual units of the system are replicated, as shown in Figure 9.1(c), then we have unit redundancy. In the general case, all the original units may be different, but the replication of any one unit involves identical units. Also, the number of units in parallel with a basic unit may differ from unit to unit. The division of a complex system into units is not unique, as long as each unit is an integral working part of the system with clearly defined inputs and outputs.

For the original (basic) system, the cost C_o and the reliability R_o can be calculated easily. We have

$$C_o = \sum_{i=1}^{n} c_i \tag{9-1}$$

and

$$R_o = \prod_{i=1}^{n} p_i \tag{9-2}$$

With system redundancy, as shown in Figure 9.1(b), the total cost and the overall system reliability are, respectively,

$$C_s = mC_o = m \sum_{i=1}^{n} c_i \tag{9-3}$$

and

$$R_s = 1 - (1 - R_o)^m$$
$$= 1 - \left[1 - \prod_{i=1}^{n} p_i \right]^m \tag{9-4}$$

For the unit redundancy configuration, shown in Figure 9.1(c), the total cost and overall system reliability can be expressed in terms of c_i and p_i as

$$C_u = \sum_{i=1}^{n} c_i m_i \tag{9-5}$$

and

$$R_u = \prod_{i=1}^{n} \left[1 - (1 - p_i)^{m_i} \right] \tag{9-6}$$

If all the units are identical, then we replace c_i by c and p_i by p and obtain, for system redundancy,

$$C_s = mnc \tag{9-7}$$

$$R_s = 1 - (1 - p^n)^m \tag{9-8}$$

and, for unit redundancy,

$$C_u = c \sum_{i=1}^{n} m_i \tag{9-9}$$

$$R_u = \prod_{i=1}^{n} [1 - (1 - p)^{m_i}] \tag{9-10}$$

If, in addition, all the m_i's are equal in the unit redundancy configuration, then we can replace m_i by m and obtain

$$C_u = cnm \tag{9-11}$$

and

$$R_u = [1 - (1 - p)^m]^n \tag{9-12}$$

In arriving at the expressions for the reliability of the various configurations, we have tacitly assumed that all the units are independent. This assumption is reasonable, and we will stick to it throughout our development of the topic.

It can be shown (see reference 33) that the system redundancy configuration is expensive and wasteful as compared to the unit redundancy configuration. Therefore, we will concentrate on the latter and look for a design that minimizes the total cost of achieving a desired improvement in the overall system reliability of the original (basic) system.

9.2.1 Cost Minimization for the Unit Redundancy Configuration

Given a basic system, as shown in Figure 9.1(a), with reliability R_o and cost C_o, we want to improve the overall system reliability to R by employing the unit redundancy configuration shown in Figure 9.1(c). The questions to be answered are (1) What should be the values of m_i so that the total cost is minimized and (2) What is the minimum cost? Thus, we want to maintain

$$R = \prod_{i=1}^{n} [1 - (1 - p_i)^{m_i}] \tag{9-13}$$

and minimize

$$C_u = \sum_{i=1}^{n} c_i m_i \tag{9-14}$$

by a suitable choice of the values m_i.

We will re-pose the problem by introducing a set of numbers α_i defined by

$$1 - (1 - p_i)^{m_i} \equiv R^{\alpha_i} \tag{9-15}$$

for $i = 1, 2, \dots, n$, and require that

$$\sum_{i=1}^{n} \alpha_i - 1 = 0 \tag{9-16}$$

so that we can satisfy the specified reliability requirement. By using Equations (9–15) and (9–16) in Equation (9–13), we can easily see that the reliability requirement is indeed satisfied. From Equation (9–15), we get

$$m_i = \frac{\ln(1 - R^{\alpha_i})}{\ln(1 - p_i)} \tag{9-17}$$

and using this in Equation (9–14), we obtain, for the total cost,

$$C_u = \sum_{i=1}^{n} \frac{c_i \ln(1 - R^{\alpha_i})}{\ln(1 - p_i)} \tag{9-18}$$

The problem boils down to minimizing C_u, a function of n variables $\alpha_1, \alpha_2, \dots, \alpha_n$, as given by Equation (9–18), subject to the restriction imposed by Equation (9–16). This is a classic optimization problem. We employ the method of Lagrangian multipliers and define a function f of n variables, α_1 through α_n, as

$$f(\alpha_1, \alpha_2, \dots, \alpha_n) \equiv \sum_{i=1}^{n} \frac{c_i \ln(1 - R^{\alpha_i})}{\ln(1 - p_i)} + k\left[\sum_{i=1}^{n} \alpha_i - 1\right] \tag{9-19}$$

where k is the Lagrangian multiplier. Then we stipulate that

$$\frac{\partial f}{\partial \alpha_i} = 0 \tag{9-20}$$

for all i. This leads to

$$\frac{c_i}{\ln(1 - p_i)} \frac{1}{(1 - R^{\alpha_i})} \frac{d}{d\alpha_i}(1 - R^{\alpha_i}) + k = 0$$

from which we obtain

$$k = \frac{c_i R^{\alpha_i} \ln R}{(1 - R^{\alpha_i}) \ln(1 - p_i)} \tag{9-21}$$

At this point, we make use of the fact that typical values of R are very close to unity, and we deduce an approximation to the value of k. We see that

$$\lim_{R \to 1} \frac{\ln R}{(1 - R^{\alpha_i})} = -\frac{1}{\alpha_i} \tag{9-22}$$

and we approximate R^{α_i} by unity to get

$$k \cong -\frac{c_i}{\alpha_i \ln(1 - p_i)} \tag{9-23}$$

Rearranging terms, we obtain

$$\alpha_i \cong - \frac{c_i}{k \ln(1 - p_i)} \tag{9-24}$$

We have n such equations for $i = 1, 2, \ldots, n$. Adding all these n equations together and using Equation (9–16), we obtain

$$\sum_{i=1}^{n} \alpha_i = 1 = - \frac{1}{k} \sum_{i=1}^{n} \frac{c_i}{\ln(1 - p_i)}$$

Finally, we deduce the value of the Lagrangian multiplier:

$$k = - \sum_{i=1}^{n} \frac{c_i}{\ln(1 - p_i)} \tag{9-25}$$

Using this value of k in Equation (9–23), we find the value of α_i required to minimize the total cost C_u in the case of unit redundancy:

$$\alpha_i = \frac{\left[\dfrac{c_i}{\ln(1 - p_i)} \right]}{\left[\displaystyle\sum_{j=1}^{n} \dfrac{c_j}{\ln(1 - p_j)} \right]} \tag{9-26}$$

for $i = 1, 2, \ldots, n$. Once the α_i values are known, the corresponding m_i values can be found from Equation (9–17). Obviously, the m_i values obtained should be rounded off to the nearest integer. Such a rounding off is not expected to cause serious problems since, in most cases, the reliability and the cost figures for units will not be known precisely.

After arriving at integer values for m_i, we can find the total cost using Equation (9–5) and the new system reliability using Equation (9–6).

■■ **Example 9–1**
Let the original system consist of three units in series, with the following reliability and cost parameters:

$$p_1 = 0.4; \; p_2 = 0.5; \; p_3 = 0.6$$

$$c_1 = 2; \; c_2 = 1; \; c_3 = 3 \text{ units of money}$$

Our objective is to introduce unit redundancy at minimum cost to achieve a new system reliability of 0.99.

For the original system,

$$C_o = 2 + 1 + 3 = 6$$

and

$$R_o = (0.4)(0.5)(0.6) = 0.12$$

Next, we calculate $c_i/[\ln(1 - p_i)]$ for $i = 1, 2,$ and 3. The results are as follows:
For $i = 1,$

$$\frac{2}{\ln(0.6)} = -3.915$$

For $i = 2,$

$$\frac{1}{\ln(0.5)} = -1.443$$

For $i = 3,$

$$\frac{3}{\ln(0.4)} = -3.274$$

Then

$$\sum_{j=1}^{3} \frac{c_j}{\ln(1 - p_j)} = -8.632$$

The α_i values can now be found. They are:

$$\alpha_1 = \frac{-3.915}{-8.632} = 0.454$$

$$\alpha_2 = \frac{-1.443}{-8.632} = 0.167$$

$$\alpha_3 = \frac{-3.274}{-8.632} = 0.379$$

The corresponding m_i values are

$$m_1 = \frac{\ln(1 - 0.99^{0.454})}{\ln(0.6)} = 10.56, \text{ or } 11 \text{ units}$$

$$m_2 = \frac{\ln(1 - 0.99^{0.167})}{\ln(0.5)} = 9.22, \text{ or } 9 \text{ units}$$

$$m_3 = \frac{\ln(1 - 0.99^{0.379})}{\ln(0.4)} = 6.08, \text{ or } 6 \text{ units}$$

Checking the reliability of the system with these values of m_i, we get

$$R_u = [1 - (0.6)^{11}][1 - (0.5)^9][1 - (0.4)^6]$$

$$= 0.990353$$

which is very close to the desired value of 0.99.
The total cost of the new system is

$$C_u = (11)(2) + (9)(1) + (6)(3)$$

$$= 49 \text{ units of money}$$

which is 8.2 times the cost of the original system.

Now suppose we had decided to employ system redundancy instead of unit redundancy. Then, in order to achieve a reliability value of 0.99, we should have m parallel branches, where m is obtained from

$$0.99 = 1 - [1 - 0.12]^m$$

Accordingly,

$$m = \frac{\ln(1 - 0.99)}{\ln(1 - 0.12)}$$

$$= 36$$

In this case, the cost will be 36 times the cost of the original system! Clearly, unit redundancy helps us achieve the required system reliability at a much lower cost.

■■

■■ **Example 9–2**
For the simple case of $n = 1$, from Equation (9–26) we see that $\alpha_1 = 1$, and from Equations (9–17) and (9–18) we get

$$m_1 = \frac{\ln(1 - R)}{\ln(1 - p)} = \frac{C_u}{c_1}$$

■■

9.3 THE ECONOMICS OF REPAIR AND MAINTENANCE

We saw earlier that proper periodic preventive maintenance increases the MTTF of components with increasing hazard functions. It also reduces the frequency of repairs in the case of repairable components. Increasing the frequency of maintenance (in effect, decreasing the time between maintenances) further reduces the frequency of repair. What, then, is the optimum value of time between maintenances? The answer depends on the relative costs of repair and maintenance.
Let

$$C_R = \text{cost of } a \text{ repair}$$

$$C_M = \text{cost of } one \text{ maintenance}$$

If C_M and C_R are about the same, the introduction of maintenance may not be economical. Fortunately, in most cases $C_M \ll C_R$, and the introduction of maintenance is justified for components with increasing hazard functions. The total cost per unit time for repairs and maintenance is

$$K = C_R f_R + C_M f_M \tag{9-27}$$

where

$$f_R = \text{frequency of repair}$$

and

$$f_M = \text{frequency of maintenance} = 1/T_M$$

To find the optimum value of T_M, we define $K_0 \equiv K/C_R$ as a function of T_M and find the value of T_M that minimizes K_0. Thus,

$$K_0 = \frac{K}{C_R} = f_R + \frac{C_M}{C_R} f_M \tag{9-28}$$

Using Equation (4–125) in Equation (9–28), we obtain

$$K_0 = \frac{1}{T_M} \left[\int_0^{T_M} L(t)\, dt + \frac{C_M}{C_R} \right] \tag{9-29}$$

or

$$T_M K_0 = \int_0^{T_M} L(t)\, dt + \frac{C_M}{C_R}$$

Differentiating with respect to T_M, we get

$$K_0 + T_M \frac{dK_0}{dT_M} = L(T_M)$$

from which it follows that

$$\frac{dK_0}{dT_M} = \frac{1}{T_M} L(T_M) - \frac{1}{T_M} K_0 \tag{9-30}$$

Equating dK_0/dT_M to zero and substituting for K_0 from Equation (9–29), we obtain the equation that needs to be solved to find the optimum value of T_M, namely,

$$T_M L(T_M) = \int_0^{T_M} L(t)\, dt + \frac{C_M}{C_R} \tag{9-31}$$

Figure 9.2 shows the variation of K_0 with T_M.

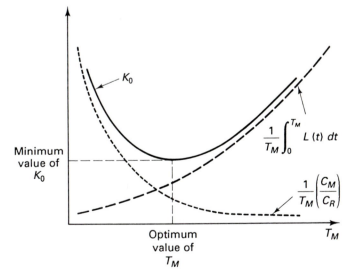

Figure 9.2   Variation of K_0 with T_M.

■■ **Example 9–3**
Consider again the component of Examples 4–22, 4–23, and 4–24. Suppose that $C_M/C_R = 0.15$. Then, using Equations (9–29) and (9–30), for $0 < T_M \leq 4$ years, we get

$$e^{T_M/4}[1 - 0.25T_M] = 0.85$$

Solving this equation, we obtain the optimum value of T_M under the stipulated conditions as 1.869 years. Thus, maintenance done at intervals of 1.869 years will minimize the total cost of maintenance and repair. In practice, this number will probably be rounded off to two years.   ■■

9.4 AVAILABILITY ANALYSIS

The productivity of a plant is measured by a combination of indices that are influenced by the magnitude, frequency, and duration of outages, as well as their costs. Availability engineering analysis is a methodology that can assist engineers in improving the productivity of a plant. This kind of analysis invariably includes cost-availability and cost-benefit trade-offs. Solutions for optimizing plant productivity are recommended on the basis of the results of availability analyses.

Conflicting requirements and constraints enter into every stage of a plant's life, from planning and design to operation, including maintenance. Thus, availability goals, reliability, maintainability, costs of outage, and life-cycle costs should be evaluated before proposing solutions to improve the productivity of a plant.

The MTTR of a component is a measure of the maintainability of that component, and the MTTF is a measure of its reliability. Often, we are faced with trading off reliability and maintainability against each other to minimize overall cost. Availability analysis offers a methodology for minimizing this cost, in addition to satisfying all the specifications.

Let us consider a repairable component with failure and repair rates of λ and μ, respectively. Then

$$\text{MTTF} = \frac{1}{\lambda} \tag{9–32}$$

$$\text{MTTR} = \frac{1}{\mu} \tag{9–33}$$

and

$$A \equiv \text{Availability} = \frac{\text{MTTF}}{\text{MTTF} + \text{MTTR}}$$

$$= \frac{\mu}{\lambda + \mu} \tag{9–34}$$

Expressing MTTR as a function of MTTF and A, we have

$$\text{MTTR} = \left(\frac{1 - A}{A}\right) \text{MTTF} \tag{9–35}$$

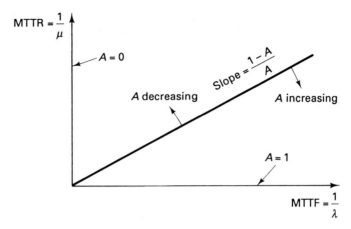

Figure 9.3   Variation of MTTR with MTTF for a fixed value of A.

For each value of A, the plot of MTTR versus MTTF will be a straight line with a slope of $(1 - A)/A$, as illustrated in Figure 9.3.

Typically, the following factors will be specified:

1. Required minimum availability level
2. Required minimum MTTF
3. Allowable maximum MTTR

The shaded area shown in Figure 9.4 is the region in which all the specifications will be satisfied. Then it is up to the designer to pick a point within this region

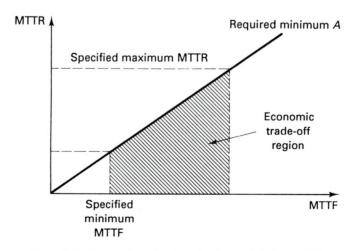

Figure 9.4   Economic trade-off region for availability analysis.

that will result in the lowest cost. A simple numerical example is used to illustrate the cost-availability trade-off calculations and how the cost-benefit approach is used in making design decisions.

■■ **Example 9–4**

A pumping system for a plant must be designed. The objective is to minimize the total present-worth revenue requirements (PWRR). Three alternatives are available, the details of which are as follows:

No. of Pumps	Percent Capacity	Total Installed Cost	Mean Repair Cost per Repair
1	100% × 1	$500,000	$1,000
2	50% × 2	$650,000	$ 900
3	50% × 3	$900,000	$ 800

Each pump, irrespective of its capacity, has a failure rate of 0.25 per year and a mean repair time of 24 hours. There is a penalty of $15,000 per hour (at full capacity) due to outages of pumps. The lifetime of the plant is estimated to be 35 years. The interest rate on borrowing is 10%, and there is an annual charge of 15% for the payback of borrowed capital funds and operation and maintenance expenditures. Thus, for each pump,

$$\text{MTTF} = \frac{1}{\lambda} = \frac{1}{0.25} = 4 \text{ years}$$

$$\text{MTTR} = 24 \text{ hr} = \frac{24}{8,760} \text{ year}$$

and

$$\text{Availability} = \frac{4}{4 + \frac{24}{8,760}} = \frac{35,040}{35,064}$$

Therefore,

$$A = 0.9993155$$

and

$$U = 0.0006845$$

With payments spread over n years, the effective present-worth factor (EPWF) for an interest rate of r per unit is

$$\text{EPWF} = \frac{(1 + r)^n - 1}{r(1 + r)^n} \tag{9–36}$$

For $r = 0.1$ and $n = 35$, EPWF $= 9.644$.

    For alternative 1, we have

$$\text{Outage hours per year} = (0.0006845)(8,760) \text{ or } (0.25)(24) = 6 \text{ hr}$$

Since an outage results in a 100% reduction in plant capacity, the annual outage cost is $6 \times 15,000 = \$90,000$. Therefore,

$$\text{PWRR} = (\text{EPWF}) \left[\left(\begin{array}{c} \text{annual} \\ \text{charge} \\ \text{rate} \end{array} \right) \left(\begin{array}{c} \text{total} \\ \text{installed} \\ \text{cost} \end{array} \right) + \left(\begin{array}{c} \text{annual} \\ \text{outage} \\ \text{cost} \end{array} \right) + \left(\begin{array}{c} \text{annual} \\ \text{repair} \\ \text{cost} \end{array} \right) \right]$$

$$= \$(9.644)[(0.15)(500,000) + 90,000 + (0.25)(1,000)]$$

$$= \$1,593,671$$

    For alternative 2, with two identical pumps, there are three possibilities to consider:

  (i) Both pumps working. Then there is no reduction in capacity and no penalty.

  (ii) One pump up and one down. Then a 50% reduction in capacity results, and

$$\text{Probability of this state} = 2AU = 0.00136806$$

(iii) Both pumps down. Then there is a 100% reduction in capacity, and

$$\text{Probability of this state} = U^2 = 469 \times 10^{-9}$$

$$\text{Annual outage cost} = \$(15,000) \left[\frac{0.00136806}{2} + 469 \times 10^{-9} \times 1 \right](8,760)$$

$$= \$(15,000)[684.03 \times 10^{-6} + 0.469 \times 10^{-6}](8,760)$$

$$= \$89,943$$

$$\text{PWRR} = (9.644)[(0.15)(650,000) + 89,943 + 2\,(0.25)(900)]$$

$$= \$1,812,040$$

    For alternative 3, there are four possibilities:

  (i) All three pumps working. Then there is no reduction in capacity.

  (ii) Only two pumps working. Then, again, there is no reduction in capacity.

  (iii) Only one pump working. Then there is a 50% reduction in capacity, and

$$\text{Probability of this state} = 3AU^2 = 1.405 \times 10^{-6}$$

  (iv) All pumps are down. Then there is a 100% reduction in capacity, and

$$\text{Probability of this state} = U^3 = 0.3207158 \times 10^{-9}$$

$$\text{Annual outage cost} = \$(15,000 \times 8,760)$$

$$\times \left[\frac{1.405 \times 10^{-6}}{2} + 0.3207158 \times 10^{-9} \right]$$

$$= \$92.35$$

$$\text{PWRR} = (9.644)[(0.15)(900,000) + 92.35 + 3(0.25)(800)]$$

$$= \$1,308,617$$

The various costs and worth-to-cost ratios are as follows:

Design	Total PWRR, $	Annual Fixed Cost, $	Annual Other Cost, $	Increase in Annual Fixed Cost, $	Improvement in Worth of Availability, $	Worth-to-Cost Ratio
1	1,593,671	75,000	90,250	········	(base case)	········
2	1,812,040	97,500	90,393	22,500	− 143	− 0.006
3	1,308,617	135,000	692	60,000	89,558	1.492

Clearly, design alternative 3 has the lowest total PWRR and the best worth-to-cost ratio. ■■

9.5 RELIABILITY REQUIREMENTS IMPOSED BY ECONOMICS

Let us consider a component with a required mission time of T, for $0 < t \leq T$. Also, let

A = benefit if the component completes the mission (equivalent to saying, "if the component does not fail in the time interval 0 to T")

B = penalty if the component does not complete the mission (equivalent to saying, "if the component fails in the time interval 0 to T")

k = average loss incurred

There are two cases of interest.

Case (i). If no renewal (repair or replacement) is possible, then

$$k = B[1 - R(T)] - A$$

$$= (B - A) - BR(T) \tag{9-37}$$

where $R(t)$ is the reliability function for the component. Typically, $B > A$, and we would like k to be negative. Therefore, we require that

$$R(T) > \frac{B - A}{B} \tag{9-38}$$

If there is a minimum benefit of γA, irrespective of the failure or success of the mission, then we replace A in Equation (9–38) by $(1 - \gamma)A$, which is the difference between the full benefit and the minimum benefit. This substitution leads to

$$R(T) > \frac{B - (1 - \gamma)A}{B} \qquad (9\text{–}39)$$

If the component has a constant hazard of λ, then

$$R(t) = e^{-\lambda t}$$

and Equation (9–39) becomes

$$e^{-\lambda T} > \frac{B - (1 - \gamma)A}{B}$$

or

$$\lambda < \left(\frac{1}{T}\right) \ln \left[\frac{B}{B - (1 - \gamma)A} \right] \qquad (9\text{–}40)$$

Case (ii).   Assuming instant renewal (ideal repair), let $H(t)$ be the renewal function, which is the expected value of the number of renewals during the time interval 0 to t. For components with a constant hazard of λ, $H(t) = \lambda t$ = expected number of failures during $(0, t)$.

With ideal repair, the average loss incurred is

$$k = BH(T) - A \qquad (9\text{–}41)$$

where T is the mission time. Now, we want k to be negative; so, replacing $H(T)$ by λT, we obtain, for the requirement,

$$A > B\lambda T \qquad (9\text{–}42)$$

or

$$\lambda < \frac{A}{BT} \qquad (9\text{–}43)$$

Introducing the factor γ, as in Equation (9–39), we get

$$\lambda < \frac{(1 - \gamma)A}{BT} \qquad (9\text{–}44)$$

Equations (9–40) and (9–44) express the reliability requirements based on economics for nonrepairable and (ideally) repairable components, respectively.

The next step is to devise a reliability test to make sure that the lots containing the components conform to the desired reliability requirements. The test involves an *acceptance number*, equal to the maximum number of failures that can be tolerated in a lot (with certain *a priori* characteristics expressed in terms of the distribution of λ) in terms of the number of failures experienced during a certain number of unit-hours of testing. In devising the test, we assume that λ is gamma distributed.

Instant renewal assumed.   Suppose the failure rate λ is known *a priori* to have a gamma distribution with parameters α and β. Then

$$f(\lambda) = \frac{\beta(\beta\lambda)^{\alpha-1}}{\Gamma(\alpha)} \, e^{-\beta\lambda} \tag{9-45}$$

The expected value of λ with this distribution is α/β. After including experimental information involving r failures and an accumulated testing (operating) time of T_Σ, we can infer the *a posteriori* distribution of λ to be

$$f(\lambda) = \frac{(\beta + T_\Sigma)^{\alpha+r}\lambda^{\alpha+r-1}}{\Gamma(\alpha + r)} \, e^{-(\beta + T_\Sigma)\lambda} \tag{9-46}$$

Note that Equation (9–46) is a gamma distribution, just like Equation (9–45), except that α and β are replaced by $(\alpha + r)$ and $(\beta + T_\Sigma)$, respectively. The *a posteriori* mean value $\hat{\lambda}$ of the failure rate with the density function given by Equation (9–46) is

$$\hat{\lambda} = \frac{\alpha + r}{\beta + T_\Sigma} \tag{9-47}$$

Replacing λ in Equation (9–44) by the expression for $\hat{\lambda}$ given by Equation (9–47), we obtain the new reliability requirement as

$$\frac{\alpha + r}{\beta + T_\Sigma} < \frac{(1 - \gamma)A}{BT} \tag{9-48}$$

or

$$r < \frac{(1 - \gamma)(\beta + T_\Sigma)A}{BT} - \alpha \tag{9-49}$$

The right-hand side of Equation (9–49) is called the *acceptance number w*. In a test involving T_Σ unit-hours, if the number of failures does not exceed w, then the lot is acceptable. Obviously, w must be a nonnegative number. Therefore,

$$\frac{(1 - \gamma)(\beta + T_\Sigma)A}{BT} \geq \alpha \tag{9-50}$$

or

$$T_\Sigma \geq \frac{\alpha BT}{(1 - \gamma)A} - \beta \tag{9-51}$$

Since T_Σ must be a positive number, we require that

$$\frac{\alpha B T}{(1 - \gamma)A} > \beta \tag{9-52}$$

or

$$\frac{\alpha}{\beta} > \frac{(1 - \gamma)A}{BT} \tag{9-53}$$

Equation (9-53) states that the *a priori* value of λ, which is equal to α/β, must not be underestimated. In other words, the reliability of the component or system must not be overestimated *a priori*.

■■   **Example 9-5**

Suppose that $B = 3A$, $\gamma = 0.7$, $\alpha = 4$, and $\beta = 10^4$. For a mission time of one month (30 days) and a reliability test conducted for a total of 60,000 unit-hours, find the acceptance number.

We have:

$$T = 30 \times 24 = 720 \text{ hours}$$

$$T_\Sigma = 60,000 \text{ hours}$$

First of all, let us check and see whether the required inequalities are satisfied. They are, from Equation (9-51),

$$6 \times 10^4 \geq \frac{4 \times 3A \times 720}{0.3 \times A} - 10^4$$

$$6 \times 10^4 \geq 1.88 \times 10^4$$

and from Equation (9-53),

$$\frac{4}{10^4} > \frac{0.3A}{3A \times 720}$$

$$4 \times 10^{-4} > 1.389 \times 10^{-4}$$

Since the inequalities are satisfied, we can now use Equation (9-49) to calculate the acceptance number:

$$w = \frac{0.3 \times [10^4 + 6 \times 10^4]A}{3A \times 720} - 4$$

$$= 5.72$$

We conclude, rather conservatively, that in a reliability test involving 60,000 unit-hours, at most 5 failures may occur, to ensure that the reliability of the components is sufficiently high to realize the expected economic benefit.   ■■

No renewal assumed.   Assuming constant-hazard components and a mission time of T, we use Equation (9–37) to give the average loss incurred, which is

$$k = B - A - Be^{-\lambda T} \tag{9-54}$$

In this equation, the only random term is $e^{-\lambda T}$, because of the randomness in λ. The average value (or expected value) of $e^{-\lambda T}$ is

$$E[e^{-\lambda T}] = \int_0^\infty e^{-\lambda T} f(\lambda) \, d\lambda \tag{9-55}$$

where $f(\lambda)$ is the *a posteriori* density function given by Equation (9–46). The result is

$$E[e^{-\lambda T}] = \left[\frac{\beta + T_\Sigma}{\beta + T_\Sigma + T} \right]^{\alpha + r} \tag{9-56}$$

Using this average value in Equation (9–54), and introducing the γ term as before, we have

$$k = B - (1 - \gamma)A - B\left[\frac{\beta + T_\Sigma}{\beta + T_\Sigma + T} \right]^{\alpha + r} \tag{9-57}$$

The requirement that k be negative leads to

$$B\left[\frac{\beta + T_\Sigma}{\beta + T_\Sigma + T} \right]^{\alpha + r} > B - (1 - \gamma)A \tag{9-58}$$

or

$$r < \frac{\log\left(\dfrac{B}{B - (1 - \gamma)A} \right)}{\log\left(\dfrac{\beta + T_\Sigma + T}{\beta + T_\Sigma} \right)} - \alpha \tag{9-59}$$

The right-hand side of Equation (9–59) is the acceptance number w for the case of no renewal.

The acceptance number w should be nonnegative. This condition requires that

$$T_\Sigma \geq \frac{T}{\left[\dfrac{B}{B - (1 - \gamma)A} \right]^{1/\alpha} - 1} - \beta \tag{9-60}$$

Moreover, T_Σ should be positive, which leads to

$$\left(\frac{\beta}{T + \beta} \right)^\alpha < \frac{B - (1 - \gamma)A}{B} \tag{9-61}$$

The left-hand side of Equation (9–61) is the *a priori* expected value of the reliability function $e^{-\lambda T}$. Therefore, Equation (9–61) states that we must not overestimate *a priori* the reliability of the system.

■■   **Example 9–6**

Let us repeat Example 9–5, but without renewal. Now, the inequalities to be satisfied are

Equation (9–60), which leads to

$$6 \times 10^4 \geq \frac{720}{\left(\dfrac{3}{3 - 0.3}\right)^{0.25} - 1} - 10^4$$

$$6 \times 10^4 \geq 1.697 \times 10^4$$

and Equation (9–61), which results in

$$\left(\frac{10^4}{720 + 10^4}\right)^4 < \left(\frac{3 - 0.3}{3}\right)$$

$$0.7572 < 0.9$$

With the required inequalities satisfied, we use Equation (9–59) to calculate the acceptance number:

$$w = \frac{\log\left(\dfrac{3}{3 - 0.3}\right)}{\log\left[\dfrac{10^4 + (6 \times 10^4) + 720}{10^4 + (6 \times 10^4)}\right]} - 4$$

$$= \frac{\log 1.111}{\log 1.0103} - 4$$

$$= 10.28 - 4$$

$$= 6.28$$

Thus, in a reliability test involving 60,000 unit-hours, at most 6 failures may occur in order for us to conclude that the reliability is sufficiently high to assure the expected economic benefit.   ■■

9.6 SUMMARY

Reliability and economics are very closely related. The final design choices for any engineering system are almost always based on the required level of reliability and the cost of failure. In this chapter, we have discussed four specific aspects of this vast topic: the economics of redundancy, the economics of repair and maintenance, availability analysis, and an analysis of reliability requirements imposed by economics for (ideally) repairable and nonrepairable components. Based on this last analysis, the concept of an acceptance number—namely, the number of failures that can be permitted in an acceptance test involving a certain number of unit-hours of testing—has been introduced. With the background acquired from this chapter, the reader will be able to better understand other economic analyses and optimization studies that dot the literature on reliability.

PROBLEMS

9.1 In the case of unit redundancy, derive an expression for the optimum cost ratio C_u/C_o if all the units have equal reliabilities and equal costs. What is the reliability improvement ratio? What can you say about the m_i values?

9.2 With unit redundancy, what is the minimum total cost C_u if all the units have equal reliabilities but unequal costs?

9.3 Given that the cost of all the units are equal, but their reliabilities are different, derive an expression for α_i for the case of optimum (minimum total cost) unit redundancy.

9.4 Repeat Example 9–1 for the case where $c_1 = 1$, $c_2 = 2$, and $c_3 = 3$ units of money. All the rest of the information is assumed the same. Compare your results with those of the example, and explain any differences.

9.5 Four components with identical reliabilities of 0.5 each are operated in a series arrangement. Their costs are 2, 3, 4, and 5 units of money, respectively. Employ the unit redundancy concept and suggest an arrangement to achieve an overall system reliability of 0.98 at minimum cost.

9.6 In Problem 9.5, assuming that system redundancy is employed, find the total cost C_s, and calculate the ratio C_s/C_u, where C_u is the minimum cost with unit redundancy.

9.7 A basic series system consists of four units with cost and reliability values as given in Figure P 9.7. Design a new system configuration incorporating the unit redundancy concept to achieve an overall system reliability of 0.995 at minimum cost. Compare your design with the one that employs the system redundancy concept.

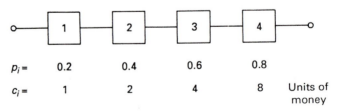

$p_i =$	0.2	0.4	0.6	0.8	
$c_i =$	1	2	4	8	Units of money

Figure P 9.7

9.8 Consider a number of identical three-state devices operated in parallel. Let the normal operation, open-mode failure, and short-mode failure probabilities of a device be p_n, q_o, and q_s, respectively. Clearly, the optimum design in this case corresponds to the maximization of parallel system reliability by adjusting the number of devices in parallel. Show that the optimum number of devices in parallel is

$$n = \frac{\ln \left[\dfrac{\ln q_o}{\ln (1 - q_s)} \right]}{\ln \left(\dfrac{1 - q_s}{q_o} \right)}$$

Calculate the numerical value for n for $q_o = 0.1$ and $q_s = 0.05$.

9.9   Repeat Example 9–3 for $C_M/C_R = 0.05$. Compare and discuss your result with the values obtained in the example.

9.10 If $L(t) = 0.125 \exp(t/8)$, and C_M/C_R stays the same as in Example 9–3, what is the new value for the optimum maintenance time interval? Discuss the differences with the value found in the example.

9.11 Evaluate the following three alternatives and arrive at a choice of one of them based on the PWRR for a pump system for a 1-GW power plant. Each pump has a separate repair crew.

Number of Pumps	% Capacity of Each Pump	Total Installed Cost	Mean Repair Cost per Repair
1	100%	$500,000	$1,000
2	60%	$700,000	$ 900
3	50%	$900,000	$ 800

The failure and repair rates for all the pumps can be assumed to be the same at 0.26 failure/year and (1/30) repair/hour. The replacement power cost is $500/MW-day, and the plant capacity factor is 0.65. Assume an interest rate of 9%, an annual charge rate of 18%, and a plant life of 35 years.

9.12 Repeat Example 9–5 with $\alpha = 4$, $\beta = 10^4$, $B = 2A$, $\gamma = 0.8$, $T = 1,000$ hours, and $T_\Sigma = 5 \times 10^4$ hours.

9.13 Repeat Example 9–6 with $\alpha = 4$, $\beta = 10^4$, $B = 2A$, $\gamma = 0.8$, and $T = 1,000$ hours. The total accumulated number of unit-hours of testing is 6×10^4 hours.

10

Accelerated Testing and Models

10.1 INTRODUCTION

Highly reliable and complex systems require extremely reliable components. The advent of such components has brought with it the difficulty of demonstrating their reliability within a reasonable period of time and a realistic sample size and budget. We simply cannot wait to accumulate failure data from field performance. Because of extremely long expected lifetimes, such a wait would make the components obsolete before the required answers are gleaned.

One approach to overcome this dilemma is to accelerate the failures of components by subjecting them, in a laboratory setting, to conditions of stress, temperature, cycles per unit time, etc., that are much more rigorous than normal. The accelerated life test data thus obtained are fitted to suitable distribution models, which are then used in conjunction with acceleration models to estimate failure rates under conditions of normal use.

Selecting appropriate levels of stress for accelerated life testing requires assumptions based on sound engineering judgment and experience with actual conditions of use. Test stresses can be applied either one at a time or in combinations. The levels of these stresses can be constant, increasing in steps, or progressively increasing. Stresses that result in a change of the state of the materials of the components and stresses that will not be experienced under normal use conditions should not be employed in accelerated life testing.

In this chapter, we introduce the reader to accelerated testing and models that involve some of the more commonly used failure distributions. Also, some well-known acceleration models and their applications are outlined. Details regarding setting up test cells and calculating the number of test cells needed for validation of the model (which is at least equal to the number of parameters to be estimated) are not discussed.

10.2 TRUE ACCELERATION

The simplest way to understand true acceleration is to consider a videocassette recorder playing a prerecorded tape in the fast-forward mode. All the events happen, but at a higher speed. In other words, the sequence and nature of the events do not change; they just happen at a faster pace. If such a statement is true of the operation of a component at higher-than-normal stress levels, then we have *true acceleration*, meaning that we have managed to accelerate failures without altering the failure mechanisms or the sequence of events. Under true acceleration, we simply have a transformation of the time scale. Such a simple time scale transformation will be applicable only over a limited range of stresses. While any single-valued, well-behaved function can be used to model true acceleration, the assumption of linearity is commonly made because of its simplicity and applicability. Under true linear acceleration, every failure time and every distribution percentile is multiplied by the same constant to obtain the corresponding values under different stress levels. We quantify the amount of acceleration by an acceleration factor

$$\mathcal{A} \equiv \frac{t_n}{t_a} \qquad (10\text{--}1)$$

where

$$t_n = \text{time to failure under normal conditions}$$

and

$$t_a = \text{time to failure under accelerated (higher stress) conditions}$$

If $f_a(t_a)$ and $f_n(t_n)$ are the density functions of the random variables t_a and t_n, respectively, then, since $t_n = \mathcal{A}t_a$, we have

$$f_n(t_n) = \left(\frac{1}{\mathcal{A}}\right) f_a(t_a)$$

$$= \left(\frac{1}{\mathcal{A}}\right) f_a\left(\frac{t_n}{\mathcal{A}}\right) \qquad (10\text{--}2)$$

Under conditions of normal use, the density function is equal to the product of the reciprocal of the acceleration factor and the density function under accelerated conditions with the time variable t_a replaced by $t_n/\mathcal{A}$. Replacing t_n by a general time variable t, we obtain

$$f_n(t) = \left(\frac{1}{\mathcal{A}}\right) f_a\left(\frac{t}{\mathcal{A}}\right) \tag{10–3}$$

The relationship between the corresponding distribution functions follows easily and is

$$F_n(t) = \left(\frac{1}{\mathcal{A}}\right) \mathcal{A} F_a\left(\frac{t}{\mathcal{A}}\right)$$

or

$$F_n(t) = F_a\left(\frac{t}{\mathcal{A}}\right) \tag{10–4}$$

Since the hazard function $\lambda(t)$ is related to the density and distribution functions by

$$\lambda(t) = \frac{f(t)}{1 - F(t)} \tag{10–5}$$

we have

$$\lambda_n(t) = \frac{f_n(t)}{1 - F_n(t)}$$

$$= \frac{(1/\mathcal{A})\, f_a(t/\mathcal{A})}{1 - F_a(t/\mathcal{A})}$$

or

$$\lambda_n(t) = \left(\frac{1}{\mathcal{A}}\right) \lambda_a\left(\frac{t}{\mathcal{A}}\right) \tag{10–6}$$

Equations (10–3), (10–4), and (10–6) are completely general and are applicable as long as the assumptions of true and linear acceleration are valid.

10.3 PHYSICAL ACCELERATION AND FAILURE DISTRIBUTIONS

Suppose that failure data at one stress condition fit a certain distribution. What will be the failure distribution under different stress conditions if the assumptions of true and linear acceleration are valid? This is the question we will answer in this section for some of the more commonly used distribution functions.

10.3.1 Exponential Distribution

Let us assume that failure data obtained under accelerated test conditions fit an exponential model with a constant failure rate of λ_a. The corresponding distribution function is

$$F_a(t) = 1 - \exp[-\lambda_a t] \tag{10-7}$$

Then, using Equation (10–4), we find that the failure distribution function under normal conditions of use is

$$F_n(t) = F_a \left(\frac{t}{\mathscr{A}} \right)$$
$$= 1 - \exp\left[-\left(\frac{\lambda_a}{\mathscr{A}} \right) t \right] \tag{10-8}$$

Since λ_a is the constant failure rate under accelerated test conditions, we can define the constant failure rate λ_n under conditions of normal use as

$$\lambda_n \equiv \frac{\lambda_a}{\mathscr{A}} \tag{10-9}$$

Clearly, the failure distribution remains exponential under conditions of normal use, with a failure rate equal to $1/\mathscr{A}$ times the failure rate under accelerated test conditions.

10.3.2 Weibull Distribution

Now let us assume that the accelerated failure data fit a Weibull distribution with a scale parameter α_a and a shape parameter β_a. Then

$$F_a(t) = 1 - \exp\left[-\left(\frac{t}{\alpha_a} \right)^{\beta_a} \right] \tag{10-10}$$

The failure distribution under conditions of normal use is

$$F_n(t) = 1 - \exp\left[-\left(\frac{t}{\alpha_a \mathscr{A}} \right)^{\beta_a} \right] \tag{10-11}$$

In terms of the scale and shape parameters α_n and β_n, respectively, under conditions of normal use, we have

$$F_n(t) = 1 - \exp\left[-\left(\frac{t}{\alpha_n} \right)^{\beta_n} \right] \tag{10-12}$$

where

$$\alpha_n = \alpha_a \mathscr{A} \tag{10-13}$$

and

$$\beta_n = \beta_a \qquad (10\text{–}14)$$

We have just shown that if the failure times at one stress level have a Weibull distribution, then, under true linear acceleration, the life distribution at any other stress level is also Weibull with the same shape parameter and a new scale parameter equal to the reciprocal of the acceleration factor times the old scale parameter.

If failure data obtained at two different stress levels do not have the same shape parameter, then either the assumption of a Weibull distribution is wrong, or we do not have true linear acceleration, or both.

The hazard function (which is the same as the failure rate) under accelerated test conditions is

$$\lambda_a(t) = \frac{f_a(t)}{1 - F_a(t)} \qquad (10\text{–}15)$$

or

$$\lambda_a(t) = \frac{\beta_a t^{\beta_a - 1}}{\alpha_a^{\beta_a}} \qquad (10\text{–}16)$$

Therefore, using Equation (10–6), we can obtain the hazard function $\lambda_n(t)$ under conditions of normal use:

$$\lambda_n(t) = \left(\frac{1}{\mathscr{A}}\right)\left[\beta_a \left(\frac{t}{\mathscr{A}}\right)^{\beta_a - 1}\right]\frac{1}{\alpha_a^{\beta_a}}$$

$$= \frac{\beta_a t^{\beta_a - 1}}{\alpha_a^{\beta_a}} \frac{1}{\mathscr{A}^{\beta_a}}$$

or

$$\lambda_n(t) = \frac{\lambda_a(t)}{\mathscr{A}^{\beta_a}} \qquad (10\text{–}17)$$

Although the failure rate changes linearly, the multiplication factor is $1/\mathscr{A}$ only for $\beta_a = 1$, in which case we revert back to the exponential distribution. For other values of the shape parameter, the multiplying factor is $(1/\mathscr{A}^{\beta_a})$, and the failure rate no longer varies inversely with $\mathscr{A}$.

10.3.3 Lognormal Distribution

Now we assume that the failure times under accelerated test conditions have a lognormal density function

$$f_a(t) = \frac{1}{t\sigma_a \sqrt{2\pi}} \exp\left[-\frac{(\ln t - \mu_a)^2}{2\sigma_a^2}\right] \qquad (10\text{–}18)$$

Assuming true linear acceleration, the corresponding density function under conditions of normal use becomes

$$f_n(t) = \left(\frac{1}{\mathcal{A}}\right) f_a \left(\frac{t}{\mathcal{A}}\right)$$

$$= \frac{1}{t\sigma_a\sqrt{2\pi}} \exp\left[-\frac{\{\ln t - (\mu_a + \ln \mathcal{A})\}^2}{2\sigma_a^2}\right] \tag{10-19}$$

Since $\mu_a = \ln T_{50a}$, where T_{50a} is the median time to failure under accelerated conditions, we can define T_{50n} as equal to $\mathcal{A}T_{50a}$, with the variance (shape parameter) remaining the same under true linear acceleration. As in the case of the Weibull distribution, true linear acceleration does not change the shape parameter; only the scale parameter is multiplied by the reciprocal of the acceleration factor $\mathcal{A}$. Moreover, if failure data obtained at two stress levels do not have the same variance, then either the assumption of lognormal distribution is not correct, or we do not have true linear acceleration, or both.

10.3.4 Gamma Distribution

Finally, let us assume that the accelerated failure times have a gamma density function

$$f_a(t) = \frac{t^{\beta_a - 1}}{\alpha_a^{\beta_a}\Gamma(\beta_a)} \exp\left(\frac{-t}{\alpha_a}\right) \tag{10-20}$$

with $\alpha_a > 0$, $\beta_a > 0$, and $t \geq 0$. The density function for the failure times under conditions of normal use will then be

$$f_n(t) = \frac{t^{\beta_n - 1}}{\alpha_n^{\beta_n}\Gamma(\beta_n)} \exp\left(\frac{-t}{\alpha_n}\right) \tag{10-21}$$

where

$$\alpha_n \equiv \mathcal{A}\alpha_a \tag{10-22}$$

and

$$\beta_n = \beta_a \tag{10-23}$$

Once again, under true linear acceleration, the shape parameter does not change, and the scale parameter is multiplied by the reciprocal of the acceleration factor.

■■ **Example 10–1**
Suppose that a certain class of electronic components tested at an elevated temperature has a constant failure rate of 20%/K. If the acceleration factor between the normal temperature and the elevated temperature is 40, then

$$\lambda_n = \frac{\lambda_a}{\mathcal{A}} = \frac{20}{40} = 0.5\%/K$$

Or, the failure rate under normal use is 5×10^{-6} hr^{-1}. Over a period of five years of normal use, the fraction of components that will fail is

$$F_n(5 \times 8{,}760) = 1 - \exp[-5 \times 10^{-6} \times 5 \times 8{,}760]$$

$$= 0.19668$$

We conclude that about 19.668% of the components will fail within five years of normal use.   ■■

■■   **Example 10–2**
Failure times of ceramic capacitors are known to follow exponential distributions. One of the easiest ways of overstressing capacitors is to operate them at a higher-than-rated voltage, maintaining the operating temperature constant. Then, the voltage acceleration factor can be assumed to be

$$\mathcal{A}_v = \left(\frac{\text{Test voltage}}{\text{Rated voltage}} \right)^n = \left[\frac{V_a}{V_r} \right]^n$$

where n is called the *power exponent*. A typical value of n for ceramic capacitors is found to be 3. For mica capacitors, $n = 5$, and for paper and film capacitors, $n = 9$. Therefore, failure rates obtained from accelerated test data at twice the rated voltage should be divided by 8 or 32 or 512 for ceramic, mica, and paper and film capacitors, respectively.

In practice, the value of n is found to be dependent on the operating temperature in a complex way. Also, since $\mathcal{A}_v = (\text{MTTF})_n/(\text{MTTF})_a$, a plot of $\ln V_a$ versus $\ln(\text{MTTF})_a$ will be a straight line with slope $-1/n$.   ■■

■■   **Example 10–3**
If temperature is used to accelerate capacitor failures, then the temperature acceleration factor is

$$\mathcal{A}_T = 2^{(T_a - T_r)/K}$$

where T_a and T_r are the accelerated and rated temperatures, in °C, respectively, and K is a constant that depends on the type of capacitor and the temperatures T_a and T_r. Typical values of K are as follows:

Material	K
Ceramic (<85°C)	8
Ceramic (85°C–125°C)	10
Ceramic (>125°C)	20
Mica	20
Paper and film	15

If mica capacitors rated at 30°C are tested at 60°C, then the failure rates obtained from accelerated tests should be divided by

$$\mathscr{A}_T = 2^{(60-30)/20} = 2^{1.5} = 2.828$$

If ceramic capacitors rated at 45°C are tested at 100°C, then

$$\mathscr{A}_T = 2^{(100-85)/10} \times 2^{(85-45)/8} = 2^{6.5} = 90.5$$

■■

■■ **Example 10–4**

Accelerated testing of solid tantalum capacitors was undertaken at 163% of rated voltage. The temperature was maintained constant at 85°C. Failure times were found to be Weibull distributed with a shape parameter of 0.5 and a scale parameter of 2.42×10^5. The acceleration factor was estimated to be 10^5. Therefore, under conditions of normal use, failure times are expected to be Weibull distributed with a shape parameter of 0.5 and a scale parameter of $2.42 \times 10^5 \times 10^5$, or 2.42×10^{10}.

The probability that these capacitors will last 1 million hours under normal use can be found as follows:

$$R(10^6) = 1 - F_n(10^6)$$

$$= \exp\left[-\left(\frac{10^6}{2.42 \times 10^{10}}\right)^{0.5}\right]$$

$$= \exp[-0.006428]$$

$$= 0.99359$$

■■

■■ **Example 10–5**

A model that has found use in studying deep-groove annular ball bearings subjected to radial loads takes the form

$$\frac{L_n}{L_a} = \left(\frac{P_a}{P_n}\right)^3$$

where

L_n = life of the ball bearings, in millions of cycles under normal load conditions

L_a = life of the ball bearings, in millions of cycles under accelerated load conditions

P_n = equivalent load under conditions of normal use

P_a = equivalent load under conditions of accelerated use

If the bearings are properly installed, failures will be due primarily to fatigue. Also, lifetimes (in number of cycles before failure) have been found to be Weibull distributed within the range of application.

From the above relationship, we obtain

$$\ln\left(\frac{1}{P_a}\right) = \left(\frac{1}{3}\right)\ln(L_a) - \left(\frac{1}{3}\right)\ln(L_n) - \ln(P_n)$$

A plot of $\ln(1/P_a)$ versus $\ln(L_a)$, which is the same as a plot of $(1/P_a)$ versus L_a on log-log paper, will be a straight line if this assumed model is applicable.   ■ ■

10.4 ACCELERATION MODELS

The development of acceleration models involves suitable assumptions regarding the relationship between the two random variables

$t_a \equiv$ continuous random variable denoting the time to failure under accelerated stress conditions

and

$t_n \equiv$ continuous random variable denoting the time to failure under conditions of normal use

We have already seen that under the assumption of true linear acceleration, the relationship between t_a and t_n is linear and is given by

$$t_a = \left(\frac{1}{\mathscr{A}}\right)t_n \qquad (10\text{-}24)$$

Also, under the assumption of true linear acceleration, exponential, Weibull, gamma, and lognormal distributions are preserved under varying stress conditions.

What happens if the relationship between t_a and t_n is not linear? We will consider two simple cases.

Case (i).   Let the accelerated failure times follow an exponential distribution; that is,

$$F_a(t_a) = 1 - \exp(-\lambda_a t_a) \qquad (10\text{-}25)$$

In addition, let

$$t_a = \left(\frac{1}{\lambda_a}\right)\left(\frac{t_n}{\alpha_n}\right)^{\beta_n}, \qquad \beta_n \neq 1 \qquad (10\text{-}26)$$

It is easy to show that the distribution of failure times under conditions of normal use is

$$F_n(t_n) = 1 - \exp\left[-\left(\frac{t_n}{\alpha_n}\right)^{\beta_n}\right] \qquad (10\text{-}27)$$

which is a Weibull distribution with a scale parameter α_n and a shape parameter β_n.

Case (ii).   Now we assume that accelerated failure times follow a Weibull distribution given by

$$F_a(t_a) = 1 - \exp\left[-\left(\frac{t_a}{\alpha_a}\right)^{\beta_a}\right] \tag{10-28}$$

with $\alpha_a > 0$, $\beta_a > 0$, $\beta_a \neq 1$, and $t_a > 0$. In addition, if

$$t_a = \alpha_a(\lambda_n t_n)^{1/\beta_a} \tag{10-29}$$

then the failure distribution under conditions of normal use becomes

$$F_n(t_n) = 1 - \exp(-\lambda_n t_n) \tag{10-30}$$

which is an exponential distribution with a constant failure rate of λ_n.

Table 10.1 summarizes the models discussed in Sections 10.3 and 10.4.

10.5 THE ARRHENIUS MODEL

Failures of many simple electronic devices and electrical insulation are due to chemical degradation processes, which are accelerated by elevated temperatures. The Arrhenius model, developed towards the end of the 19th century to describe the reaction rates of chemical processes, has found application in accelerated life-testing technology.

The Arrhenius model is applicable if

(i) the most significant stresses are thermal,
(ii) at any temperature, lifetimes follow a lognormal distribution,
(iii) the standard deviation of the natural logarithm of the failure times is independent of temperature, and
(iv) the mean value of the natural logarithm of failure times as a function of temperature is expressed by the Arrhenius relation

$$\mu\left(\frac{1}{T}\right) = A + (B)\left(\frac{1}{T}\right) \tag{10-31}$$

where the parameters A and B depend on the characteristics of the material being tested and the test methods.

The median time to failure T_{50_1} at temperature T_1 K is

$$T_{50_1} = \exp\left[\mu\left(\frac{1}{T_1}\right)\right] = e^A e^{(B/T_1)} \tag{10-32}$$

TABLE 10.1 ACCELERATION MODELS

Failure density functions	Relationship between t_a and t_n	Density function $f_a(t_a)$ under accelerated test conditions	Density function $f_n(t_n)$ under conditions of normal use
Exponential/Exponential	$t_a = \dfrac{1}{\mathcal{A}}\, t_n$	$\lambda_a e^{-\lambda_a t_a}$	$\left(\dfrac{\lambda_a}{\mathcal{A}}\right) e^{-(\lambda_a/\mathcal{A})t_n}$
Weibull/Weibull	$t_a = \dfrac{1}{\mathcal{A}}\, t_n$	$\dfrac{\beta_a t_a^{\beta_a-1}}{\alpha_a^{\beta_a}}\, e^{-(t_a/\alpha_a)^{\beta_a}}$	$\dfrac{\beta_a t_n^{\beta_a-1}}{(\mathcal{A}\alpha_a)^{\beta_a}}\, e^{-(t_n/(\mathcal{A}\alpha_a))^{\beta_a}}$
Lognormal/Lognormal	$t_a = \dfrac{1}{\mathcal{A}}\, t_n$	$\dfrac{1}{t_a \sigma_a \sqrt{2\pi}}\exp\left[-\dfrac{(\ln t_a - \mu_a)^2}{2\sigma_a^2}\right]$	$\dfrac{1}{t_n \sigma_a \sqrt{2\pi}}\exp\left[-\dfrac{\{\ln t_n - (\mu_a + \ln \mathcal{A})\}^2}{2\sigma_a^2}\right]$
Gamma/Gamma	$t_a = \dfrac{1}{\mathcal{A}}\, t_n$	$\dfrac{t_a^{\beta_a-1}}{\alpha_a^{\beta_a}\,\Gamma(\beta_a)}\, e^{-t_a/\alpha_a}$	$\dfrac{t_n^{\beta_a-1}}{(\mathcal{A}\alpha_a)^{\beta_a}\,\Gamma(\beta_a)}\, e^{-t_n/(\mathcal{A}\alpha_a)}$
Exponential/Weibull	$t_a = \left(\dfrac{1}{\lambda_a}\right)\dfrac{(t_n)^{\beta_n}}{(\alpha_n)}\,;\ \beta_n \neq 1$	$\lambda_a\, e^{-\lambda_a t_a}$	$\dfrac{\beta_n t_n^{\beta_n-1}}{\alpha_n^{\beta_n}}\, e^{-(t_n/\alpha_n)^{\beta_n}}$
Weibull/Exponential	$t_a = \alpha_a(\lambda_n t_n)^{1/\beta_a}\,;\ \beta_a \neq 1$	$\dfrac{\beta_a t_a^{\beta_a-1}}{\alpha_a^{\beta_a}}\, e^{-(t_a/\alpha_a)^{\beta_a}}$	$\lambda_n\, e^{-\lambda_n t_n}$

Similarly, the median time to failure at temperature T_2 K is

$$T_{50_2} = e^A e^{(B/T_2)} \tag{10–33}$$

Assuming true acceleration, the acceleration factor for conducting the tests at temperature T_2 K instead of T_1 K, for $T_2 > T_1$, can be calculated as

$$\mathscr{A} = \frac{T_{50_1}}{T_{50_2}} = \exp\left[B\left(\frac{1}{T_1} - \frac{1}{T_2}\right)\right] \tag{10–34}$$

Once the parameter B is known, the acceleration factor between any two temperatures can easily be calculated using Equation (10–34). Conversely,

$$B = (\ln \mathscr{A})\left[\frac{1}{T_1} - \frac{1}{T_2}\right]^{-1} \tag{10–35}$$

or

$$B = \left[\ln\left(\frac{T_{50_1}}{T_{50_2}}\right)\right]\left[\frac{1}{T_1} - \frac{1}{T_2}\right]^{-1} \tag{10–36}$$

Equation (10–36) provides a convenient way to estimate the parameter B from failure data obtained by testing at two different temperatures.

■■   **Example 10–6**
With $B = 5,800$, if the normal-use and accelerated testing temperatures are 25°C and 65°C, respectively, then $T_1 = 298$ K, $T_2 = 338$ K, and

$$\ln \mathscr{A} = \left(\frac{1}{T_1} - \frac{1}{T_2}\right) 5,800 = (3.97125 \times 10^{-4})5,800$$

$$= 2.303$$

The corresponding acceleration factor is equal to 10.
   If we double the acceleration temperature to 130°C, then the new acceleration factor is

$$\mathscr{A} = \exp\left[\left(\frac{1}{298} - \frac{1}{403}\right)5,800\right] = 159.34$$

Thus, doubling the higher temperature increases the acceleration factor sixteenfold.
   If the value of B is doubled, then for $T_1 = 25$°C and $T_2 = 65$°C, $\mathscr{A}$ is equal to 100. For $T_1 = 25$°C and $T_2 = 130$°C, the new value of $\mathscr{A}$ is 25,389!
   It is possible to develop a lookup table for acceleration factors for different values of T_1, T_2, and B. The constant B is sometimes expressed as $\Delta H/k$, where k is the Boltzmann constant, equal to 8.6168×10^{-5} ev/K. The value of $1/k$ is 11,605. ■■

10.6 THE EYRING MODEL

Often, more than thermal stresses are involved in the failure of components. The Arrhenius model cannot handle multiple stresses. The Eyring model offers a general approach to handling these additional stresses. Specifically, for the case of thermal stress and n additional stresses, the median time to failure is assumed to be of the form

$$T_{50} = e^A T^\alpha e^{(B/T)} \prod_{i=1}^{n} \exp[\{k_{1i} + (k_{2i}/T)\}S_i] \qquad (10\text{–}37)$$

where S_1, S_2, ... , S_n are the stresses placed on the component in addition to thermal stress. With $\alpha = 0$ and no additional stresses, the Eyring model reduces to the Arrhenius model.

The Eyring model has three parameters in the temperature term, and two parameters are added for each additional stress to be considered. Thus, the application of this model becomes quite difficult when several stresses are involved, since the number of stress cells should at least be equal to the number of parameters to be estimated. In reality, we require more stress cells to test the adequacy of the model.

For the special case of two stresses, thermal and one other,

$$T_{50} = e^A T^\alpha e^{(B/T)} \exp[\{k_{11} + (k_{21}/T)\}S_1] \qquad (10\text{–}38)$$

■■ Example 10–7

Let the two stresses acting on a component be temperature and the natural logarithm of applied voltage, $\ln V$. Also, let $\alpha = 0$ and $k_{21} = 0$. Then

$$T_{50} = C e^{(B/T)} V^{k_{11}}$$

where

$$C \equiv e^A$$

This model has only three parameters to be evaluated. Depending on the way various physical quantities contribute to the degradation of materials and the eventual failure of the component, the model can take on many different forms. If, in addition, the test temperature is held constant, then

$$T_{50} \propto V^{k_{11}}$$

An alternative form that can be used is

$$(\text{MTTF})_n = \left[\frac{V_a}{V_n}\right]^k (\text{MTTF})_a$$

where $(\text{MTTF})_n$ corresponds to operation at the normal-use voltage V_n and $(\text{MTTF})_a$ corresponds to accelerated testing at voltage V_a. It is easy to see that the value of k can be experimentally determined by conducting a series of constant-voltage tests. A plot of $\ln V_a$ versus $\ln(\text{MTTF})_a$ yields a straight line with a slope of $-(1/k)$.

■■

10.7 SUMMARY

This chapter has presented a brief discussion of the need for and the models used in accelerated testing of components. True acceleration is equivalent to a linear change in the time scale, and under such conditions, exponential, Weibull, log-normal, and gamma distributions are preserved under different stresses. Certain special relationships between t_a and t_n result in an exponential model becoming Weibull and vice versa. Two other popular models, the Arrhenius and Eyring models, were also presented in their elementary forms. Information on the important topic of designing experiments to arrive at the parameters employed in these models can be found in the literature.

PROBLEMS

10.1 If the times to failure under normal and accelerated conditions are related by

$$t_n = \mathcal{A}(t_a + t_0)$$

where t_0 is a constant, find the relationships between
  (*i*) $f_n(t)$ and $f_a(t)$,
  (*ii*) $F_n(t)$ and $F_a(t)$, and
  (*iii*) $\lambda_n(t)$ and $\lambda_a(t)$.

10.2 Under true acceleration and a linear acceleration factor $\mathcal{A}$, derive the relationship between $(MTTF)_n$ and $(MTTF)_a$. Justify your steps.

10.3 For the situation considered in Problem 10.1, find the relationship between $(MTTF)_n$ and $(MTTF)_a$.

10.4 Find the relationship between the median times to failure T_{50a} and T_{50n} under accelerated and normal-use conditions, assuming true linear acceleration and exponential failure distributions.

10.5 Repeat Problem 10.4 for the case of Weibull distributions.

10.6 Find the ratios of test voltage to rated voltage necessary in a constant-temperature accelerated testing of (*i*) ceramic, (*ii*) mica, and (*iii*) paper capacitors to realize a tenfold decrease in the value of the MTTF. Assume exponential distributions of failure times.

10.7 Using the values given in Example 10–3, calculate the temperature acceleration factor for ceramic capacitors rated at 50°C and tested at 95°C. Repeat this exercise for a rated temperature of 80°C and testing temperature of 130°C.

10.8 Accelerated testing of an electronic component at an elevated temperature resulted in failure times with exponential distribution and a median time to failure of 2,773 hours. Assuming an acceleration factor of 25, find (*i*) the normal-use failure rate, (*ii*) the median time to failure under normal use, and (*iii*) the fraction of a group of similar components expected to fail during their useful lifetime of two years.

10.9 Under normal operating conditions, only 1% of a certain type of components fail during their first year of operation. In order to obtain more failure data, an experiment is devised for the accelerated testing of these components. The accelerated test results indicate an MTTF of 4,500 hours. Assuming true acceleration and exponential distributions, estimate the acceleration factor.

10.10 Certain components under study exhibit exponential failure distribution with an MTTF of 10,000 hours. What acceleration factor should one strive for in order to obtain a median time to failure of no more than 100 hours under accelerated test conditions?

10.11 Failure times obtained from accelerated testing of a certain type of component are found to be gamma distributed with a mean value of 25 hours and a variance of 62.5 hours^2. Assuming true acceleration and an acceleration factor of 8, calculate the percentage of failures that can be expected under normal-use conditions during the first 200 hours of operation.

10.12 Repeat Problem 10.11 if the mean value of accelerated failure data is 23.72 hours and all the other values remain the same.

10.13 Calculate the Arrhenius acceleration factors for $B = 11,605$ and 7,000 at a testing temperature of 105°C. The normal-use temperature is 65°C.

10.14 For the Arrhenius model, derive the relationship between $\ln \mathcal{A}$ and $[(1/T_1) - (1/T_2)]$, where $\mathcal{A}$ is the acceleration factor and T_1 and T_2 are the lower and higher temperatures, respectively. For $B = 6,000$ and $T_1 = 25°C$, plot $\ln \mathcal{A}$ versus $[(1/T_1) - (1/T_2)]$ for values of T_2 ranging from 55°C to 105°C. Comment on the plot obtained.

10.15 The parameter B in Equation (10–35) is sometimes expressed in the form

$$B = \frac{\Delta H}{k}$$

where ΔH is a constant and k is the Boltzmann constant, the value of which is 8.6168 $\times 10^{-5}$ ev/K. Calculate the acceleration factors for $\Delta H = 0.5$ and
(*i*) $T_1 = 25°C$, $T_2 = 75°C$,
(*ii*) $T_1 = 75°C$, $T_2 = 125°C$, and
(*iii*) $T_1 = 125°C$, $T_2 = 155°C$.

10.16 Repeat Problem 10.15 for $\Delta H = 1.0$.

10.17 An acceleration factor of 10 was realized by elevating the operating temperature of a device from 65°C to 105°C. Assuming an Arrhenius model, calculate the acceleration factor that can be expected by raising the test temperature to 125°C.

10.18 Random samples of capacitors were tested at two different temperatures, with the following results:

Testing at 85°C: $T_{50} = 40,000$ hours

Testing at 125°C: $T_{50} = 250$ hours

Assuming an Arrhenius acceleration model, estimate the value of the constant B. What is the corresponding value of ΔH?

Fundamentals
of Matrices

A.1 BASIC DEFINITIONS AND NOTATION

A matrix is a rectangular array of elements, with m rows and n columns arranged in the following manner:

$$\mathbf{A} \equiv \begin{bmatrix} a_{11} & a_{12} & a_{13} & \ldots & a_{1n} \\ a_{21} & a_{22} & a_{23} & \ldots & a_{2n} \\ \vdots & \vdots & \vdots & \vdots & \vdots \\ a_{m1} & a_{m2} & a_{m3} & \ldots & a_{mn} \end{bmatrix} \equiv [a_{ij}] \qquad \text{(A–1)}$$

Either a boldface uppercase letter (in this case, $\mathbf{A}$) or bracketed notation (here, $[a_{ij}]$) is used to represent a matrix. The notation a_{ij} denotes the (ij)th element, or the element located in ith row and jth column. One easy way to remember the sequence is to memorize the verse "eye row, jay column."

For a real matrix, a_{ij} is real for all i and j; a_{ij} is complex for all i and j for a complex matrix. The dimension of the matrix is $m \times n$ (said "m by n") when we have m rows and n columns. If $m = n$, we have a square matrix, and a_{ii} is then called the ith diagonal element. If $n = 1$, we have a column matrix (or column vector), and if $m = 1$, we have a row matrix (or row vector). Row and column matrices (vectors) are denoted by boldface lowercase letters.

The compactness of matrix notation can be seen by considering a set of m simultaneous equations in n variables:

$$
\begin{aligned}
a_{11}x_1 + a_{12}x_2 + \ldots + a_{1n}x_n &= y_1 \\
a_{21}x_1 + a_{22}x_2 + \ldots + a_{2n}x_n &= y_2 \\
\vdots \qquad \vdots \qquad \vdots \qquad \vdots \qquad \vdots & \\
a_{m1}x_1 + a_{m2}x_2 + \ldots + a_{mn}x_n &= y_m
\end{aligned}
\tag{A-2}
$$

These m equations can be represented using summation notation as

$$
\sum_{j=1}^{n} a_{ij}x_j = y_i, \quad i = 1, 2, \ldots, m
\tag{A-3}
$$

If we adopt vector-matrix notation, Equation (A–3) takes on the extremely compact form

$$
\mathbf{Ax} = \mathbf{y}
\tag{A-4}
$$

in which $\mathbf{A}$ is the $m \times n$ matrix given in Equation (A–1), $\mathbf{x}$ is the $n \times 1$ column vector

$$
\mathbf{x} =
\begin{bmatrix}
x_1 \\
x_2 \\
\vdots \\
x_n
\end{bmatrix}
\tag{A-5}
$$

and $\mathbf{y}$ is the $m \times 1$ column vector

$$
\mathbf{y} =
\begin{bmatrix}
y_1 \\
y_2 \\
\vdots \\
y_m
\end{bmatrix}
\tag{A-6}
$$

Compactness and streamlining of notation are not the only reasons for using matrices; matrices also help us to apply the rich theory of linear algebra in a systematic way to analyze complex problems.

A.2 TYPES OF MATRICES

There are many different types of matrices, including the following:

1. *Zero matrix* or *null matrix*, **0**: All of the elements of this matrix are zero, and the matrix can have any dimension.

2. *Diagonal matrix*: square matrix with all nondiagonal elements zero.

3. *Identity matrix* **I**: diagonal matrix with $a_{ii} = 1$ for all i.

4. *Real* or *complex* or *imaginary* matrix.

5. *Upper triangular matrix*: square matrix in which all the elements *below* the diagonal are zero. A *strictly upper triangular matrix* has its diagonal elements zero also.

6. *Lower triangular matrix*: square matrix in which all the elements *above* the diagonal are zero. A *strictly lower triangular matrix* has its diagonal elements zero also.

7. The *transpose* $\mathbf{A}^T$ of an $m \times n$ matrix **A** is the $n \times m$ matrix obtained by interchanging the rows and columns of **A**. Thus,

$$\text{if } \mathbf{A} = [a_{ij}], \text{ then } \mathbf{A}^T = [a_{ji}] \qquad (A–7)$$

8. A *symmetric matrix* is a square matrix in which $a_{ij} = a_{ji}$. Obviously, if **A** is symmetric, then $\mathbf{A} = \mathbf{A}^T$.

9. A matrix is said to be *skew symmetric* if it is square and $a_{ij} = -a_{ji}$. In a skew symmetric matrix, all the diagonal elements are zero and $\mathbf{A} = -\mathbf{A}^T$.

10. The *conjugate* $\overline{\mathbf{A}}$ of a matrix **A** is obtained by replacing each element by its complex conjugate. Thus, if $\overline{a}_{ij}$ is the complex conjugate of a_{ij}, then

$$\mathbf{A} = [a_{ij}]; \overline{\mathbf{A}} = [\overline{a}_{ij}] \qquad (A–8)$$

For a real matrix, $\mathbf{A} = \overline{\mathbf{A}}$, and for an imaginary matrix, $\mathbf{A} = -\overline{\mathbf{A}}$.

11. The *associate matrix* of a matrix **A** is **A**'s transposed conjugate $(\overline{\mathbf{A}})^T$.

12. A matrix **A** is said to be *Hermitian* if it is equal to its associate. In the case of real matrices, symmetric and Hermitian mean the same thing.

13. A matrix is said to be *skew Hermitian* if it is equal to the negative of its associate, i.e., if $\mathbf{A} = -(\overline{\mathbf{A}})^T$.

14. A *Metzler matrix* is a real square matrix in which all nondiagonal elements are nonnegative; that is, $a_{ij} \geq 0$ for all $i \neq j$.

15. Consider a set of n functions of n variables, and let

$$\mathbf{f} = \begin{bmatrix} f_1(x_1, x_2, \ldots, x_n) \\ f_2(x_1, x_2, \ldots, x_n) \\ \vdots \\ f_n(x_1, x_2, \ldots, x_n) \end{bmatrix} \qquad (A–9)$$

Suppose that all the partial derivatives of all these functions exist at $\mathbf{x} = \bar{\mathbf{x}}$, where

$$\bar{\mathbf{x}} = [\bar{x}_1, \bar{x}_2, \ldots, \bar{x}_n]^T \qquad (A–10)$$

Then the *Jacobian matrix* **F** of **f** at $\mathbf{x} = \bar{\mathbf{x}}$ is defined as follows:

$$
\mathbf{F} = \begin{bmatrix}
\left(\dfrac{\partial f_1}{\partial x_1}\right) & \left(\dfrac{\partial f_1}{\partial x_2}\right) & \cdots & \left(\dfrac{\partial f_1}{\partial x_n}\right) \\[2ex]
\left(\dfrac{\partial f_2}{\partial x_1}\right) & \left(\dfrac{\partial f_2}{\partial x_2}\right) & \cdots & \left(\dfrac{\partial f_2}{\partial x_n}\right) \\[2ex]
\vdots & \vdots & \vdots & \vdots \\[2ex]
\left(\dfrac{\partial f_n}{\partial x_1}\right) & \left(\dfrac{\partial f_n}{\partial x_2}\right) & \cdots & \left(\dfrac{\partial f_n}{\partial x_n}\right)
\end{bmatrix}_{\mathbf{x} = \bar{\mathbf{x}}}
\tag{A-11}
$$

16. A *block diagonal* or *quasidiagonal matrix* has the form

$$
\mathbf{A} = \begin{bmatrix}
\mathbf{A}_1 & & & \\
& \mathbf{A}_2 & & \\
& & \ddots & \\
& & & \mathbf{A}_k
\end{bmatrix}
\tag{A-12}
$$

or

$$
\mathbf{A} = \mathrm{diag}[\mathbf{A}_1 \quad \mathbf{A}_2 \quad \cdots \quad \mathbf{A}_k]
\tag{A-13}
$$

The matrices $\mathbf{A}_1$ through $\mathbf{A}_k$ need not all be of the same size. If all of them are one by one, then **A** becomes a diagonal matrix.

17. A *companion matrix* has ones along an off-diagonal and zeros everywhere else, except along the bottom row *or* the last column. The following matrices illustrate the two alternatives:

$$
\begin{bmatrix}
0 & 1 & 0 & \cdots & 0 & 0 \\
0 & 0 & 1 & \cdots & 0 & 0 \\
\vdots & \vdots & \vdots & \vdots & \vdots & \vdots \\
0 & 0 & 0 & \cdots & 0 & 1 \\
-a_0 & -a_1 & -a_2 & \cdots & -a_{n-2} & -a_{n-1}
\end{bmatrix}
$$

$$
\begin{bmatrix}
0 & 0 & \cdots & 0 & -a_0 \\
1 & 0 & \cdots & 0 & -a_1 \\
0 & 1 & \cdots & 0 & -a_2 \\
\vdots & \vdots & \vdots & \vdots & \vdots \\
0 & 0 & \cdots & 1 & -a_{n-1}
\end{bmatrix}
$$

A.3 MATRIX OPERATIONS

1. The addition or subtraction of two matrices **A** and **B** of the same dimensions is defined as follows:

$$
\mathbf{C} = \mathbf{A} \pm \mathbf{B} \text{ means } c_{ij} = a_{ij} \pm b_{ij}
\tag{A-14}
$$

2. Equality can exist only between matrices of the same dimensions. Thus,

$$\mathbf{A} = \mathbf{B} \text{ means } a_{ij} = b_{ij} \tag{A-15}$$

3. Commutative law:

$$\mathbf{A} + \mathbf{B} = \mathbf{B} + \mathbf{A} \tag{A-16}$$

$$\mathbf{A} - \mathbf{B} = \mathbf{A} + (-\mathbf{B}) = -\mathbf{B} + \mathbf{A} \tag{A-17}$$

4. Associative law:

$$\mathbf{A} + \mathbf{B} + \mathbf{C} = \mathbf{A} + (\mathbf{B} + \mathbf{C}) = (\mathbf{A} + \mathbf{B}) + \mathbf{C} \tag{A-18}$$

5. Scalar multiplication:

For any real or complex α, if $\mathbf{A} = [a_{ij}]$, then

$$\alpha\mathbf{A} = [\alpha a_{ij}] = \mathbf{A}\alpha \tag{A-19}$$

Note that *every* element is multiplied by α.

6. Distributive law:

$$\alpha(\mathbf{A} + \mathbf{B}) = \alpha\mathbf{A} + \alpha\mathbf{B} \tag{A-20}$$

7. Matrix multiplication: Two matrices are said to be *conformable* if the number of columns in the first matrix is equal to the number of rows in the second matrix. Matrix multiplication is defined only for two conformable matrices, as follows:

$$\underset{m \times n}{\mathbf{A}} \quad \times \quad \underset{n \times p}{\mathbf{B}} \quad = \quad \underset{m \times p}{\mathbf{C}} \tag{A-21}$$

$$c_{ij} = \sum_{k=1}^{n} a_{ik}b_{kj} \tag{A-22}$$

In other words, the first column of $\mathbf{C}$ is equal to the matrix $\mathbf{A}$ times the first column of $\mathbf{B}$, and so on. Moreover,

$$\underset{m \times n}{\mathbf{A}} \quad \times \quad \underset{n \times n}{\mathbf{I}} \quad = \mathbf{A}\mathbf{I} = \underset{m \times n}{\mathbf{A}} \quad ; \quad \underset{m \times m}{\mathbf{I}} \quad \times \quad \underset{m \times n}{\mathbf{A}} \quad = \mathbf{I}\mathbf{A} = \underset{m \times n}{\mathbf{A}} \tag{A-23}$$

We have the associative law of matrix multiplication:

$$\mathbf{A}\mathbf{B}\mathbf{C} = \mathbf{A}(\mathbf{B}\mathbf{C}) = (\mathbf{A}\mathbf{B})\mathbf{C} \tag{A-24}$$

The commutative law is not valid in general for matrix multiplication, even if both products are defined. Thus,

$$\mathbf{A}\mathbf{B} \neq \mathbf{B}\mathbf{A}, \text{ in general} \tag{A-25}$$

The distributive law is applicable, and we have

$$\mathbf{A}(\mathbf{B} + \mathbf{C}) = \mathbf{AB} + \mathbf{AC} \qquad (A\text{--}26)$$

$$(\mathbf{B} + \mathbf{C})\mathbf{A} = \mathbf{BA} + \mathbf{CA} \qquad (A\text{--}27)$$

Also, $\mathbf{AB} = \mathbf{0}$ does not necessarily mean that either $\mathbf{A} = \mathbf{0}$ or $\mathbf{B} = \mathbf{0}$.

8. The *inner product* (or *dot product*) of a real $n \times 1$ vector $\mathbf{c}$ and a real $n \times 1$ vector $\mathbf{d}$ is

$$\langle \mathbf{c}, \mathbf{d} \rangle = \sum_{i=1}^{n} c_i d_i = \mathbf{c}^T \mathbf{d} = \mathbf{d}^T \mathbf{c} \qquad (A\text{--}28)$$

The inner product is a scalar quantity.

9. The *outer product* of two real vectors $\mathbf{c}$ and $\mathbf{d}$ is an $n \times n$ matrix

$$\mathbf{c} \rangle\langle \mathbf{d} = \mathbf{cd}^T \qquad (A\text{--}29)$$

10. Reversal rule:

$$(\mathbf{AB})^T = \mathbf{B}^T \mathbf{A}^T \qquad (A\text{--}30)$$

$$(\mathbf{ABC})^T = \mathbf{C}^T \mathbf{B}^T \mathbf{A}^T \qquad (A\text{--}31)$$

11. Differentiation:

$$\dot{\mathbf{A}}(t) = \frac{d}{dt} \mathbf{A}(t) = [\dot{a}_{ij}(t)] \qquad (A\text{--}32)$$

Note that each element is differentiated.

12. Integration:

$$\int \mathbf{A}(t)\, dt = \left[\int a_{ij}(t)\, dt \right] \qquad (A\text{--}33)$$

13. The *trace* of a matrix is the sum of all the diagonal elements of the matrix:

$$\text{Trace of } \mathbf{A} = \text{Tr}(\mathbf{A}) = \sum_{i=1}^{n} a_{ii} \qquad (A\text{--}34)$$

$$\text{Tr}(\mathbf{A} \pm \mathbf{B}) = \text{Tr}(\mathbf{A}) \pm \text{Tr}(\mathbf{B}) \qquad (A\text{--}35)$$

$$\text{Tr}(\mathbf{AB}) = \text{Tr}(\mathbf{BA}) \qquad (A\text{--}36)$$

$$\text{Tr}(\mathbf{A}^T) = \text{Tr}(\mathbf{A}) \qquad (A\text{--}37)$$

$$\text{Tr}(\mathbf{AB}) = \text{Tr}(\mathbf{B}^T \mathbf{A}^T) \qquad (A\text{--}38)$$

14. The *determinant* of a matrix $\mathbf{A}$ is defined for square matrices only and

$$\text{determinant of } \mathbf{A} = |\mathbf{A}| = \det \mathbf{A} = \sum_{j=1}^{n} a_{ij} C_{ij} = \sum_{i=1}^{n} a_{ij} C_{ij} \qquad (A\text{--}39)$$

where C_{ij} is called the *cofactor* of the element a_{ij}. The two rightmost expressions in Equation (A–39) are the result of applying Laplace's expansion with respect to column j and row i, respectively.

The cofactor itself is the signed minor of a_{ij}, where

$$\text{minor of } a_{ij} = M_{ij} = \text{determinant of the array formed by deleting}$$
$$\text{the } i\text{th row and the } j\text{th column from } \mathbf{A}.$$

(A–40)

Thus, we have

$$\text{cofactor of } a_{ij} = C_{ij} = (-1)^{i+j} M_{ij} \qquad \text{(A–41)}$$

If all the elements in a row or column are zero, then det $\mathbf{A} = 0$. Following are some useful properties of det $\mathbf{A}$:

(*i*) If any two rows (columns) are proportional, then $|\mathbf{A}| = 0$.
(*ii*) If a row (column) is a linear combination of any number of other rows (columns), then $|\mathbf{A}| = 0$.
(*iii*) Interchanging any two rows (columns) of a matrix changes the sign of $|\mathbf{A}|$.
(*iv*) Multiplying all the elements of a row (column) of a matrix by a scalar α yields a matrix whose determinant is $\alpha |\mathbf{A}|$.
(*v*) Any multiple of a row (column) can be added to any other row (column) without changing the determinant of $\mathbf{A}$.

Rank of a matrix $\mathbf{A}$:

$$\text{rank } (\mathbf{A}) = r_A = \text{size of the largest nonzero determinant} \qquad \text{(A–42)}$$
$$\text{that can be formed from } \mathbf{A}$$

If $\mathbf{A}$ is $m \times n$, then maximum value of r_A = smaller of m and n (A–43)

If $\mathbf{A}$ is $n \times n$ and $r_A = n$, then $\mathbf{A}$ is called *nonsingular* (A–44)

If $\mathbf{C} = \mathbf{AB}$, then $0 \le r_C \le \min\{r_A, r_B\}$ (A–45)

Taking the derivative of both sides of Equation (A–39) with respect to a_{ij}, we see that

$$\frac{\partial |\mathbf{A}|}{\partial a_{ij}} = C_{ij} \qquad \text{(A–46)}$$

If $\mathbf{A}$ is a function of t, then $|\mathbf{A}|$ is also a function of t, and $d|\mathbf{A}|/dt$ can be expressed as a sum of n separate determinants, the first of which has its first row (column) differentiated, the second of which has its second row (column) differentiated, and so on through all the n rows (columns).

15. If $\mathbf{A}$ is a triangular matrix, then

$$|\mathbf{A}| = \prod_{i=1}^{n} a_{ii} = \text{the product of all diagonal elements of } \mathbf{A} \quad \text{(A–47)}$$

16. The determinant of the product of two matrices is equal to the product of the determinants of the two individual matrices. Thus,

$$|\mathbf{AB}| = \det(\mathbf{AB}) = |\mathbf{A}\,\|\,\mathbf{B}| = (\det \mathbf{A})(\det \mathbf{B}) \qquad \text{(A–48)}$$

where both $\mathbf{A}$ and $\mathbf{B}$ are $n \times n$ square matrices.

17. Determinant of the transpose of a matrix:

$$|\mathbf{A}^T| = |\mathbf{A}| \qquad \text{(A–49)}$$

18. Matrix inversion: For any square matrix $\mathbf{A}$, the inverse $\mathbf{A}^{-1}$ is defined as follows:

$$\mathbf{A}^{-1}\mathbf{A} = \mathbf{A}\mathbf{A}^{-1} = \mathbf{I} \qquad \text{(A–50)}$$

The inverse can be found using

$$\mathbf{A}^{-1} = \frac{1}{|\mathbf{A}|}\,\text{adj}\,\mathbf{A} \qquad \text{(A–51)}$$

where the adjoint of $\mathbf{A}$ is

$$\text{adj}\,\mathbf{A} = [C_{ij}]^T \qquad \text{(A–52)}$$

Also,

$$(\mathbf{ABC} \ldots \mathbf{W})^{-1} = \mathbf{W}^{-1} \ldots \mathbf{B}^{-1}\mathbf{A}^{-1} \qquad \text{(A–53)}$$

19. The transpose of the inverse of a matrix is equal to the inverse of the transpose. That is,

$$(\mathbf{A}^{-1})^T = (\mathbf{A}^T)^{-1} \qquad \text{(A–54)}$$

20. As a special case of Equation (A–53), if $\mathbf{A}$ and $\mathbf{B}$ are two square matrices of the same order with inverses $\mathbf{A}^{-1}$ and $\mathbf{B}^{-1}$, respectively, the

$$(\mathbf{AB})^{-1} = \mathbf{B}^{-1}\mathbf{A}^{-1} \qquad \text{(A–55)}$$

21. Previna's theorem:

$$\begin{vmatrix} \mathbf{D}_1 & \vdots & \mathbf{D}_2 \\ \cdots & \vdots & \cdots \\ \mathbf{D}_3 & \vdots & \mathbf{D}_4 \end{vmatrix} = |\mathbf{D}_1\,\|\,\mathbf{D}_4 - \mathbf{D}_3\mathbf{D}_1^{-1}\mathbf{D}_2| \qquad \text{(A–56)}$$

22. Inverse of a matrix by partitioning:

Let the matrix $\mathbf{A}$ be $n \times n$, and let $(p + q) = n$. Partition $\mathbf{A}$ as shown into four submatrices, $\mathbf{A}_{11}$, $\mathbf{A}_{12}$, $\mathbf{A}_{21}$, and $\mathbf{A}_{22}$. Then the inverse of $\mathbf{A}$ can be found as follows. If

$$\mathbf{A} = \begin{array}{c} p \times p \\ q \times p \end{array}\begin{bmatrix} \mathbf{A}_{11} & \vdots & \mathbf{A}_{12} \\ \cdots & \vdots & \cdots \\ \mathbf{A}_{21} & \vdots & \mathbf{A}_{22} \end{bmatrix}\begin{array}{c} p \times q \\ q \times q \end{array} \qquad \text{(A–57)}$$

Then

$$\mathbf{A}^{-1} = \begin{bmatrix} \mathbf{A}_{11}^{-1} + (\mathbf{A}_{11}^{-1}\mathbf{A}_{12})\mathbf{E}^{-1}(\mathbf{A}_{21}\mathbf{A}_{11}^{-1}) & -(\mathbf{A}_{11}^{-1}\mathbf{A}_{12})\mathbf{E}^{-1} \\ \\ -\mathbf{E}^{-1}(\mathbf{A}_{21}\mathbf{A}_{11}^{-1}) & \mathbf{E}^{-1} \end{bmatrix} \quad \text{(A-58)}$$

where

$$\mathbf{E} \equiv \mathbf{A}_{22} - \mathbf{A}_{21}(\mathbf{A}_{11}^{-1}\mathbf{A}_{12}) \quad \text{(A-59)}$$

23. Elementary operations:

The following three operations on a matrix are called *elementary operations*:
   (*i*) Interchange of two rows or columns.
   (*ii*) Multiplication of all elements in a row (column) by a constant $\alpha \neq 0$.
   (*iii*) Multiplication of one row (column) by a constant α and addition of the result to another row (column), element by element.

These elementary operations can be performed successively on a matrix $\mathbf{A}$ to reduce it to the identity matrix $\mathbf{I}$. The same operations performed on $\mathbf{I}$ yield the inverse $\mathbf{A}^{-1}$ of $\mathbf{A}$. Also, by performing a sequence of elementary row and column operations, any matrix of rank r can be reduced to one of the Hermite normal forms

$$\mathbf{I}_r; \; [\mathbf{I}_r \; \vdots \; \mathbf{0}]; \; \begin{bmatrix} \mathbf{I}_r \\ \cdots \\ \mathbf{0} \end{bmatrix}; \; \text{and} \; \begin{bmatrix} \mathbf{I}_r & \vdots & \mathbf{0} \\ \cdots & \vdots & \cdots \\ \mathbf{0} & \vdots & \mathbf{0} \end{bmatrix}$$

where $\mathbf{I}_r$ is an identity matrix of dimension r.

A.4 MORE TYPES OF MATRICES

1. *Orthogonal matrix* (also called *normal matrix*):

A real matrix $\mathbf{A}$ is orthogonal if its inverse is equal to its transpose. Thus,

$$\mathbf{A}\mathbf{A}^T = \mathbf{A}^T\mathbf{A} = \mathbf{I} \quad \text{(A-60)}$$

or

$$\mathbf{A}^{-1} = \mathbf{A}^T \quad \text{(A-61)}$$

2. *Unitary matrix*:

A complex matrix $\mathbf{A}$ is unitary if

$$(\overline{\mathbf{A}})^T\mathbf{A} = \mathbf{A}(\overline{\mathbf{A}})^T = \mathbf{I} \quad \text{(A-62)}$$

Unitary matrix with real elements is an orthogonal matrix.

3. We have already seen the adjoint matrix

$$\text{adj } \mathbf{A} = [C_{ij}]^T \quad \text{(A-63)}$$

and the inverse matrix

$$\mathbf{A}^{-1} = \left(\frac{1}{|\mathbf{A}|}\right) \text{adj } \mathbf{A} \qquad (A\text{-}64)$$

for a square $\mathbf{A}$.

4. A matrix $\mathbf{A}$ is said to be *singular* if its determinant is zero. In this case, its inverse does not exist. Nonsingular matrices have nonzero determinants, and their inverses exist.

5. If a matrix is its own inverse, it is called an *involutary matrix*. Thus,

$$\text{If } \mathbf{A} \text{ is involutary, then } \mathbf{AA} = \mathbf{I} \qquad (A\text{-}65)$$

Boolean Algebra

Boolean algebra is employed in problems involving binary variables. A binary variable has only two values, denoted by "1" and "0," or "A" and "$\overline{A}$," or "true" and "false," or "high" and "low," or "switch closed" and "switch open," among other things. Since the two states can be captured in functional propositions, Boolean algebra is sometimes also called propositional calculus. In algebra involving binary states, the plus sign, "+," is used to denote the "or" function, and the multiplication sign, "·," is used to denote the "and" function. These two signs are called *logical sum* and *logical* product, respectively. Naturally, the + and · signs used in this context will not follow conventional arithmetic rules. With this background, the following theorems are assembled here for easy reference:

$$1 \cdot 1 = 1 \qquad (B-1)$$

$$1 + 1 = 1 \qquad (B-2)$$

$$1 \cdot 0 = 0 \qquad (B-3)$$

$$1 + 0 = 1 \qquad (B-4)$$

Let A, B, and C be Boolean variables. Then

$$A \cdot 1 = A \qquad (B-5)$$

$$A + A = A \qquad (B-6)$$

$$A \cdot 0 = 0 \qquad \text{(B–7)}$$

$$A + 0 = A \qquad \text{(B–8)}$$

$$A \cdot A = A \qquad \text{(B–9)}$$

$$A + 1 = 1 \qquad \text{(B–10)}$$

$$A + \overline{A} = 1 \qquad \text{(B–11)}$$

$$A \cdot \overline{A} = 0 \qquad \text{(B–12)}$$

$$A + AB = A \qquad \text{(B–13)}$$

$$A(A + B) = A \qquad \text{(B–14)}$$

Associative law: $\qquad (A + B) + C = A + (B + C) \qquad$ (B–15)

Associative law: $\qquad (AB)C = A(BC) \qquad$ (B–16)

Commutative law: $\qquad A + B = B + A \qquad$ (B–17)

Commutative law: $\qquad A \cdot B = B \cdot A \qquad$ (B–18)

Distributive law: $\qquad A(B + C) = AB + AC \qquad$ (B–19)

Distributive law: $\qquad A + BC = (A + B)(A + C) \qquad$ (B–20)

Double complement: $\qquad \overline{\overline{A}} = A \qquad$ (B–21)

De Morgan's law: $\qquad \overline{A + B} = \overline{A}\,\overline{B} \qquad$ (B–22)

De Morgan's law: $\qquad \overline{AB} = \overline{A} + \overline{B} \qquad$ (B–23)

$$A + \overline{A}B = A + B \qquad \text{(B–24)}$$

$$A(\overline{A} + B) = AB \qquad \text{(B–25)}$$

$$(A + B)(\overline{A} + C) = AC + \overline{A}B \qquad \text{(B–26)}$$

$$(AC + B\overline{C}) = \overline{A}C + \overline{B}\,\overline{C} \qquad \text{(B–27)}$$

The Laplace Transform

C.1 BASIC DEFINITIONS AND NOTATION

The Laplace transform is a tool, much like logarithms, which enables us to solve complex problems by simpler procedures. It transforms a function of a real variable, typically, time t, into a function of a complex variable

$$s = \sigma + j\omega, \quad \sigma > 0 \qquad \text{(C--1)}$$

In other words, the transformation takes us from the time domain to the complex frequency domain. The Laplace transform of $f(t)$ will be denoted $F(s)$.

Our interest in Laplace transforms lies in the fact that, by using them, we can transform a linear differential equation with constant coefficients into an algebraic equation in s in the complex frequency domain. This enables us to solve linear differential equations with constant coefficients in a systematic and convenient way, and we can handle abrupt inputs and initial conditions easily. Thus,

$$\mathscr{L}[f(t)] = F(s) \equiv \int_0^\infty f(t)\, e^{-st}\, dt \qquad \text{(C--2)}$$

where

$$e^{-st} = e^{-\sigma t} e^{-j\omega t} \qquad \text{(C--3)}$$

and

$$|e^{-st}| = e^{-\sigma t} \tag{C-4}$$

By choosing a large enough σ, we can ensure the convergence of this integral. For all the functions of interest to us, we will assume that the Laplace transform exists.

C.2 LAPLACE TRANSFORMS OF SOME COMMON FUNCTIONS

1. *Unit-step function U(t):*

$$\mathcal{L}[U(t)] = \int_0^\infty e^{-st}\, dt = \frac{1}{s} \tag{C-5}$$

2. *Step function KU(t):*

$$\mathcal{L}[KU(t)] = K\mathcal{L}[U(t)] = (K/s) \tag{C-6}$$

3. *Exponential function e^{-at}, for a real and $\sigma > -a$:*

$$\mathcal{L}[e^{-at}] = \int_0^\infty e^{-(s+a)t}\, dt = \frac{1}{s+a} \tag{C-7}$$

As $a \to 0$, $f(t) \to U(t)$ and $F(s) \to (1/s)$.

4. *Unit ramp function $f(t) = t$ for $t > 0$:*

$$\mathcal{L}[t] = \int_0^\infty te^{-st}\, dt = \frac{1}{s^2} \tag{C-8}$$

Evaluation of this integral requires integration by parts and the use of $\lim_{t \to \infty} te^{-st} = 0$.

5. *Unit impulse function $\delta(t)$:*
   Assuming that the impulse occurs at $t = 0^+$ instead of at $t = 0$, by using the sampling property of the delta function, we get

$$\mathcal{L}[\delta(t)] = \int_0^\infty \delta(t)e^{-st}\, dt$$
$$= e^{-st}|_{t=0^+} = 1 \tag{C-9}$$

If we take the lower limit of integration as 0^- instead of 0, we can assume that the impulse occurs at $t = 0$.

6. Complex exponential function $\exp[j\omega t]$:
   Replacing a by $-j\omega$ in Equation (C–7), we get

$$\mathcal{L}[e^{j\omega t}] = \frac{1}{s - j\omega} \tag{C-10}$$

Since

$$\frac{1}{s - j\omega} = \frac{s}{s^2 + \omega^2} + j\left(\frac{\omega}{s^2 + \omega^2}\right)$$

and $e^{j\omega t} = \cos \omega t + j \sin \omega t$
we conclude that

$$\mathcal{L}[\cos \omega t] = \frac{s}{s^2 + \omega^2} \tag{C-11}$$

and

$$\mathcal{L}[\sin \omega t] = \frac{\omega}{s^2 + \omega^2} \tag{C-12}$$

C.3 PROPERTIES OF LAPLACE TRANSFORMS

Laplace transforms have the following properties:

1. Linearity:

$$\mathcal{L}[af(t)] = aF(s) \tag{C-13}$$

2. Superposition:

$$\mathcal{L}[af(t) \pm bg(t)] = aF(s) \pm bG(s) \tag{C-14}$$

3. Exponential attenuation and translation in the s domain:

$$\mathcal{L}[f(t)e^{\mp at}] = F(s \pm a) \tag{C-15}$$

4. Multiplication by time and complex differentiation:

$$\mathcal{L}[tf(t)] = -\frac{d}{ds}F(s) \tag{C-16}$$

$$\mathcal{L}[(-1)^n t^n f(t)] = \frac{d^n}{ds^n}F(s) \tag{C-17}$$

5. Differentiation in the time domain:

$$\mathcal{L}\left[\frac{d^n f(t)}{dt^n}\right] = s^n F(s) - s^{n-1}f(0) - s^{n-2}f^{(1)}(0) - \dots - f^{(n-1)}(0) \tag{C-18}$$

where superscripts in parentheses indicate the order of the time derivative. If there are any discontinuities, they are supposed to occur at $t = 0^+$. All functions evaluated at time $t = 0$ imply evaluation at $t = 0^-$, even though it is not explicitly stated so.

6. Integration in the time domain:

$$\mathscr{L}\left[\int_0^t f(\xi)\,d\xi\right] = \left(\frac{1}{s}\right)F(s) \tag{C-19}$$

$$\mathscr{L}\left[\underbrace{\int_0^t \cdots \int_0^t}_{n \text{ times}} f(\xi)\,d\xi\right] = \left[\frac{1}{s^n}\right]F(s) \tag{C-20}$$

7. Convolution in the time domain:

$$\mathscr{L}\left[\int_0^t f(\xi)g(t-\xi)\,d\xi\right] = F(s)G(s) \tag{C-21}$$

8. Translation in time for $a > 0$:

$$\mathscr{L}[f(t-a)U(t-a)] = e^{-as}F(s) \tag{C-22}$$

9. Integration in the complex domain:

$$\mathscr{L}\left[\left(\frac{1}{t}\right)f(t)\right] = \int_s^\infty F(\xi)\,d\xi \tag{C-23}$$

10. Initial-value theorem:

$$\lim_{s\to\infty}\,[sF(s)] = f(0^+) \tag{C-24}$$

11. Final-value theorem:

$$\lim_{s\to 0}\,[sF(s)] = f(\infty) \tag{C-25}$$

C.4 SOME COMMONLY OCCURRING TRANSFORMS

The following are Laplace transforms that occur commonly in analysis:

$$\mathscr{L}\,[k] = \frac{k}{s} \tag{C-26}$$

$$\mathscr{L}[t^n] = \frac{n!}{s^{n+1}}, \quad n = 0, 1, 2, 3, \ldots \tag{C-27}$$

$$\mathscr{L}[e^{-at}] = \frac{1}{s+a} \tag{C-28}$$

$$\mathscr{L}[te^{-at}] = \frac{1}{(s+a)^2} \tag{C-29}$$

$$\mathscr{L}\left[\frac{t^{n-1}e^{-at}}{(n-1)!}\right] = \frac{1}{(s+a)^n}, \quad n = 1, 2, 3, \ldots \tag{C-30}$$

$$\mathcal{L}[e^{-at}\sin\omega t] = \frac{\omega}{(s+a)^2 + \omega^2} \tag{C–31}$$

$$\mathcal{L}[e^{-at}\cos\omega t] = \frac{s+a}{(s+a)^2 + \omega^2} \tag{C–32}$$

$$\mathcal{L}\left[e^{-at}\left\{\frac{k_2 - ak_1}{\omega}\sin\omega t + k_1\cos\omega t\right\}\right] = \frac{k_1 s + k_2}{(s+a)^2 + \omega^2} \tag{C–33}$$

$$\mathcal{L}[2ke^{-at}\cos(\omega t + \phi)] = \frac{ke^{j\phi}}{s+a-j\omega} + \frac{ke^{-j\phi}}{s+a+j\omega} \tag{C–34}$$

$$\mathcal{L}[\sinh\omega t] = \frac{\omega}{s^2 - \omega^2} \tag{C–35}$$

$$\mathcal{L}[\cosh\omega t] = \frac{s}{s^2 - \omega^2} \tag{C–36}$$

$$\mathcal{L}[e^{-at}\sinh\omega t] = \frac{\omega}{(s+a)^2 - \omega^2} \tag{C–37}$$

$$\mathcal{L}[e^{-at}\cosh\omega t] = \frac{s+a}{(s+a)^2 - \omega^2} \tag{C–38}$$

$$\mathcal{L}\left[\frac{e^{-at} - e^{-bt}}{b-a}\right] = \frac{1}{(s+a)(s+b)}, \quad a \neq b \tag{C–39}$$

$$\mathcal{L}\left[\frac{ae^{-at} - be^{-bt}}{a-b}\right] = \frac{s}{(s+a)(s+b)}, \quad a \neq b \tag{C–40}$$

More transforms can be found in any standard book of mathematical tables.

C.5 INVERSE LAPLACE TRANSFORM

The inverse Laplace transform, denoted by $\mathcal{L}^{-1}[F(s)]$, is the time function $f(t)$. That is,

$$f(t) = \mathcal{L}^{-1}[F(s)] \tag{C–41}$$

In general, $\mathcal{L}^{-1}$ is found using complex variable theory in terms of the inversion integral in the equation

$$f(t) = \frac{1}{2\pi j}\lim_{T\to\infty}\int_{a-jT}^{a+jT} e^{sT}F(s)\,ds \tag{C–42}$$

where a is chosen such that all the poles (singular points) of $F(s)$ lie to the left of the vertical line described by the equation $\mathcal{R}e(s) = a$ in the complex plane.

Often, the transform $F(s)$ can be written as the ratio of two polynomials $N(s)$ and $D(s)$. In such cases, the method of partial-fraction expansion can be conveniently employed to find $f(t)$.

C.6 RATIONAL FUNCTIONS AND PARTIAL-FRACTION EXPANSIONS

Let us consider the transform $F(s)$ expressed as a ratio of two polynomials:

$$F(s) = \frac{N(s)}{D(s)} = \frac{b_m s^m + b_{m-1} s^{m-1} + \ldots + b_1 s + b_0}{s^n + a_{n-1} s^{n-1} + \ldots + a_1 s + a_0} \qquad \text{(C–43)}$$

Such functions are called *rational functions*. If $m \le n$, we have a proper rational function, and if $m < n$, we have a *strictly proper rational function*. The *poles* of $F(s)$ are the values of s for which $D(s) = 0$, and the *zeros* of $F(s)$ are the values of s for which $N(s) = 0$. Assuming that there are no pole-zero cancellations (meaning that $N(s) \ne 0$ at any of the poles), let $s_1, s_2, \ldots, s_n$ be the roots of $D(s) = 0$. Then

$$D(s) = (s - s_1)(s - s_2)(s - s_3) \ldots (s - s_n) \qquad \text{(C–44)}$$

If $F(s)$ is strictly proper, we can employ the method of partial-fraction expansion to express $F(s)$ as the sum of simpler transforms. Then we can use a table of inverse Laplace transforms to find $f(t)$. Several cases of interest are discussed next.

Distinct poles.   If all the n roots of $D(s) = 0$ are distinct, we can express the strictly proper rational function $F(s)$ as

$$F(s) = \frac{A_1}{s - s_1} + \frac{A_2}{s - s_2} + \ldots + \frac{A_n}{s - s_n} \qquad \text{(C–45)}$$

where $A_1, A_2, \ldots, A_n$ are constants to be evaluated. It is easy to see that

$$A_i = \lim_{s \to s_i} [(s - s_i)F(s)] \text{ for } i = 1, 2, \ldots, n \qquad \text{(C–46)}$$

The corresponding inverse transform is

$$f(t) = A_1 e^{s_1 t} + A_2 e^{s_2 t} + \ldots + A_n e^{s_n t} \qquad \text{(C–47)}$$

Repeated poles.   If s_i is a pole that is repeated m times, then, corresponding to s_i, we include the following terms in the partial-fraction expansion:

$$\frac{A_{i1}}{s - s_i} + \frac{A_{i2}}{(s - s_i)^2} + \ldots + \frac{A_{im}}{(s - s_i)^m}$$

The constants A_{ip}, for $1 \leq p \leq m$, are found using

$$A_{ip} = \lim_{s \to s_i} \left[\frac{1}{(m-p)!} \frac{d^{m-p}}{ds^{m-p}} \{(s - s_i)^m F(s)\} \right] \qquad \text{(C–48)}$$

Complex poles.   If $D(s)$ has no imaginary terms, then complex poles must occur in conjugate pairs. As an example, let $s_1 = (-a + j\omega)$ and $s_2 = (-a - j\omega)$. Then, corresponding to these two poles, we will have, in the partial-fraction expansion,

$$\frac{A_1}{s + a - j\omega}$$

and

$$\frac{A_2}{s + a + j\omega}$$

Now we have to allow the coefficients A_1 and A_2 to be complex. They are found from

$$A_1 = \lim_{s \to (-a + j\omega)} [(s + a - j\omega)F(s)] \qquad \text{(C–49)}$$

and

$$A_2 = \lim_{s \to (-a - j\omega)} [(s + a + j\omega)F(s)] \qquad \text{(C–50)}$$

We can easily show that A_1 and A_2 are complex conjugates of each other. Therefore, if

$$A_1 = Ke^{j\phi} \qquad \text{(C–51)}$$

then

$$A_2 = Ke^{-j\phi} \qquad \text{(C–52)}$$

The corresponding inverse transform is $[2Ke^{-at} \cos(\omega t + \phi)]$ for $t > 0$.
    Alternatively, since

$$(s + a - j\omega)(s + a + j\omega) = (s + a)^2 + \omega^2 \qquad \text{(C–53)}$$

corresponding to the complex pole pair, we include the term $(B_1 S + B_2)/[(s + a)^2 + \omega^2]$ in the partial-fraction expansion. This term can be modified to

$$B_1 \frac{s + a}{(s + a)^2 + \omega^2} + \left(\frac{B_2 - aB_1}{\omega} \right) \left[\frac{\omega}{(s + a)^2 + \omega^2} \right]$$

which, when inverted, yields

$$B_1 e^{-at} \cos \omega t + B_3 e^{-at} \sin \omega t \qquad \text{(C–54)}$$

where

$$B_3 = \frac{B_2 - aB_1}{\omega} \qquad (C\text{--}55)$$

Using trigonometric identities, we can express Equation (C–54) in the form

$$2Ke^{-at} \cos(\omega t + \phi) \text{ for } t > 0$$

where

$$K \equiv \left(\frac{1}{2}\right) \sqrt{B_1^2 + B_3^2} \qquad (C\text{--}56)$$

and

$$\phi \equiv - \tan^{-1}\left(\frac{B_3}{B_1}\right) \qquad (C\text{--}57)$$

If $F(s)$ is *not* strictly proper, $m = n$, and we divide the numerator by the denominator and write

$$F(s) = A + F_1(s) \qquad (C\text{--}58)$$

where $F_1(s)$ *is* strictly proper. Then

$$f(t) = A\delta(t) + \mathcal{L}^{-1}[F_1(s)] \qquad (C\text{--}59)$$

and the inversion of $F_1(s)$ can be handled by the techniques described earlier.

C.7 SOLUTION OF AN INTEGRO-DIFFERENTIAL EQUATION USING LAPLACE TRANSFORMS

The first step in finding the solution to an integro-differential equation is to transform it via Laplace transform and use all the initial conditions as required. This yields an algebraic equation in the variable s. Next, we manipulate and solve the algebraic equation for the transform of the solution. Evaluation of the corresponding inverse transform gives us the solution as a function of time. Often, this procedure is simpler and less time consuming than the classical approach. Appendix D discusses differential equations and their solutions.

D

Differential Equations

Differential equations are used in almost all branches of science and engineering. Although there are many types of differential equations, we will focus only on the type that we need in the context of the material presented in this book.

D.1 DEFINITIONS

Let us consider a function $y(t)$ defined over $t_0 \leq t \leq t_1$, and let us assume that the first n derivatives of this function exist. Also, let $a_n(t)$, $a_{n-1}(t)$, $a_{n-2}(t)$, ... , $a_1(t)$, $a_0(t)$, and $g(t)$ be continuous functions defined over the same time interval. Then

$$a_n(t) \frac{d^n y(t)}{dt^n} + a_{n-1}(t) \frac{d^{n-1}y(t)}{dt^{n-1}} + \ldots + a_1(t) \frac{dy(t)}{dt} + a_0(t)y(t) = g(t)$$

(D–1)

is called a *nonhomogeneous linear differential equation of the nth order*. The equation is linear, since all the derivatives occur linearly (in their first power). The function $g(t)$ on the right-hand side is called the *driving* (or *forcing*) function. If it is zero, the equation becomes a homogeneous linear differential equation.

The functions $a_0(t)$ through $a_n(t)$ are the coefficients of the differential equation. If they are constant, then we have a linear differential equation with constant coefficients. Equation (D–1) has a unique solution, and it can be found if the n initial conditions

$$y(0), \left.\frac{dy(t)}{dt}\right|_{t=0}, \ldots, \left.\frac{d^{n-1}y(t)}{dt^{n-1}}\right|_{t=0}$$

are known.

D.2 LINEAR DIFFERENTIAL EQUATION WITH CONSTANT COEFFICIENTS

Without any loss of generality, we can say that a nonhomogeneous linear differential equation with constant coefficients is of the form

$$\frac{d^n y(t)}{dt^n} + a_{n-1}\frac{d^{n-1}y(t)}{dt^{n-1}} + \ldots + a_1 \frac{dy(t)}{dt} + a_0 y(t) = g(t) \qquad \text{(D–2)}$$

The collection of all solutions to this equation is the collection of all functions of the form

$$y(t) = \bar{y}(t) + z(t) \qquad \text{(D–3)}$$

where $z(t)$ is a solution to the corresponding homogeneous equation and $\bar{y}(t)$ is a known solution (called the *particular integral*) to the nonhomogeneous differential equation.

The corresponding homogeneous equation, obtained by setting $g(t) = 0$, has n linearly independent solutions, $z_1(t), z_2(t), \ldots, z_n(t)$. Any solution $z(t)$ of the homogeneous equation can be expressed as a linear combination of the $z_i(t)$; that is,

$$z(t) = c_1 z_1(t) + c_2 z_2(t) + \ldots + c_n z_n(t) \qquad \text{(D–4)}$$

for some constants $c_1, c_2, \ldots, c_n$.

For homogeneous linear differential equations with constant coefficients, we postulate a solution of the form

$$z(t) = e^{\alpha t} \qquad \text{(D–5)}$$

for some constant α. Substituting Equation (D–5) into Equation (D–2) with $g(t) = 0$ yields

$$\alpha^n + a_{n-1}\alpha^{n-1} + \ldots + a_1\alpha + a_0 = 0 \qquad \text{(D–6)}$$

The left-hand side of Equation (D–6) is called the *characteristic polynomial* of the differential equation. It has n roots, called *characteristic values*. Let these roots be α_i, for $i = 1, 2, \ldots, n$. Then

$$z(t) = c_1 e^{\alpha_1 t} + c_2 e^{\alpha_2 t} + \ldots + c_n e^{\alpha_n t} \qquad \text{(D–7)}$$

If the roots are distinct, n different solutions are obtained, corresponding to the n degrees of freedom inherent in the original equation.

Several techniques are available for finding $\bar{y}(t)$. The n integration constants, c_1 through c_n, are found by using the n known initial conditions. Of course, the process could get cumbersome. However, the use of Laplace transforms opens up a systematic and convenient avenue to finding the solution.

D.3 SOME SIMPLE EXAMPLES

D.3.1 First-Order Differential Equation

Consider the first-order differential equation

$$\frac{dy}{dt} + ky = e^{-at} \tag{D–8}$$

for $t > 0$ and $k \neq a$, with initial condition: $y(0) = 1$. The characteristic equation for this differential equation is $\alpha + k = 0$, or $\alpha = -k$, and $z(t) = z_1(t) = e^{-kt}$. Using D as the differential operator, the Equation (D–9) can be written as $(D + k)y = e^{-at}$.

The particular integral is

$$\bar{y}(t) = \frac{e^{-at}}{D + k} = \frac{e^{-at}}{-a + k}$$

The validity of $\bar{y}(t)$ can easily be verified. The general solution is

$$y(t) = c_1 e^{-kt} + \frac{e^{-at}}{k - a}$$

We find c_1 using the initial condition at $t = 0$:

$$y(0) = 1 = c_1 + \frac{1}{k - a}$$

or

$$c_1 = 1 - \frac{1}{k - a}$$

Finally, the solution of Equation (D–9) can be written as

$$y(t) = e^{-kt} + \frac{1}{k - a} [e^{-at} - e^{-kt}] \tag{D–9}$$

Next, we will solve the same equation by the application of Laplace transforms. Laplace transforming Equation (D–8), we get

$$sY(s) - 1 + kY(s) = \frac{1}{s + a}$$

or

$$Y(s) = \frac{s + a + 1}{(s + k)(s + a)}$$

Employing the partial-fractions technique, we express $Y(s)$ as

$$Y(s) = \frac{A_1}{s + k} + \frac{A_2}{s + a}$$

We can then find

$$A_1 = \frac{a + 1 - k}{a - k}$$

and

$$A_2 = \frac{1}{k - a}$$

Rearranging $Y(s)$, we have

$$Y(s) = \frac{1}{s + k} + \frac{1}{a - k}\frac{1}{s + k} + \frac{1}{k - a}\frac{1}{s + a}$$

After taking the inverse Laplace transform of $Y(s)$, we obtain

$$y(t) = e^{-kt} + \frac{1}{k - a}[e^{-at} - e^{-kt}]$$

which is the same solution we got using classical techniques.

D.3.2 Second-Order Differential Equation

For the second-order differential equation

$$\frac{d^2y}{dt^2} + y = 2t \tag{D–10}$$

with initial conditions $y(0) = 2$, $y'(0) = -1$, the characteristic equation is

$$\alpha^2 + 1 = 0$$

so

$$\alpha = +j \text{ or } -j$$

and

$$z(t) = k_1 z_1(t) + k_2 z_2(t)$$
$$= k_1 e^{jt} + k_2 e^{-jt}$$

The particular integral is

$$\bar{y}(t) = \frac{2t}{D^2 + 1}$$

By inspection, it is easy to see and verify that

$$\bar{y}(t) = 2t$$

so the general solution to Equation (D–10) is

$$y(t) = 2t + k_1 e^{jt} + k_2 e^{-jt}$$

Using the two known initial conditions, we have

$$k_1 + k_2 = 2$$

$$k_1 - k_2 = j3$$

Solving for k_1 and k_2, we get

$$k_1 = \frac{2 + j3}{2}$$

$$k_2 = \frac{2 - j3}{2}$$

As expected, k_1 and k_2 are complex conjugates of each other. The general solution now becomes

$$y(t) = 2t + \left[1 + j\left(\frac{3}{2}\right)\right] e^{jt} + \left[1 - j\left(\frac{3}{2}\right)\right] e^{-jt}$$

By using Euler's identities, we can express the solution as

$$y(t) = 2t + 2 \cos t - 3 \sin t$$

Next, let us solve Equation (D–10) using Laplace transforms. We obtain

$$s^2 Y(s) - 2s + 1 + Y(s) = \frac{2}{s^2}$$

Solving for $Y(s)$ gives

$$Y(s) = \frac{2s^3 - s^2 + 2}{s^2(s^2 + 1)}$$

Once again, using partial-fraction techniques, we see that

$$Y(s) = \frac{2s}{s^2 + 1} - \frac{3}{s^2 + 1} + \frac{2}{s^2}$$

Taking the inverse Laplace transforms of each of these terms, we get

$$y(t) = 2 \cos t - 3 \sin t + 2t$$

as before.

D.3.3 Simultaneous Differential Equations

Suppose we want to find $x(t)$ and $y(t)$ solving

$$\begin{bmatrix} \dot{x}(t) \\ \dot{y}(t) \end{bmatrix} = \begin{bmatrix} 1 & -2 \\ -1 & 1 \end{bmatrix} \begin{bmatrix} x(t) \\ y(t) \end{bmatrix}$$

subject to the initial conditions $x(0) = 1$ and $y(0) = 2$. (The dot refers to the derivative with respect to time.) We first write the two simultaneous equations individually as

$$\dot{x}(t) = x(t) - 2y(t)$$

$$\dot{y}(t) = -x(t) + y(t)$$

Taking the Laplace transform of these two equations and incorporating the initial conditions into the analysis, we have

$$sX(s) - 1 = X(s) - 2Y(s)$$

$$sY(s) - 2 = -X(s) + Y(s)$$

Solving these equations for $X(s)$ and $Y(s)$, we obtain

$$X(s) = \frac{s - 5}{(s - s_1)(s - s_2)} = \left(\frac{A_1}{s - s_1} \right) + \left(\frac{A_2}{s - s_2} \right)$$

$$Y(s) = \frac{2s - 3}{(s - s_1)(s - s_2)} = \left(\frac{A_3}{s - s_1} \right) + \left(\frac{A_4}{s - s_2} \right)$$

where

$$s_1 = (1 + \sqrt{2}), \, s_2 = (1 - \sqrt{2}), \, A_1 = \frac{1 - 2\sqrt{2}}{2},$$

$$A_2 = \frac{1 + 2\sqrt{2}}{2}, \, A_3 = \frac{4 - \sqrt{2}}{4}, \, A_4 = \frac{4 + \sqrt{2}}{4}$$

After taking the inverse Laplace transforms, we obtain the solution as

$$x(t) = \left(\frac{1 - 2\sqrt{2}}{2} \right) e^{(1 + \sqrt{2})t} + \left(\frac{1 + 2\sqrt{2}}{2} \right) e^{(1 - \sqrt{2})t}$$

$$y(t) = \left(\frac{4 - \sqrt{2}}{4} \right) e^{(1 + \sqrt{2})t} + \left(\frac{4 + \sqrt{2}}{4} \right) e^{(1 - \sqrt{2})t}$$

_____

Some Useful Derivatives
and Integrals

E.1 DERIVATIVES

$$\frac{d}{dx} a^{u(x)} = a^u \ln a \frac{du}{dx} \tag{E-1}$$

$$\frac{d}{dx} [u(x)]^{v(x)} = vu^{v-1} \frac{du}{dx} + u^v \ln u \frac{dv}{dx} \tag{E-2}$$

$$\frac{d}{dx} \int_{u(x)}^{v(x)} f(\xi, x) \, d\xi = \int_{u(x)}^{v(x)} \frac{d}{dx} f(\xi, x) \, d\xi + f(v,x) \frac{dv}{dx} - f(u,x) \frac{du}{dx} \tag{E-3}$$

E.2 INDEFINITE INTEGRALS

$$\int u \, dv = uv - \int v \, du \quad \text{(integration by parts)} \tag{E-4}$$

$$\int \ln x \, dx = x \ln x - x \tag{E-5}$$

$$\int a^u \, du = \frac{a^u}{\ln a}, \quad a > 0 \text{ and } a \neq 1 \tag{E-6}$$

$$\int x^n e^{ax} dx = \frac{e^{ax}}{a}\left[x^n - \frac{nx^{n-1}}{a} + \frac{n(n-1)x^{n-2}}{a^2} - \cdots + \frac{(-1)^n n!}{a^n}\right] \quad (E-7)$$

for positive integer values of n

$$\int xe^{-x^2} dx = -\frac{1}{2}e^{-x^2} \quad (E-8)$$

$$\int x^n e^{-x^2} dx = -\frac{1}{2}x^{n-1}e^{-x^2} + \left(\frac{n-1}{2}\right)\int x^{n-2}e^{-x^2} dx \quad (E-9)$$

E.3 DEFINITE INTEGRALS

$$\int_0^\infty e^{-\alpha x} dx = \frac{1}{\alpha} \text{ for } \alpha > 0 \quad (E-10)$$

$$\int_0^\infty x^n e^{-\alpha x} dx = \frac{\Gamma(n+1)}{\alpha^{n+1}}$$

$$= \frac{n!}{\alpha^{n+1}} \text{ for } \alpha > 0 \text{ and } n = \text{integer} > 0 \quad (E-11)$$

$$\int_{-\infty}^\infty e^{-\alpha x^2} dx = 2\int_0^\infty e^{-\alpha x^2} dx$$

$$= \sqrt{\frac{\pi}{\alpha}} \text{ for } \alpha > 0 \quad (E-12)$$

$$\int_0^\infty x^n e^{-\alpha x^2} dx = \frac{\Gamma\left(\frac{n+1}{2}\right)}{2\alpha^{(n+1)/2}} \quad (E-13)$$

$$\int_0^\infty e^{-x}\ln x \, dx = -\gamma \quad (E-14)$$

where

$$\gamma = \lim_{n\to\infty}\left[\left\{1 + \frac{1}{2} + \frac{1}{3} + \cdots + \frac{1}{n}\right\} - \ln n\right]$$

$$= 0.577215665 \quad (E-15)$$

(Euler-Mascheroni constant)

$$\int_0^\infty e^{-x^2}\ln x \, dx = -\frac{\sqrt{\pi}}{4}(\gamma + 2\ln 2) \quad (E-16)$$

The Beta Function

The *beta function* is defined as

$$B(\gamma, \beta) = \int_0^1 \xi^{\gamma - 1}(1 - \xi)^{\beta - 1} \, d\xi \qquad \text{(F–1)}$$

for $\gamma > 0$ and $\beta > 0$. The beta function can be expressed in terms of the gamma function (see Appendix G) as

$$B(\gamma, \beta) = \frac{\Gamma(\gamma)\Gamma(\beta)}{\Gamma(\gamma + \beta)} \qquad \text{(F–2)}$$

for $\gamma > 0$ and $\beta > 0$. An alternative form of Equation (F–2) is

$$B(\gamma, \beta) = \frac{\Gamma(\gamma + 1)\Gamma(\beta + 1)}{\Gamma(\gamma + \beta + 2)} \qquad \text{(F–3)}$$

for $\gamma > -1$ and $\beta > -1$.

It is easily seen that

$$B(\gamma, \beta) = B(\beta, \gamma) \qquad \text{(F–4)}$$

Extensions to negative values of γ and β are accomplished using

$$\Gamma(n) = \frac{\Gamma(n + 1)}{n} \qquad \text{(F–5)}$$

for $n < 0$.

For $\gamma > 0$ and $\beta > 0$, we can express the beta function as

$$B(\gamma, \beta) = \int_0^\infty \frac{\xi^{\gamma-1}}{(1 + \xi)^{\gamma+\beta}} \, d\xi \tag{F–6}$$

or

$$B(\gamma, \beta) = \int_0^{\pi/2} 2(\sin \xi)^{2\gamma-1}(\cos \xi)^{2\beta-1} \, d\xi \tag{F–7}$$

The Gamma Function

The complete *gamma function*, also known as the *factorial function*, is defined as

$$\Gamma(x) = \int_0^\infty t^{x-1} e^{-t} \, dt, \quad x > 0 \tag{G-1}$$

For integer values of x,

$$\Gamma(x + 1) = x! = x\Gamma(x) \tag{G-2}$$

$$\Gamma(x) = (x - 1)! \tag{G-3}$$

$$\Gamma(x) = (x - 1)\Gamma(x - 1) \tag{G-4}$$

For negative values of x, we use

$$\Gamma(x) = \frac{\Gamma(x + 1)}{x} \tag{G-5}$$

repeatedly, if necessary, to find the values of the gamma function.
For very large positive values of x,

$$\Gamma(x + 1) = x! = x^x e^{-x} \sqrt{2\pi x} \left[1 + \frac{1}{12x} + \frac{1}{288x^2} - \cdots \right] \tag{G-6}$$

or

$$\Gamma(x + 1) = x! \cong x^x e^{-x} \sqrt{2\pi x} \tag{G-7}$$

This approximation is known as *Sterling's formula for large factorials.*
Some special values of the gamma function are the following:

$$\Gamma(1) = 1.0 \tag{G-8}$$

$$\Gamma(2) = 1.0 \tag{G-9}$$

$$\Gamma(1/2) = \int_0^\infty \frac{e^{-t}}{\sqrt{t}} \, dt = \sqrt{\pi} \tag{G-10}$$

Numerical values of the gamma function from 1 to 2 by hundredths are tabulated in the following table.

GAMMA FUNCTION

$$\Gamma(x) = \int_0^\infty t^{x-1}e^{-t}dt \qquad \text{for } 1 \leqq x \leqq 2$$

[For other values use the formula $\Gamma(x + 1) = x\Gamma(x)$]

x	$\Gamma(x)$	x	$\Gamma(x)$
1.00	1.00000	1.50	.88623
1.01	.99433	1.51	.88659
1.02	.98884	1.52	.88704
1.03	.98355	1.53	.88757
1.04	.97844	1.54	.88818
1.05	.97350	1.55	.88887
1.06	.96874	1.56	.88964
1.07	.96415	1.57	.89049
1.08	.95973	1.58	.89142
1.09	.95546	1.59	.89243
1.10	.95135	1.60	.89352
1.11	.94740	1.61	.89468
1.12	.94359	1.62	.89592
1.13	.93993	1.63	.89724
1.14	.93642	1.64	.89864
1.15	.93304	1.65	.90012
1.16	.92980	1.66	.90167
1.17	.92670	1.67	.90330
1.18	.92373	1.68	.90500
1.19	.92089	1.69	.90678
1.20	.91817	1.70	.90864
1.21	.91558	1.71	.91057
1.22	.91311	1.72	.91258
1.23	.91075	1.73	.91467
1.24	.90852	1.74	.91683
1.25	.90640	1.75	.91906
1.26	.90440	1.76	.92137
1.27	.90250	1.77	.92376
1.28	.90072	1.78	.92623
1.29	.89904	1.79	.92877
1.30	.89747	1.80	.93138
1.31	.89600	1.81	.93408
1.32	.89464	1.82	.93685
1.33	.89338	1.83	.93969
1.34	.89222	1.84	.94261
1.35	.89115	1.85	.94561
1.36	.89018	1.86	.94869
1.37	.88931	1.87	.95184
1.38	.88854	1.88	.95507
1.39	.88785	1.89	.95838
1.40	.88726	1.90	.96177
1.41	.88676	1.91	.96523
1.42	.88636	1.92	.96877
1.43	.88604	1.93	.97240
1.44	.88581	1.94	.97610
1.45	.88566	1.95	.97988
1.46	.88560	1.96	.98374
1.47	.88563	1.97	.98768
1.48	.88575	1.98	.99171
1.49	.88595	1.99	.99581
1.50	.88623	2.00	1.00000

Reprinted with permission from Schaum's Outline Series, McGraw-Hill Publishing Company, Spiegel, *Mathematical Handbook*, © 1968.

Tables of Values for *n* Factorial

n	$n!$
0	1 (by definition)
1	1
2	2
3	6
4	24
5	120
6	720
7	5040
8	40,320
9	362,880
10	3,628,800
11	39,916,800
12	479,001,600
13	6,227,020,800
14	87,178,291,200
15	1,307,674,368,000
16	20,922,789,888,000
17	355,687,428,096,000
18	6,402,373,705,728,000
19	121,645,100,408,832,000
20	2,432,902,008,176,640,000
21	51,090,942,171,709,440,000
22	1,124,000,727,777,607,680,000
23	25,852,016,738,884,976,640,000
24	620,448,401,733,239,439,360,000
25	15,511,210,043,330,985,984,000,000
26	403,291,461,126,605,635,584,000,000
27	10,888,869,450,418,352,160,768,000,000
28	304,888,344,611,713,860,501,504,000,000
29	8,841,761,993,739,701,954,543,616,000,000
30	265,252,859,812,191,058,636,308,480,000,000
31	8.22284×10^{33}
32	2.63131×10^{35}
33	8.68332×10^{36}
34	2.95233×10^{38}
35	1.03331×10^{40}
36	3.71993×10^{41}
37	1.37638×10^{43}
38	5.23023×10^{44}
39	2.03979×10^{46}

n	$n!$
40	8.15915×10^{47}
41	3.34525×10^{49}
42	1.40501×10^{51}
43	6.04153×10^{52}
44	2.65827×10^{54}
45	1.19622×10^{56}
46	5.50262×10^{57}
47	2.58623×10^{59}
48	1.24139×10^{61}
49	6.08282×10^{62}
50	3.04141×10^{64}
51	1.55112×10^{66}
52	8.06582×10^{67}
53	4.27488×10^{69}
54	2.30844×10^{71}
55	1.26964×10^{73}
56	7.10999×10^{74}
57	4.05269×10^{76}
58	2.35056×10^{78}
59	1.38683×10^{80}
60	8.32099×10^{81}
61	5.07580×10^{83}
62	3.14700×10^{85}
63	1.98261×10^{87}
64	1.26887×10^{89}
65	8.24765×10^{90}
66	5.44345×10^{92}
67	3.64711×10^{94}
68	2.48004×10^{96}
69	1.71122×10^{98}
70	1.19786×10^{100}
71	8.50479×10^{101}
72	6.12345×10^{103}
73	4.47012×10^{105}
74	3.30789×10^{107}
75	2.48091×10^{109}
76	1.88549×10^{111}
77	1.45183×10^{113}
78	1.13243×10^{115}
79	8.94618×10^{116}

n	$n!$
80	7.15695×10^{118}
81	5.79713×10^{120}
82	4.75364×10^{122}
83	3.94552×10^{124}
84	3.31424×10^{126}
85	2.81710×10^{128}
86	2.42271×10^{130}
87	2.10776×10^{132}
88	1.85483×10^{134}
89	1.65080×10^{136}
90	1.48572×10^{138}
91	1.35200×10^{140}
92	1.24384×10^{142}
93	1.15677×10^{144}
94	1.08737×10^{146}
95	1.03300×10^{148}
96	9.91678×10^{149}
97	9.61928×10^{151}
98	9.42689×10^{153}
99	9.33262×10^{155}
100	9.33262×10^{157}

Reprinted with permission from Schaum's Outline Series, McGraw-Hill Publishing Company, Spiegel, *Mathematical Handbook*, © 1968

Binomial Coefficients

BINOMIAL COEFFICIENTS

$$\binom{n}{k} = \frac{n!}{k!(n-k)!} = \frac{n(n-1)\cdots(n-k+1)}{k!} = \binom{n}{n-k}, \; 0! = 1$$

k \ n	0	1	2	3	4	5	6	7	8	9
1	1	1								
2	1	2	1							
3	1	3	3	1						
4	1	4	6	4	1					
5	1	5	10	10	5	1				
6	1	6	15	20	15	6	1			
7	1	7	21	35	35	21	7	1		
8	1	8	28	56	70	56	28	8	1	
9	1	9	36	84	126	126	84	36	9	1
10	1	10	45	120	210	252	210	120	45	10
11	1	11	55	165	330	462	462	330	165	55
12	1	12	66	220	495	792	924	792	495	220
13	1	13	78	286	715	1287	1716	1716	1287	715
14	1	14	91	364	1001	2002	3003	3432	3003	2002
15	1	15	105	455	1365	3003	5005	6435	6435	5005
16	1	16	120	560	1820	4368	8008	11440	12870	11440
17	1	17	136	680	2380	6188	12376	19448	24310	24310
18	1	18	153	816	3060	8568	18564	31824	43758	48620
19	1	19	171	969	3876	11628	27132	50388	75582	92378
20	1	20	190	1140	4845	15504	38760	77520	125970	167960
21	1	21	210	1330	5985	20349	54264	116280	203490	293930
22	1	22	231	1540	7315	26334	74613	170544	319770	497420
23	1	23	253	1771	8855	33649	100947	245157	490314	817190
24	1	24	276	2024	10626	42504	134596	346104	735471	1307504
25	1	25	300	2300	12650	53130	177100	480700	1081575	2042975
26	1	26	325	2600	14950	65780	230230	657800	1562275	3124550
27	1	27	351	2925	17550	80730	296010	888030	2220075	4686825
28	1	28	378	3276	20475	98280	376740	1184040	3108105	6906900
29	1	29	406	3654	23751	118755	475020	1560780	4292145	10015005
30	1	30	435	4060	27405	142506	593775	2035800	5852925	14307150

Note that each number is the sum of two numbers in the row above; one of these numbers is in the same column and the other is in the preceding column [e.g., $56 = 35 + 21$]. The arrangement is often called *Pascal's triangle*.

For $k > 15$, use the fact that $\begin{pmatrix} n \\ k \end{pmatrix} = \begin{pmatrix} n \\ n - k \end{pmatrix}$.

n \ k	10	11	12	13	14	15
10	1					
11	11	1				
12	66	12	1			
13	286	78	13	1		
14	1001	364	91	14	1	
15	3003	1365	455	105	15	1
16	8008	4368	1820	560	120	16
17	19448	12376	6188	2380	680	136
18	43758	31824	18564	8568	3060	816
19	92378	75582	50388	27132	11628	3876
20	184756	167960	125970	77520	38760	15504
21	352716	352716	293930	203490	116280	54264
22	646646	705432	646646	497420	319770	170544
23	1144066	1352078	1352078	1144066	817190	490314
24	1961256	2496144	2704156	2496144	1961256	1307504
25	3268760	4457400	5200300	5200300	4457400	3268760
26	5311735	7726160	9657700	10400600	9657700	7726160
27	8436285	13037895	17383860	20058300	20058300	17383860
28	13123110	21474180	30421755	37442160	40116600	37442160
29	20030010	34597290	51895935	67863915	77558760	77558760
30	30045015	54627300	86493225	119759850	145422675	155117520

Tables of Values for the Exponential Functions e^x and e^{-x}

TABLE OF VALUES FOR THE EXPONENTIAL FUNCTION e^x

x	0	1	2	3	4	5	6	7	8	9
.0	1.0000	1.0101	1.0202	1.0305	1.0408	1.0513	1.0618	1.0725	1.0833	1.0942
.1	1.1052	1.1163	1.1275	1.1388	1.1503	1.1618	1.1735	1.1853	1.1972	1.2092
.2	1.2214	1.2337	1.2461	1.2586	1.2712	1.2840	1.2969	1.3100	1.3231	1.3364
.3	1.3499	1.3634	1.3771	1.3910	1.4049	1.4191	1.4333	1.4477	1.4623	1.4770
.4	1.4918	1.5068	1.5220	1.5373	1.5527	1.5683	1.5841	1.6000	1.6161	1.6323
.5	1.6487	1.6653	1.6820	1.6989	1.7160	1.7333	1.7507	1.7683	1.7860	1.8040
.6	1.8221	1.8404	1.8589	1.8776	1.8965	1.9155	1.9348	1.9542	1.9739	1.9937
.7	2.0138	2.0340	2.0544	2.0751	2.0959	2.1170	2.1383	2.1598	2.1815	2.2034
.8	2.2255	2.2479	2.2705	2.2933	2.3164	2.3396	2.3632	2.3869	2.4109	2.4351
.9	2.4596	2.4843	2.5093	2.5345	2.5600	2.5857	2.6117	2.6379	2.6645	2.6912
1.0	2.7183	2.7456	2.7732	2.8011	2.8292	2.8577	2.8864	2.9154	2.9447	2.9743
1.1	3.0042	3.0344	3.0649	3.0957	3.1268	3.1582	3.1899	3.2220	3.2544	3.2871
1.2	3.3201	3.3535	3.3872	3.4212	3.4556	3.4903	3.5254	3.5609	3.5966	3.6328
1.3	3.6693	3.7062	3.7434	3.7810	3.8190	3.8574	3.8962	3.9354	3.9749	4.0149
1.4	4.0552	4.0960	4.1371	4.1787	4.2207	4.2631	4.3060	4.3492	4.3929	4.4371
1.5	4.4817	4.5267	4.5722	4.6182	4.6646	4.7115	4.7588	4.8066	4.8550	4.9037
1.6	4.9530	5.0028	5.0531	5.1039	5.1552	5.2070	5.2593	5.3122	5.3656	5.4195
1.7	5.4739	5.5290	5.5845	5.6407	5.6973	5.7546	5.8124	5.8709	5.9299	5.9895
1.8	6.0496	6.1104	6.1719	6.2339	6.2965	6.3598	6.4237	6.4883	6.5535	6.6194
1.9	6.6859	6.7531	6.8210	6.8895	6.9588	7.0287	7.0993	7.1707	7.2427	7.3155
2.0	7.3891	7.4633	7.5383	7.6141	7.6906	7.7679	7.8460	7.9248	8.0045	8.0849
2.1	8.1662	8.2482	8.3311	8.4149	8.4994	8.5849	8.6711	8.7583	8.8463	8.9352
2.2	9.0250	9.1157	9.2073	9.2999	9.3933	9.4877	9.5831	9.6794	9.7767	9.8749
2.3	9.9742	10.074	10.176	10.278	10.381	10.486	10.591	10.697	10.805	10.913
2.4	11.023	11.134	11.246	11.359	11.473	11.588	11.705	11.822	11.941	12.061
2.5	12.182	12.305	12.429	12.554	12.680	12.807	12.936	13.066	13.197	13.330
2.6	13.464	13.599	13.736	13.874	14.013	14.154	14.296	14.440	14.585	14.732
2.7	14.880	15.029	15.180	15.333	15.487	15.643	15.800	15.959	16.119	16.281
2.8	16.445	16.610	16.777	16.945	17.116	17.288	17.462	17.637	17.814	17.993
2.9	18.174	18.357	18.541	18.728	18.916	19.106	19.298	19.492	19.688	19.886
3.0	20.086	20.287	20.491	20.697	20.905	21.115	21.328	21.542	21.758	21.977
3.1	22.198	22.421	22.646	22.874	23.104	23.336	23.571	23.807	24.047	24.288
3.2	24.533	24.779	25.028	25.280	25.534	25.790	26.050	26.311	26.576	26.843
3.3	27.113	27.385	27.660	27.938	28.219	28.503	28.789	29.079	29.371	29.666
3.4	29.964	30.265	30.569	30.877	31.187	31.500	31.817	32.137	32.460	32.786
3.5	33.115	33.448	33.784	34.124	34.467	34.813	35.163	35.517	35.874	36.234
3.6	36.598	36.966	37.338	37.713	38.092	38.475	38.861	39.252	39.646	40.045
3.7	40.447	40.854	41.264	41.679	42.098	42.521	42.948	43.380	43.816	44.256
3.8	44.701	45.150	45.604	46.063	46.525	46.993	47.465	47.942	48.424	48.911
3.9	49.402	49.899	50.400	50.907	51.419	51.935	52.457	52.985	53.517	54.055
4.	54.598	60.340	66.686	73.700	81.451	90.017	99.484	109.95	121.51	134.29
5.	148.41	164.02	181.27	200.34	221.41	244.69	270.43	298.87	330.30	365.04
6.	403.43	445.86	492.75	544.57	601.85	665.14	735.10	812.41	897.85	992.27
7.	1096.6	1212.0	1339.4	1480.3	1636.0	1808.0	1998.2	2208.3	2440.6	2697.3
8.	2981.0	3294.5	3641.0	4023.9	4447.1	4914.8	5431.7	6002.9	6634.2	7332.0
9.	8103.1	8955.3	9897.1	10938	12088	13360	14765	16318	18034	19930
10.	22026									

Reprinted with permission from Schaum's Outline Series, McGraw-Hill Publishing Company, Spiegel, *Mathematical Handbook*, © 1968

TABLE OF VALUES FOR THE EXPONENTIAL FUNCTION e^{-x}

x	0	1	2	3	4	5	6	7	8	9
.0	1.00000	.99005	.98020	.97045	.96079	.95123	.94176	.93239	.92312	.91393
.1	.90484	.89583	.88692	.87810	.86936	.86071	.85214	.84366	.83527	.82696
.2	.81873	.81058	.80252	.79453	.78663	.77880	.77105	.76338	.75578	.74826
.3	.74082	.73345	.72615	.71892	.71177	.70469	.69768	.69073	.68386	.67706
.4	.67032	.66365	.65705	.65051	.64404	.63763	.63128	.62500	.61878	.61263
.5	.60653	.60050	.59452	.58860	.58275	.57695	.57121	.56553	.55990	.55433
.6	.54881	.54335	.53794	.53259	.52729	.52205	.51685	.51171	.50662	.50158
.7	.49659	.49164	.48675	.48191	.47711	.47237	.46767	.46301	.45841	.45384
.8	.44933	.44486	.44043	.43605	.43171	.42741	.42316	.41895	.41478	.41066
.9	.40657	.40252	.39852	.39455	.39063	.38674	.38289	.37908	.37531	.37158
1.0	.36788	.36422	.36060	.35701	.35345	.34994	.34646	.34301	.33960	.33622
1.1	.33287	.32956	.32628	.32303	.31982	.31664	.31349	.31037	.30728	.30422
1.2	.30119	.29820	.29523	.29229	.28938	.28650	.28365	.28083	.27804	.27527
1.3	.27253	.26982	.26714	.26448	.26185	.25924	.25666	.25411	.25158	.24908
1.4	.24660	.24414	.24171	.23931	.23693	.23457	.23224	.22993	.22764	.22537
1.5	.22313	.22091	.21871	.21654	.21438	.21225	.21014	.20805	.20598	.20393
1.6	.20190	.19989	.19790	.19593	.19398	.19205	.19014	.18825	.18637	.18452
1.7	.18268	.18087	.17907	.17728	.17552	.17377	.17204	.17033	.16864	.16696
1.8	.16530	.16365	.16203	.16041	.15882	.15724	.15567	.15412	.15259	.15107
1.9	.14957	.14808	.14661	.14515	.14370	.14227	.14086	.13946	.13807	.13670
2.0	.13534	.13399	.13266	.13134	.13003	.12873	.12745	.12619	.12493	.12369
2.1	.12246	.12124	.12003	.11884	.11765	.11648	.11533	.11418	.11304	.11192
2.2	.11080	.10970	.10861	.10753	.10646	.10540	.10435	.10331	.10228	.10127
2.3	.10026	.09926	.09827	.09730	.09633	.09537	.09442	.09348	.09255	.09163
2.4	.09072	.08982	.08892	.08804	.08716	.08629	.08543	.08458	.08374	.08291
2.5	.08208	.08127	.08046	.07966	.07887	.07808	.07730	.07654	.07577	.07502
2.6	.07427	.07353	.07280	.07208	.07136	.07065	.06995	.06925	.06856	.06788
2.7	.06721	.06654	.06587	.06522	.06457	.06393	.06329	.06266	.06204	.06142
2.8	.06081	.06020	.05961	.05901	.05843	.05784	.05727	.05670	.05613	.05558
2.9	.05502	.05448	.05393	.05340	.05287	.05234	.05182	.05130	.05079	.05029
3.0	.04979	.04929	.04880	.04832	.04783	.04736	.04689	.04642	.04596	.04550
3.1	.04505	.04460	.04416	.04372	.04328	.04285	.04243	.04200	.04159	.04117
3.2	.04076	.04036	.03996	.03956	.03916	.03877	.03839	.03801	.03763	.03725
3.3	.03688	.03652	.03615	.03579	.03544	.03508	.03474	.03439	.03405	.03371
3.4	.03337	.03304	.03271	.03239	.03206	.03175	.03143	.03112	.03081	.03050
3.5	.03020	.02990	.02960	.02930	.02901	.02872	.02844	.02816	.02788	.02760
3.6	.02732	.02705	.02678	.02652	.02625	.02599	.02573	.02548	.02522	.02497
3.7	.02472	.02448	.02423	.02399	.02375	.02352	.02328	.02305	.02282	.02260
3.8	.02237	.02215	.02193	.02171	.02149	.02128	.02107	.02086	.02065	.02045
3.9	.02024	.02004	.01984	.01964	.01945	.01925	.01906	.01887	.01869	.01850
4.	.018316	.016573	.014996	.013569	.012277	.011109	.010052	$.0^2$90953	$.0^2$82297	$.0^2$74466
5.	$.0^2$67379	$.0^2$60967	$.0^2$55166	$.0^2$49916	$.0^2$45166	$.0^2$40868	$.0^2$36979	$.0^2$33460	$.0^2$30276	$.0^2$27394
6.	$.0^2$24788	$.0^2$22429	$.0^2$20294	$.0^2$18363	$.0^2$16616	$.0^2$15034	$.0^2$13604	$.0^2$12309	$.0^2$11138	$.0^2$10078
7.	$.0^3$91188	$.0^3$82510	$.0^3$74659	$.0^3$67554	$.0^3$61125	$.0^3$55308	$.0^3$50045	$.0^3$45283	$.0^3$40973	$.0^3$37074
8.	$.0^3$33546	$.0^3$30354	$.0^3$27465	$.0^3$24852	$.0^3$22487	$.0^3$20347	$.0^3$18411	$.0^3$16659	$.0^3$15073	$.0^3$13639
9.	$.0^3$12341	$.0^3$11167	$.0^3$10104	$.0^4$91424	$.0^4$82724	$.0^4$74852	$.0^4$67729	$.0^4$61283	$.0^4$55452	$.0^4$50175
10.	$.0^4$45400									

Reprinted with permission from Schaum's Outline Series, McGraw-Hill Publishing Company, Spiegel, *Mathematical Handbook*, © 1968

K

Standard Normal
Curve Areas

This table gives areas under the standard normal distribution ϕ between 0 and $t \geq 0$ in steps of 0.01.

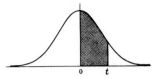

t	0	1	2	3	4	5	6	7	8	9
0.0	.0000	.0040	.0080	.0120	.0160	.0199	.0239	.0279	.0319	.0359
0.1	.0398	.0438	.0478	.0517	.0557	.0596	.0636	.0675	.0714	.0754
0.2	.0793	.0832	.0871	.0910	.0948	.0987	.1026	.1064	.1103	.1141
0.3	.1179	.1217	.1255	.1293	.1331	.1368	.1406	.1443	.1480	.1517
0.4	.1554	.1591	.1628	.1664	.1700	.1736	.1772	.1808	.1844	.1879
0.5	.1915	.1950	.1985	.2019	.2054	.2088	.2123	.2157	.2190	.2224
0.6	.2258	.2291	.2324	.2357	.2389	.2422	.2454	.2486	.2518	.2549
0.7	.2580	.2612	.2642	.2673	.2704	.2734	.2764	.2794	.2823	.2852
0.8	.2881	.2910	.2939	.2967	.2996	.3023	.3051	.3078	.3106	.3133
0.9	.3159	.3186	.3212	.3238	.3264	.3289	.3315	.3340	.3365	.3389
1.0	.3413	.3438	.3461	.3485	.3508	.3531	.3554	.3577	.3599	.3621
1.1	.3643	.3665	.3686	.3708	.3729	.3749	.3770	.3790	.3810	.3830
1.2	.3849	.3869	.3888	.3907	.3925	.3944	.3962	.3980	.3997	.4015
1.3	.4032	.4049	.4066	.4082	.4099	.4115	.4131	.4147	.4162	.4177
1.4	.4192	.4207	.4222	.4236	.4251	.4265	.4279	.4292	.4306	.4319
1.5	.4332	.4345	.4357	.4370	.4382	.4394	.4406	.4418	.4429	.4441
1.6	.4452	.4463	.4474	.4484	.4495	.4505	.4515	.4525	.4535	.4545
1.7	.4554	.4564	.4573	.4582	.4591	.4599	.4608	.4616	.4625	.4633
1.8	.4641	.4649	.4656	.4664	.4671	.4678	.4686	.4693	.4699	.4706
1.9	.4713	.4719	.4726	.4732	.4738	.4744	.4750	.4756	.4761	.4767
2.0	.4772	.4778	.4783	.4788	.4793	.4798	.4803	.4808	.4812	.4817
2.1	.4821	.4826	.4830	.4834	.4838	.4842	.4846	.4850	.4854	.4857
2.2	.4861	.4864	.4868	.4871	.4875	.4878	.4881	.4884	.4887	.4890
2.3	.4893	.4896	.4898	.4901	.4904	.4906	.4909	.4911	.4913	.4916
2.4	.4918	.4920	.4922	.4925	.4927	.4929	.4931	.4932	.4934	.4936
2.5	.4938	.4940	.4941	.4943	.4945	.4946	.4948	.4949	.4951	.4952
2.6	.4953	.4955	.4956	.4957	.4959	.4960	.4961	.4962	.4963	.4964
2.7	.4965	.4966	.4967	.4968	.4969	.4970	.4971	.4972	.4973	.4974
2.8	.4974	.4975	.4976	.4977	.4977	.4978	.4979	.4979	.4980	.4981
2.9	.4981	.4982	.4982	.4983	.4984	.4984	.4985	.4985	.4986	.4986
3.0	.4987	.4987	.4987	.4988	.4988	.4989	.4989	.4989	.4990	.4990
3.1	.4990	.4991	.4991	.4991	.4992	.4992	.4992	.4992	.4993	.4993
3.2	.4993	.4993	.4994	.4994	.4994	.4994	.4994	.4995	.4995	.4995
3.3	.4995	.4995	.4995	.4996	.4996	.4996	.4996	.4996	.4996	.4997
3.4	.4997	.4997	.4997	.4997	.4997	.4997	.4997	.4997	.4997	.4998
3.5	.4998	.4998	.4998	.4998	.4998	.4998	.4998	.4998	.4998	.4998
3.6	.4998	.4998	.4999	.4999	.4999	.4999	.4999	.4999	.4999	.4999
3.7	.4999	.4999	.4999	.4999	.4999	.4999	.4999	.4999	.4999	.4999
3.8	.4999	.4999	.4999	.4999	.4999	.4999	.4999	.4999	.4999	.4999
3.9	.5000	.5000	.5000	.5000	.5000	.5000	.5000	.5000	.5000	.5000

Reprinted with permission from Schaum's Outline Series, McGraw-Hill Publishing Company, Lipschutz, *Probability*, © 1968

Standard Normal Curve Ordinates

This table gives values $\phi(t)$ of the standard normal distribution ϕ at $t \geqq 0$ in steps of 0.01.

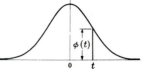

t	0	1	2	3	4	5	6	7	8	9
0.0	.3989	.3989	.3989	.3988	.3986	.3984	.3982	.3980	.3977	.3973
0.1	.3970	.3965	.3961	.3956	.3951	.3945	.3939	.3932	.3925	.3918
0.2	.3910	.3902	.3894	.3885	.3876	.3867	.3857	.3847	.3836	.3825
0.3	.3814	.3802	.3790	.3778	.3765	.3752	.3739	.3725	.3712	.3697
0.4	.3683	.3668	.3653	.3637	.3621	.3605	.3589	.3572	.3555	.3538
0.5	.3521	.3503	.3485	.3467	.3448	.3429	.3410	.3391	.3372	.3352
0.6	.3332	.3312	.3292	.3271	.3251	.3230	.3209	.3187	.3166	.3144
0.7	.3123	.3101	.3079	.3056	.3034	.3011	.2989	.2966	.2943	.2920
0.8	.2897	.2874	.2850	.2827	.2803	.2780	.2756	.2732	.2709	.2685
0.9	.2661	.2637	.2613	.2589	.2565	.2541	.2516	.2492	.2468	.2444
1.0	.2420	.2396	.2371	.2347	.2323	.2299	.2275	.2251	.2227	.2203
1.1	.2179	.2155	.2131	.2107	.2083	.2059	.2036	.2012	.1989	.1965
1.2	.1942	.1919	.1895	.1872	.1849	.1826	.1804	.1781	.1758	.1736
1.3	.1714	.1691	.1669	.1647	.1626	.1604	.1582	.1561	.1539	.1518
1.4	.1497	.1476	.1456	.1435	.1415	.1394	.1374	.1354	.1334	.1315
1.5	.1295	.1276	.1257	.1238	.1219	.1200	.1182	.1163	.1145	.1127
1.6	.1109	.1092	.1074	.1057	.1040	.1023	.1006	.0989	.0973	.0957
1.7	.0940	.0925	.0909	.0893	.0878	.0863	.0848	.0833	.0818	.0804
1.8	.0790	.0775	.0761	.0748	.0734	.0721	.0707	.0694	.0681	.0669
1.9	.0656	.0644	.0632	.0620	.0608	.0596	.0584	.0573	.0562	.0551
2.0	.0540	.0529	.0519	.0508	.0498	.0488	.0478	.0468	.0459	.0449
2.1	.0440	.0431	.0422	.0413	.0404	.0396	.0387	.0379	.0371	.0363
2.2	.0355	.0347	.0339	.0332	.0325	.0317	.0310	.0303	.0297	.0290
2.3	.0283	.0277	.0270	.0264	.0258	.0252	.0246	.0241	.0235	.0229
2.4	.0224	.0219	.0213	.0208	.0203	.0198	.0194	.0189	.0184	.0180
2.5	.0175	.0171	.0167	.0163	.0158	.0154	.0151	.0147	.0143	.0139
2.6	.0136	.0132	.0129	.0126	.0122	.0119	.0116	.0113	.0110	.0107
2.7	.0104	.0101	.0099	.0096	.0093	.0091	.0088	.0086	.0084	.0081
2.8	.0079	.0077	.0075	.0073	.0071	.0039	.0067	.0065	.0063	.0061
2.9	.0060	.0058	.0056	.0055	.0053	.0051	.0050	.0048	.0047	.0046
3.0	.0044	.0043	.0042	.0040	.0039	.0038	.0037	.0036	.0035	.0034
3.1	.0033	.0032	.0031	.0030	.0029	.0028	.0027	.0026	.0025	.0025
3.2	.0024	.0023	.0022	.0022	.0021	.0020	.0020	.0019	.0018	.0018
3.3	.0017	.0017	.0016	.0016	.0015	.0015	.0014	.0014	.0013	.0013
3.4	.0012	.0012	.0012	.0011	.0011	.0010	.0010	.0010	.0009	.0009
3.5	.0009	.0008	.0008	.0008	.0008	.0007	.0007	.0007	.0007	.0006
3.6	.0006	.0006	.0006	.0005	.0005	.0005	.0005	.0005	.0005	.0004
3.7	.0004	.0004	.0004	.0004	.0004	.0004	.0003	.0003	.0003	.0003
3.8	.0003	.0003	.0003	.0003	.0003	.0002	.0002	.0002	.0002	.0002
3.9	.0002	.0002	.0002	.0002	.0002	.0002	.0002	.0002	.0001	.0001

Reprinted with permission from Schaum's Outline Series, McGraw-Hill Publishing Company, Lipschutz, *Probability*, © 1968

References

ABRAMOWITZ, M., and I. A. STEGUN, EDS. *Handbook of Mathematical Functions with Formulas, Graphs, and Mathematical Tables*. Washington: National Bureau of Standards, Applied Mathematics Series, 55, 1966.

ALDER, H. L., and E. B. ROESSLER. *Introduction to Probability and Statistics*. 4th ed. San Francisco: W. H. Freeman & Company, 1968.

ARSENAULT, J. E., and J. A. ROBERTS, EDS. *Reliability and Maintainability of Electronic Systems*. Potomac, Maryland: Computer Science Press, 1980.

BARLOW, R. E. "Mathematical Theory of Reliability: A Historical Perspective," *IEEE Transactions on Reliability* R-33(1): 16–20, 1984.

BILLINTON, R. *Power System Reliability Evaluation*. New York: Gordon and Breach, 1970.

BILLINTON, R., R. J. RINGLEE, and A. J. WOOD. *Power-System Reliability Calculations*. Cambridge, Massachusetts: MIT Press, 1973.

BILLINTON, R., and R. N. ALLAN. *Reliability Evaluation of Engineering Systems: Concepts and Techniques*. New York: Plenum, 1983.

BILLINTON, R., and R. N. ALLAN. *Reliability Evaluation of Power Systems*. Boston: Pitman/Plenum, 1984.

BOLZ, R. E., and G. L. TUVE, EDS. *CRC Handbook of Tables for Applied Engineering Science*. 2d ed. Boca Raton, Florida: CRC Press, 1973.

BONIS, A. J. *Reliability Notes*. Prepared for Oklahoma State University/American Society for Quality Control Training Course, Section D, 1981.

BREIPOHL, A. M. *Probabilistic Systems Analysis*. New York: John Wiley, 1970.

CÁTUNEANU, V. M., and A. MIHALACHE. "Some Aspects of the Relationship between Reliability and Efficiency." *Revue Roumaine des Sciences Techniques, Serie Electrotechnique et Energetique* 27(3):337–342, 1982.

CHANG, N. E., "Application of Availability Analysis Techniques to Improve Power Plant Productivity," *IEEE paper No. 82 JPGC 609-6*, presented at the 1982 IEEE/ASME/ASCE Joint Power Generation Conference, Denver, Colorado, October 1982.

CLOSE, C. M., and D. K. FREDERICK. *Modeling and Analysis of Dynamic Systems* (Chapter 12). Boston, Massachusetts: Houghton Mifflin Company, 1978.

DESHMUKH, R. G., and R. RAMAKUMAR. "Reliability Analysis of Combined Wind Electric and Conventional Generation Systems." *Solar Energy* 28(4):345–352, 1982.

DHILLON, B. S., and C. SINGH. *Engineering Reliability—New Techniques and Applications*. New York: Wiley-Interscience, 1981.

DHILLON, B. S. *Power System Reliability, Safety and Management*. Ann Arbor, Michigan: Ann Arbor Science, 1983.

DHILLON, B. S. *Reliability Engineering in System Design and Operation*. New York: Van Nostrand Reinhold Company, 1983.

DOYLE, E. A. JR. "How Parts Fail," *IEEE Spectrum (special issue on reliability)* 18 (10):36–43, 1981.

DRAKE, A. W. *Fundamentals of Applied Probability Theory*. New York: McGraw-Hill, 1967.

DUDLEY, R. H., K. K. HEKIMIAN, and H. LAITIN. "The Economics of Reliability." *Annals of Reliability and Maintainability, 1970*. Vol. 9: *Assurance Technology Spinoffs,* Ninth Reliability and Maintainability Conference, Detroit, Michigan, pp. 451–470, 1970.

ENDRENYI, J. *Reliability Modeling in Electric Power Systems*. New York: John Wiley, 1978.

GREEN, A. E., and A. J. BOURNE. *Reliability Technology*. New York: Wiley-Interscience, 1972.

GYLLENHALL, P. R., and J. E. ROBINSON. "A Reliability-Cost Optimization Procedure." *Proceedings of the Fifth National Symposium on Reliability and Quality Control in Electronics*, Philadelphia, pp. 43–54, 1959.

HECHT, H., "Economic Formulation of Reliability Objectives." *Proceedings of the 1971 Annual Symposium on Reliability*. Washington, pp. 280–284, 1971.

IEEE Tutorial Course Text. *Probability Analysis of Power System Reliability*. IEEE Publication No. 71 M 30-PWR, 1971.

IEEE Tutorial Course Text. *Power System Reliability Evaluation*. IEEE Publication No. 82 EHO 195-8-PWR, 1982.

KEMENY, J. G., and J. L. SNELL. *Finite Markov Chains*. Princeton, New Jersey: Van Nostrand, 1960.

LEWIS, E. E. *Introduction to Reliability Engineering*. New York: John Wiley, 1987.

LIPSCHUTZ, S. *Theory and Problems on Probability* (Schaum's Outline Series). New York: McGraw-Hill, 1968.

LUENBERGER, D. G. *Introduction to Dynamic Systems—Theory, Models and Applications*. New York: John Wiley & Sons, 1979.

MILLER, I., and J. E. FREUND. *Probability and Statistics for Engineers*. Englewood Cliffs, New Jersey: Prentice-Hall, 1965.

MOSKOWITZ, F., and J. B. MCLEAN. "Some Reliability Aspects of System Design." *IRE Transactions on Reliability and Quality Control* RQC-8:7–35, 1956.

NELSON, W., "Analysis of Accelerated Life Test Data—Part I: The Arrhenius Model and Graphical Methods." *IEEE Transactions on Electrical Insulation* EI-6 (4):165–181, 1971.

O'CONNOR, P. D. T. *Practical Reliability Engineering*. 2d ed. New York: John Wiley, 1985.

OLKIN, I., L. J. GLESER, and C. DERMAN. *Probability Models and Applications*. New York: Macmillan, 1980.

PAPOULIS, A. *Probability, Random Variables, and Stochastic Processes*. New York: McGraw-Hill, 1965.

PEARSON, K., ED. *Tables of The Incomplete Γ-Function*. Cambridge, England: Cambridge University Press and the Office of Biometrika, 1951.

RAMAKUMAR, R. *An Introduction to Engineering Systems Reliability Evaluation*. Notes prepared for an intensive one-day short course, Oklahoma State University, Stillwater, Oklahoma, 1985.

RAMAKUMAR, R., and M. F. MCNITT-GRAY. "Wind Power," in *Standard Handbook for Electrical Engineers*, 12th ed., edited by D. G. Fink and H. W. Beaty. New York: McGraw-Hill Publishing Company, pp. 11–15 to 11–22, 1987.

RAMAKUMAR, R., and I. ABOUZAHR. "Continued Collection and Analysis of Insolation Data Using the PSO/OSU Insolation Monitoring Station in Stillwater," *Final Report*, prepared for Public Service Company of Oklahoma, Tulsa, Oklahoma, 1991.

SCHNEEWEISS, W. G. "Computing Failure Frequency, MTBF & MTTR via Mixed Products of Availabilities and Unavailabilities." *IEEE Transactions on Reliability*, R-30 (4):362–363, 1981.

SHOOMAN, M. L. *Probabilistic Reliability: an Engineering Approach*. 2d ed. Malabar, Florida: R. E. Krieger Publishing Company, 1990.

SIEWIOREK, D. P., and R. S. SWARZ. *The Theory and Practice of Reliable System Design*. Bedford, Massachusetts: Digital Press, 1982.

SINGH, C., and R. BILLINTON. *System Reliability Modeling and Evaluation*. London: Hutchinson, 1977.

SPIEGEL, M. R. *Theory and Problems of Statistics* (Schaum's Outline Series). New York: McGraw-Hill, 1961.

SPIEGEL, M. R., *Mathematical Handbook of Formulas and Tables* (Schaum's Outline Series). New York: McGraw-Hill, 1968.

TOBIAS, P. A., and D. TRINDADE. *Applied Reliability*. New York: Van Nostrand, 1986.

TRIVEDI, K. S. *Probability and Statistics with Reliability, Queuing, and Computer Science Applications*. Englewood Cliffs, New Jersey: Prentice-Hall, 1982.

YURKOWSKY, W., et al. "Accelerated Testing Technology, Volume II, Handbook of Accelerated Life Testing Methods." *Technical Report No. RADC-TR-67-420*, prepared for Rome Air Development Center by Hughes Aircraft Company, 1967.

Answers to Problems

CHAPTER 2

2.1 (a) $\frac{3}{8}$ (b) $\frac{3}{8}$ (c) $\frac{3}{8}$ (d) $\frac{7}{8}$ (e) $\frac{1}{2}$

2.2 $P(X) = \frac{1}{32}, \frac{12}{32}, \frac{11}{32}, \frac{5}{32}, \frac{2}{32}$, and $\frac{1}{32}$ for $X = 0, 1, 2, 3, 4$, and 5, respectively.

2.3 (*i*) 52   (*ii*) 6

2.4 (b) $\frac{7}{16}$ (c) $\frac{7}{16}$ (d) $\frac{2}{7}$ (e) $\frac{1}{4}$ (f) $\frac{1}{2}$

2.5 105,336

2.6 240

2.7 33,600

2.8 16,800

2.9 0.027136

2.10 40,732,184,857

2.11 0.88

2.12 (*i*) 144   (*ii*) 126   (*iii*) 504

2.13 (*i*) 0.003375   (*ii*) 0.385875

2.14 0.0333

2.16 (*i*) 0.6205   (*ii*) 0.1395   (*iii*) 0.8295   (*iv*) 0.1705

2.17 (*i*) $\frac{3}{16}$ (*ii*) $\frac{23}{32}$ (*iii*) $\frac{5}{16}$ (*iv*) $\frac{1}{4}$ (*v*) $\frac{1}{4}$

2.18 (*i*) 0.63   (*ii*) 0.7875

2.19 (*i*) no   (*ii*) 0.18   (*iii*) no   (*iv*) 0.2   (*v*) 0.82

2.20 (*i*) $\frac{56}{120}$ (*ii*) $\frac{56}{120}$ (*iii*) $\frac{64}{120}$; 1.2$\underline{2}$

2.21 0.70968

2.22 (*i*) 0.38   (*ii*) 0.3871   (*iii*) 0.31579, 0.52632, 0.15789

2.23 0.1, 0.4, 0.1, 0.2, 0.2 for X = 2, 6, 10, 11, 15, where X = total number of dollars drawn.

2.24 $K = \frac{1}{30}$; $\frac{10}{3}$, $\frac{186}{270}$, 0.82999, -0.59259, 1.51852

2.25 $(n + 1)/2$; $[(n^2 - 1)/12]$

2.26 $\frac{11}{9}$

2.27 (*i*) 0.9, 0.09, 0.009, 0.001 for N = 1, 2, 3, and >3, respectively.
(*ii*) \$44.4 million

2.28 $F(x) = \displaystyle\sum_{k=0}^{\infty} \frac{a^k e^{-a}}{k!} u(x - k)$, where $u \equiv$ unit step function

2.29 2.5, 1.25, 1.118, 0, 2.5

2.30 (*i*) $\frac{1}{6}$ (*ii*) $\frac{1}{2}$, $\frac{5}{12}$

2.31 (*i*) $F(t) = t^3$ (*ii*) 0.623   (*iii*) 0.6, 0.0375

2.32 (a) $k = \dfrac{\pi}{2(t_2 - t_1)}$

(b) $F(t) = \dfrac{1}{2}\left[1 - \cos\left(\dfrac{t - t_1}{t_2 - t_1} \right) \pi \right]$

2.33 (a) $K = 1/t_1$

(b) $F(t) = \begin{cases} 0 & \text{for } t < (t_0 - t_1) \\[2mm] \dfrac{1}{2t_1^2}(t - t_0 + t_1)^2 & (t_0 - t_1) \leq t < t_0 \\[2mm] \dfrac{1}{2} + \left(\dfrac{t - t_0}{t_1} \right) - \dfrac{(t - t_0)^2}{2t_1^2} & t_0 \leq t < (t_0 + t_1) \\[2mm] 0 & (t_0 + t_1) \leq t \end{cases}$

2.35 (*ii*) $\bar{F}(t) = \exp(-\lambda t)$

CHAPTER 3

3.1 yes; $R(t) = \exp\left[-\dfrac{kt^2}{2} \right]$; $Q(t) = 1 - R(t)$; $f(t) = kt \exp\left[-\dfrac{kt^2}{2} \right]$

3.3 Kt^m

3.4 (*i*) 0.095   (*ii*) 0.6321

3.5 (*i*) $b^{n-1} = a/(n-1)$

   (*ii*) $R(t) = 1 - \left(\dfrac{a}{1-n}\right)[(b+t)^{1-n} - b^{1-n}]$

3.6 $R(t) = \exp\left[-a\left\{C_1 t + \dfrac{C_2}{b}(1 - e^{-bt}) + \dfrac{C_3}{c}(1 - e^{-ct})\right\}\right]$

   $f(t) = \lambda(t)R(t)$

3.7 (*i*) $Q_T(t) = \alpha Q_e(t) + (1-\alpha)\,Q_n(t)$

   (*ii*) $\lambda_T(t) = \dfrac{\alpha f_e(t) + (1-\alpha)\,f_n(t)}{1 - \{\alpha Q_e(t) + (1-\alpha)Q_n(t)\}}$

3.8 (a)

$$f(t) = \begin{cases} k\left(1 - \dfrac{t}{t_c}\right)\exp\left\{-kt + \dfrac{kt^2}{2t_c}\right\} & \text{for } 0 < t < t_c \\[2mm] 0 & \text{elsewhere} \end{cases}$$

$$Q(t) = \begin{cases} 0 & \text{for } t \le 0 \\[2mm] 1 - \exp\left[-kt + \dfrac{kt^2}{2t_c}\right] & \text{for } 0 < t < t_c \\[2mm] 1 - \exp\left[-\dfrac{kt_c}{2}\right] & \text{for } t > t_c \end{cases}$$

   (b) $1 - \exp\left[-\dfrac{kt_c}{2}\right]$

3.9 0.826875

3.10 (*i*) yes   (*ii*) $(2b)/(1 - bt)$ for $0 \le t < (1/b)$ (*iii*) $1/(3b)$

3.11 $R(t) = \exp[-2 \times 10^{-9}t]$; 0.999982; 5×10^8 hr; 5.708 years

3.12 $R(t) = \exp[-4 \times 10^{-6}t]$; 0.96557; 250,000 hr; 25 hr

3.13 $an(1 - at)^{n-1}$; $1 - (1 - at)^n$; $(an)/(1 - at)$ for $0 \le t < (1/a)$; MTTF $= 1/[a(n+1)]$; $t_m = (1/a)[1 - (0.5)^{1/n}]$

3.15 $R(t) = \exp[-\Lambda(t)]$

   $Q(t) = 1 - R(t)$

   $\lambda(t) = \dfrac{d}{dt}\Lambda(t)$

   $f(t) = \left[\dfrac{d}{dt}\Lambda(t)\right]\exp[-\Lambda(t)]$

3.16 $\Lambda(t) = (kt^2)/2$; $(k/2)(t_2 + t_1)$; $(kT)/2$

3.17 (a) $[1/(1 + kT)]^n$ (b) $k/(1 + kt)$ (c) $-(1/T)\ln[1/(1 + kT)]$

3.18 $(\ln 2)/\lambda$

3.20 (*i*) 0.9085   (*ii*) 0.93935

CHAPTER 4

4.1 (*i*) 0.273414   (*ii*) 0.59673

4.2 (**a**) 0.605   (**b**) 0.395   (**c**) 0.07562

4.3 6

4.4 zero skewness

4.5 option 2; expected loss of load values are 1.005 and 0.5262 MW.

4.6 0.02408

4.7 (*i*) 0.5787   (*ii*) 5   (*iii*) 2.04   (*iv*) 0.59799

4.8 (*i*) 0.181   (*ii*) 0.1637   (*iii*) 0.01725   (*iv*) 0.2707

4.10 (**a**) 0.27067   (**b**) 0.07468

4.11 (*i*) 0.0613   (*ii*) 0.08033   (*iii*) 0.36789

4.12 (**a**) 0.1353   (**b**) 0.18394   (**c**) 0.0803

4.13 0.41981

4.14 (*i*) 0.14957   (*ii*) 0.17098   (*iii*) 0.56625

4.15 (*i*) 0.27067   (*ii*) 0.13534   (*iii*) 0.14287

4.16 3

4.17 (**a**) 0.184737   (**b**) (*i*) 2.680643   (**b**) (*ii*) 0.319357

4.18 0.201; 1% decrease

4.20 $t_f = -\dfrac{\ln(1 - 0.01p_f)}{\lambda}$

4.21 (**a**) 0.99633   (**b**) 0.0036749   (**c**) 4,761,905 hr; 3,300,701 hr   (**d**) 2,051,347 hr

4.22 7063.4

4.23 once in 20 days

4.24 0.01499

4.25 once in 19 days

4.26 (*i*) 0.0228   (*ii*) 0.6826   (*iii*) 0.1587   (*iv*) 0.0026   (*v*) $(\mu + 1.28\sigma)$

4.27 (*i*) 0.0446   (*ii*) 0.685   (*iii*) 0.659

4.28 0.26%

4.29 41.25 months

4.30 0.98889; 3.8989 mm

4.31 a.   (*i*) 1   (*ii*) 0.2   (*iii*) 0   (*iv*) 0.4   **b.** $\frac{1}{3}$ **c.** 0.5

4.32 (**a**) $t_1 = \mu - \sqrt{3}\sigma$ and $t_2 = \mu + \sqrt{3}\sigma$ (**b**) 1,000 ± 5% ohms

4.33 MTTF $= \sqrt{\dfrac{\pi}{2k}}$; $T_{50} = \sqrt{\dfrac{2\ln 2}{k}}$

4.34 (*i*) 0.00995   (*ii*) 0.001159   (*iii*) 17.725 weeks; 16.651 weeks

4.35 $k = (\pi/2)\,\lambda_0^2$

4.36 $Q_c(t) = 1 - \exp\left[-\frac{k}{2}(2tT + t^2)\right]$; it has memory

4.37 0.632, does not depend; 0.36788, 0.13534, 0.04979, 0.01832

4.39 $Q(t) = 1 - \exp\left[-\frac{t^\beta}{\theta}\right]$

$\lambda(t) = \left(\frac{1}{\theta}\right)\beta t^{\beta-1}$

$R(t) = \exp\left[-\frac{t^\beta}{\theta}\right]$

$f(t) = \frac{\beta t^{\beta-1}}{\theta}\exp\left[-\frac{t^\beta}{\theta}\right]$

4.40 (*i*) design II   (*ii*) depends on whether the $500 increase is worth the slight improvement in reliability. If yes, then I; if not, choose II.

4.41 0.31442; 6, 18, 4.243

4.42 (*i*) 0.919699   (*ii*) 0.676676

4.44 (*i*) 0.1812   (*ii*) 0.69

4.45 $T_{50} = e^\mu$; 0.0636; 17.75%

4.46 63,800 cycles; 0.0017

4.47 MTTF $= T_{50}e^{\sigma^2/2}$; var$[t] = T_{50}^2 e^{\sigma^2}[e^{\sigma^2} - 1]$

4.48 $A = \dfrac{\Gamma(\gamma + \beta + 2)}{\Gamma(\gamma + 1)\,\Gamma(\beta + 1)}$

4.50 $F(t) = 1 - \exp[-t^a]$ for $t \geq 0$; scale parameter $= 1$ and shape parameter $= a$.

4.51 mean $= -0.5772157$; median $= -0.3665$

4.56 $f(t) = \frac{\beta(t - t_0)^{\beta-1}}{\alpha^\beta}\exp\left[-\left(\frac{t - t_0}{\alpha}\right)^\beta\right]$ for $t \geq t_0$

$Q(t) = 1 - \exp\left[-\left(\frac{t - t_0}{\alpha}\right)^\beta\right]$ for $t \geq t_0$

MTTF $= \left[\alpha\Gamma\left(1 + \frac{1}{\beta}\right)\right] + t_0$

Variance does not change.

4.57 $Q_c(t) = 1 - \exp\left[-\lambda_0 t + kt(T - T_0) - \frac{kt^2}{2}\right]$

4.58 $f(t) = \left(K + K\alpha t + \frac{K}{2}\alpha^2 t^2\right)\exp\left[-\left(Kt + K\alpha\frac{t^2}{2} + \frac{K\alpha^2}{6}t^3\right)\right]$

$R(t) = \exp\left[-\left(Kt + K\alpha\frac{t^2}{2} + \frac{K\alpha^2}{6}t^3\right)\right]$

$Q(t) = 1 - R(t)$

4.59 0.01705 W; 0.02685 W

4.60 (*i*) 0.90421   (*ii*) 0.853354

4.61 (a) 2.22 months   **(b)** 3.941 months

4.62 $\lambda(t) = \dfrac{1}{5 - t}$ for $0 \le t < 5$; MTTF $= 4.5$ yr

4.63 10 years versus 3.9633 years; 20 years versus 3.9633 years

4.64 0.1107 failure/year

4.65 0.1219

4.66 $1 - \dfrac{10^{-7(k-1)}}{(k-1)!} e^{-10^{-7}t}$

$$\times \left[t^{k-1} + \frac{k-1}{10^{-7}} t^{k-2} + \frac{(k-1)(k-2)}{10^{-14}} t^{k-3} + \cdots + \frac{(k-1)!}{10^{-7(k-1)}} \right]$$

4.67 mean $= (\beta\gamma + y_m)$, where γ is Euler's constant.

4.69 $F_Y(y) = 1 - \left(\dfrac{1}{\beta^m} \right) (y_1 - y)^m$ valid for $|y_1 - y| \ll \beta$

$$F_Z(z) = \left(\frac{1}{\beta^m} \right) (z - y_1)^m \qquad \text{valid for } |z - y_1| \ll \beta$$

4.70 mean $= \beta\Gamma\left(1 + \dfrac{1}{m} \right) + y_1$

variance $= \beta^2 \left[\Gamma\left(1 + \dfrac{2}{m} \right) - \Gamma^2\left(1 + \dfrac{1}{m} \right) \right]$

4.73 $\gamma = 0.025$; $\beta = 8.225$

CHAPTER 5

5.1 0.9487; 0.9740; 0.9826; 0.9869; 0.9895

5.2 $[R(t)]^{1-n}$

5.3 $1 + Q(t)$

5.4 4; no

5.6 (a) 0.96333   **(b)** 0.512 by trial and error

5.7 $pp_1^2 (-pp_1 + 2)$; 0.978775

5.8 0.981204

5.9 0.988475; 0.9996295

5.10 $p[3p - p^2 - 6p^3 + 9p^4 - 5p^5 + p^6]$; 0.8993619

5.11 0.975

5.12 I: $p^2(4 - 4p + p^2)$; II: $p^2(2 - p^2)$

5.13 0.9234

5.14 0.99144

5.15 (*i*) 0.995   (*ii*) 0.9905   (*iii*) 0.6666

5.16 0.951792

5.17 $2p^2 + 2p^3 - 5p^4 + 2p^5$

5.18 $2p^2 + 2p^3 - 5p^4 + 2p^5$

5.19 $2q^5 - 5q^4 + 2q^3 + 2q^2 = Q_s = (1 - R_s)$

5.20 $2q^2 + 2q^3 - 5q^4 + 2q^5$

5.21 $R_s = R_1R_2R_3 + R_1R_2Q_3R_4 + R_1Q_2R_3 + R_1Q_2Q_3R_4R_5$
$\qquad\quad + Q_1R_2R_3R_4 + Q_1R_2R_3Q_4R_5 + Q_1R_2Q_3R_4$

$\qquad Q_s = R_1R_2Q_3Q_4 + R_1Q_2Q_3R_4Q_5 + R_1Q_2Q_3Q_4 + Q_1R_2R_3Q_4Q_5 + Q_1R_2Q_3Q_4 + Q_1Q_2$

5.22 $R_aR_b + R_bR_c + R_cR_d - R_bR_c[R_a + R_d]; 3p^2 - 2p^3$

5.24 $R_1R_2R_3(1 - R_2) + (2R_1 - R_1^2)(2R_3 - R_3^2)R_2$

5.26 $R_aR_b + R_cR_d(R_b + R_e) - R_bR_cR_d(R_a + R_e)$

5.27 $ab + cd + adf + ceb$

5.28 ab, cd, adf, ceb

5.29 (a) $abc + d$ (c) $\frac{67}{256}$

5.30 (ii) 0.332   (iii) (0.35185 − 0.332)

5.32 $(a + bde)f$

5.34 (i) 1,792/16,384   (ii) 1,072/16,384   (iii) 67/1,024

5.35 (i) 0.99877317   (ii) 0.99886786

5.36 $q_o < q_s$

5.37 (i) 3.94%   (ii) 18.725

5.38 (a) 0.99570735; 0.99663265   (b) 0.99663265; 0.99570735

5.40 0.369619

5.41 (i) 1.002497   (ii) 1.002503

5.42 $\dfrac{1 - Q^n(t)}{1 - Q(t)}$; 1.4; 1.56; 1.624

5.43 (a) $\dfrac{1 + Q_1(t) + Q_1^2(t)}{3Q_1^2(t)}$;

$\qquad$ (b) $\dfrac{1 + Q_1(t) + Q_1^2(t) + \cdots + Q_1^{n-1}(t)}{nQ_1^{n-1}(t)}$

5.44 $(e^{-\lambda_1 t})(2e^{-\lambda_2 t} - e^{-2\lambda_2 t})[3e^{-kt^2} - 2e^{-3kt^2/2}]$

5.45 $1/(2\lambda)$

5.46   (i) $T_d = -\dfrac{\ln p_d}{\lambda}$

$\qquad$ (ii) $T_d = -\dfrac{\ln[1 - \sqrt{1 - p_d}]}{\lambda}$

$\qquad$ (iii) $\dfrac{\ln[1 - \sqrt{1 - p_d}]}{\ln p_d}$;   4.9343, 3,6079,
$\qquad\qquad\qquad\qquad\qquad\quad$ 3.0143, 2.6565

5.49 9

5.50 $e^{-3\lambda t} + 2e^{-2\lambda t}; \dfrac{4}{3\lambda}$

5.51 $p^2 + 3p^3 - 4p^4 + p^5; 0.7/\lambda$

5.52 58.33 %/K

5.53 $1.217/\lambda$

5.54 $\dfrac{1}{2}\sqrt{\dfrac{\pi}{k/2}} + \dfrac{1}{\lambda_0} - \dfrac{1}{2}\sqrt{\dfrac{\pi}{k/2}}\, e^{\lambda_0^2/(2k)}\, \text{erfc}\left(\dfrac{\lambda_0}{\sqrt{2k}}\right)$

5.55 $e^{-(\lambda_1+\lambda_2)t}\left[e^{-\lambda_4 t} + e^{-\lambda_3 t} - e^{-(\lambda_3+\lambda_4)t}\right]$

5.56 $1 - 8 \times 10^{-7}$

5.57 $e^{-\lambda t} + P_{fss}\lambda t e^{-\lambda t}$

5.58 0.8050

5.60 $e^{-(\lambda_1+\lambda_2)t} + \left[e^{-\lambda_1 t} - e^{-(\lambda_1+\lambda_2)t}\right] + \left[e^{-\lambda_2 t} - e^{-(\lambda_1+\lambda_2)t}\right]$

$$+ (\lambda_2 e^{-\lambda_3 t})\left[\frac{1}{(\lambda_2+\lambda_{3s}-\lambda_3)}\{1 - e^{-(\lambda_2+\lambda_{3s}-\lambda_3)t}\}\right.$$

$$\left. - \frac{1}{(\lambda_1+\lambda_2+\lambda_{3s}-\lambda_3)}\{1 - e^{-(\lambda_1+\lambda_2+\lambda_{3s}-\lambda_3)t}\}\right]$$

$$+ (\lambda_1 e^{-\lambda_3 t})\left[\frac{1}{(\lambda_1+\lambda_{3s}-\lambda_3)}\{1 - e^{-(\lambda_1+\lambda_{3s}-\lambda_3)t}\}\right.$$

$$\left. - \frac{1}{(\lambda_1+\lambda_2+\lambda_{3s}-\lambda_3)}\{1 - e^{-(\lambda_1+\lambda_2+\lambda_{3s}-\lambda_3)t}\}\right]$$

5.61 $N - 1$

5.62 5; 5.5

5.63 (*i*) 0.459205   (*ii*) 0.896088   (*iii*) 0.751967

5.64 (*i*) 0.664127   (*ii*) 0.978150   (*iii*) 0.768863   (*iv*) 0.849794

5.65 $[4e^{-\lambda t} - 6e^{-2\lambda t} + 4e^{-3\lambda t} - e^{-4\lambda t}]$

5.66 $q_{SA} = \sqrt{\dfrac{[1 - (1 - q_{SAB})(1 - q_{SCA}q_{SBC})][1 - (1 - q_{SCA})(1 - q_{SAB}q_{SBC})]}{1 - (1 - q_{SBC})(1 - q_{SCA}q_{SAB})}}$

and so on.

CHAPTER 6

6.1 (**b**) $\left(\frac{29}{54}\right)$ (**c**) $\left(\frac{1}{25}\right)$ [8   3   14]

6.2 Periodic Markov chain with a period of 2.

6.3 $\left(\frac{1}{21}\right)$ [9   4   8]

6.5 $(\frac{1}{5})$ [2 1 2]

6.6 $(\frac{75}{180})$, $(\frac{50}{180})$, $(\frac{55}{180})$

6.8 $\frac{16}{9}$

6.9 (*ii*) $\frac{9}{5}$ (*iii*) $\frac{7}{5}$ (*iv*) $\frac{11}{20}$, $\frac{9}{20}$ (*v*) $\frac{13}{20}$, $\frac{7}{20}$

6.10 (b) $\frac{5}{12}$ (c) $\frac{2}{7}$, $\frac{3}{7}$, $\frac{2}{7}$

6.11 0.483, 0.276, 0.241

6.12 (*i*) $\frac{1}{2}$, $\frac{1}{6}$, $\frac{1}{3}$ (*ii*) 5 years

6.13 (*ii*) $\frac{9}{16}$ (*iii*) $\frac{2}{5}$

6.15 $$\begin{bmatrix} P_1'(t) \\ P_2'(t) \\ P_3'(t) \end{bmatrix} = \begin{bmatrix} -(\lambda_1 + \lambda_2) & 0 & \mu_2 \\ \lambda_1 & 0 & \mu_1 \\ \lambda_2 & 0 & -(\mu_1 + \mu_2) \end{bmatrix} \begin{bmatrix} P_1(t) \\ P_2(t) \\ P_3(t) \end{bmatrix}$$

6.16 $$\begin{bmatrix} P_1'(t) \\ P_2'(t) \\ P_3'(t) \\ P_4'(t) \end{bmatrix} = \begin{bmatrix} -(\lambda + \mu_1) & \mu & \lambda_1 & 0 \\ \lambda & -(\mu + \mu_r) & 0 & \lambda_2 \\ \mu_1 & \mu_r & -(\lambda_1 + \lambda_s) & \mu_s \\ 0 & 0 & \lambda_s & -(\lambda_2 + \mu_s) \end{bmatrix} \begin{bmatrix} P_1(t) \\ P_2(t) \\ P_3(t) \\ P_4(t) \end{bmatrix}$$

6.17 $$\begin{bmatrix} -(\lambda_1 + \lambda_2 + \lambda_3 + \lambda_4) & 0 & 0 & 0 & 0 \\ \lambda_1 & -(\mu_1 + \mu_3) & 0 & 0 & 0 \\ \lambda_4 & \mu_1 & -\mu_2 & 0 & 0 \\ \lambda_3 & 0 & \mu_2 & -\mu_4 & 0 \\ \lambda_2 & \mu_3 & 0 & \mu_4 & 0 \end{bmatrix}$$

6.19 $e^{-(\lambda_1 + \lambda_2)t}$; $\dfrac{\lambda_1}{\lambda_1 + \lambda_2 - \lambda_{20}} e^{-\lambda_{20}t} - \dfrac{\lambda_1 e^{-(\lambda_1 + \lambda_2)t}}{\lambda_1 + \lambda_2 - \lambda_{20}}$; $\dfrac{\lambda_2 e^{-\lambda_{10}t}}{\lambda_1 + \lambda_2 - \lambda_{10}}$

$$- \frac{\lambda_2}{\lambda_1 + \lambda_2 - \lambda_{10}} e^{-(\lambda_1 + \lambda_2)t}; \left(1 - \Sigma \text{ of the other three} \right)$$

6.20 $e^{-(\lambda_1 + \lambda_2)t}$; $e^{-\lambda_2 t} - e^{-(\lambda_1 + \lambda_2)t}$; $e^{-\lambda_1 t} - e^{-(\lambda_1 + \lambda_2)t}$; $(1 - \Sigma$ of the other three)

6.22 $$P_g(t) = e^{-(\lambda_{gf} + \lambda_{gb})t}$$

$$P_f(t) = \frac{\lambda_{gf}}{\lambda_{gf} + \lambda_{gb} - \lambda_{fb}} [e^{-\lambda_{fb}t} - e^{-(\lambda_{gf} + \lambda_{gb})t}]$$

$$P_b(t) = 1 - P_g(t) - P_f(t)$$

6.23 $P_1(t) = (0.75) e^{-0.8t} + 0.25$

$P_2(t) = (0.75) (1 - e^{-0.8t})$

$P_{10} = 0.25$

$P_{20} = 0.75$

6.24 0.6, 0.4; $\frac{15}{27}$, $\frac{12}{27}$

6.25 $\left(1 + \dfrac{w}{d} \right)^{-4}$;

$$\frac{4w}{d}\left(1 + \frac{w}{d}\right)^{-4};$$

$$6\left(\frac{w}{d}\right)^2\left(1 + \frac{w}{d}\right)^{-4};$$

$$4\left(\frac{w}{d}\right)^3\left(1 + \frac{w}{d}\right)^{-4};$$

$$\left(\frac{w}{d}\right)^4\left(1 + \frac{w}{d}\right)^{-4}$$

CHAPTER 7

7.1

$1.2071\,e^{-0.5858t} - 0.2071\,e^{-3.4142t}$	$-0.7071e^{-0.5858t} + 0.7071e^{-3.4142t}$
$0.3536\,\{e^{-0.5858t} - e^{-3.4142t}\}$	$-0.2071e^{-0.5858t} + 1.2071e^{-3.4142t}$

or

$1 - t^2 + (\frac{8}{6})t^3 - \cdots$	$-2t + 4t^2 - (\frac{28}{6})t^3 + \cdots$
$t - 2t^2 + (\frac{14}{6})t^3 - \cdots$	$1 - 4t + 7t^2 - (\frac{48}{6})t^3 + \cdots$

7.2 $A(t) = 0.986486 + 0.013514\,e^{-0.20274t}$

0.9869466; 0.986502; 1.35135%

7.5 $R(t) = 4e^{-1.5t} - 3e^{-2t}$

7.6 $(\mu\gamma)/\mathcal{D}$; $(6\lambda\gamma)/\mathcal{D}$; $(6\lambda\mu)/\mathcal{D}$

where $\mathcal{D} \equiv 6\lambda(\mu + \gamma) + \mu\gamma$

7.7 (*i*) 5.47645×10^{-4} (*ii*) 0.19989 failure/yr   (*iii*) 1 day

7.8 $P_{Li} = \beta_i\,e$ for $i = 1, 2, 3, \ldots, n$

Mean duration $= e$ day for load levels L_1 through L_n and $(1 - e)$ day for L_0.

7.9 (*ii*) $P_i = \dfrac{\lambda_i}{\mu_i}\,P_0$ for i $= 1, 2, \ldots, n$, where $P_0 = \dfrac{1}{1 + \sum\limits_{i=1}^{n}\dfrac{\lambda_i}{\mu_i}}$

(*iii*) P_0 (*iv*) $1 - P_0$

(*v*) MTTF $= \dfrac{1}{\sum\limits_{i=1}^{n}\lambda_i}$; MTTR $= \dfrac{\sum\limits_{i=1}^{n}\dfrac{\lambda_i}{\mu_i}}{\sum\limits_{i=1}^{n}\lambda_i}$ (*vi*) $P_0\sum\limits_{i=1}^{n}\lambda_i$

7.10 $\dfrac{\mu(\lambda_1 + \lambda_2)}{\lambda_2\mu + \lambda_1\mu + (\lambda_1 + \lambda_3)\lambda_2}$; 0.834725, 0.165275

7.11 $P_1(t) = \dfrac{1}{s_1 - s_2}[(s_1 + \lambda + \mu)e^{s_1t} - (s_2 + \lambda + \mu)e^{s_2t}]$

$$P_2(t) = \frac{2\lambda}{s_1 - s_2} [e^{s_1 t} - e^{s_2 t}]$$

$$P_3(t) = 1 + \frac{1}{s_1 - s_2} [s_2 e^{s_1 t} - s_1 e^{s_2 t}]$$

where $s_1, s_2 = \dfrac{-(3\lambda + \mu) \pm \sqrt{\lambda^2 + 6\lambda\mu + \mu^2}}{2}$

7.12 $\dfrac{3}{2\lambda} + \dfrac{\mu}{2\lambda^2}$

7.13 $\dfrac{3\lambda + \mu}{\lambda_1(\lambda + \mu) + 2\lambda^2}$

7.16 (*ii*) 2941.176 gph   (*iii*) 2941.176 gph

7.17   (*i*) $A^3 + 3A^2 U + 3AU^2$

(*ii*) $[3\lambda A^3 + 3A^2 U(2\lambda - \mu) + 3AU^2(\lambda - 2\mu)]$

(*iii*) $A^3 - 3A^2 + 3A$ (*iv*) $3\lambda A^3 - 6\lambda A^2 + 3\lambda A$ (*v*) U^3 (*vi*) $3\mu U^3$
1.176×10^{-5} hr^{-1}; 85,033.33 hr; 0.680272 hr

7.18 (a) $A^5 + 5A^4 U + 10A^3 U^2$; $U^5 + 5U^4 A + 10U^3 A^2$

(b) $5\lambda A^5 + 5A^4 U(4\lambda - \mu) + 10A^3 U^2(3\lambda - 2\mu)$; $5\mu U^5 + 5U^4 A(4\mu - \lambda) + 10U^3 A^2$
$(3\mu - 2\lambda)$

(c) 2.910897×10^{-3} (d) 2.9109×10^{-3}; 295.50454

7.19 $P_1 = \dfrac{\mu_o \mu_s}{\mathscr{D}}$, $P_2 = \dfrac{\lambda_o \mu_s}{\mathscr{D}}$, $P_3 = \dfrac{\lambda_s \mu_o}{\mathscr{D}}$, where

$\mathscr{D} \equiv \mu_o \mu_s + \lambda_o \mu_s + \lambda_s \mu_o$; $P_1(\lambda_o + \lambda_s)$, $P_2 \mu_o$, $P_3 \mu_s$; $[1/(\lambda_o + \lambda_s)]$, $(1/\mu_o)$, $(1/\mu_s)$

7.20 $\dfrac{\mu_o + \lambda_o}{\mu_o \lambda_s}$; 500.5 hr

7.21 $\dfrac{1}{\lambda_o + \lambda_s}$; $\dfrac{\lambda_o}{\lambda_o + \lambda_s}$, $\dfrac{\lambda_s}{\lambda_o + \lambda_s}$

7.22 $R(t) = \left(\dfrac{1}{\lambda_a - \lambda_b} \right) [\lambda_a e^{-\lambda_b t} - \lambda_b e^{-\lambda_a t}]$; MTTF $= \dfrac{1}{\lambda_a} + \dfrac{1}{\lambda_b}$

7.23 $R(t) = e^{-\lambda t} (1 + \lambda t)$; MTTF $= 2/\lambda$

7.24 (*iii*) $R(t) = e^{-\lambda_a t} + \left(\dfrac{\lambda_a}{\lambda_a + \lambda_{bs} - \lambda_b} \right) [e^{-\lambda_b t} - e^{-(\lambda_a + \lambda_{bs})t}]$

(*iv*) MTTF $= \dfrac{1}{\lambda_a} + \left(\dfrac{\lambda_a}{\lambda_a + \lambda_{bs} - \lambda_b} \right) \left[\dfrac{1}{\lambda_b} - \dfrac{1}{\lambda_a + \lambda_{bs}} \right]$

7.25 $R(t) = e^{-\lambda t} \left[1 + \dfrac{\lambda}{\lambda_{bs}} \right] - \left(\dfrac{\lambda}{\lambda_{bs}} \right) e^{-(\lambda + \lambda_{bs})t}$

MTTF $= \dfrac{1}{\lambda} + \dfrac{1}{\lambda_{bs}} - \dfrac{\lambda}{\lambda_{bs}(\lambda + \lambda_{bs})}$

7.26 $R(t) = e^{-\lambda_c t} + \dfrac{(1 - p)\lambda_c}{\lambda_c - \lambda_D} [e^{-\lambda_D t} - e^{-\lambda_c t}]$

$$\text{MTTF} = \frac{1}{\lambda_c} + \frac{1 - p}{\lambda_D}$$

7.27 $R(t) = e^{-\lambda t} [1 + (1 - p)\lambda t]$; $\text{MTTF} = (2 - p)/\lambda$

7.29 $\text{MTTF} = \dfrac{11\lambda^2 + 4\lambda\mu + \mu^2}{6\lambda^3}$

7.31 *(i)* $\dfrac{\alpha\mu^2}{\mathcal{D}}, \dfrac{n\lambda\mu(n\lambda + \mu)}{\mathcal{D}}, \dfrac{n\lambda\alpha\mu}{\mathcal{D}}, \dfrac{\alpha n^2\lambda^2}{\mathcal{D}}$

where $\mathcal{D} \equiv (\alpha\mu^2 + n^2\lambda^2\mu + n\lambda\mu^2 + \alpha n^2\lambda^2 + \alpha n\lambda\mu)$

(iii) $\dfrac{1}{n\lambda}, \dfrac{1}{\alpha}, \dfrac{1}{n\lambda + \mu}, \dfrac{1}{\mu}$

(0.892061, 0.00981267, 0.0892061, 0.00892061);

(0.00892061, 0.00981267, 0.009812671, 0.000892061) hr^{-1};

(100, 1, 9.091, 10) hr

7.32 $\dfrac{2\mu^3}{\mathcal{D}}, \dfrac{6\lambda\mu^2}{\mathcal{D}}, \dfrac{6\lambda^2\mu}{\mathcal{D}}, \dfrac{3\lambda^3}{\mathcal{D}}$, where $\mathcal{D} \equiv 2(\lambda + \mu)^3 + \lambda^3$

7.34 $\lambda_{12} = \lambda + \lambda_M$, $\lambda_{21} = \dfrac{\mu\mu_M(\lambda + \lambda_M)}{\mu\lambda_M + \lambda\mu_M}$

7.35 $\lambda_{12} = \lambda$; $\lambda_{21} = \left(\dfrac{\gamma\mu}{\gamma + \mu}\right)$

As $\gamma \rightarrow \infty$, the model becomes a simple binary model with failure and repair rates of λ and μ, respectively.

7.37 Transition rates: $4\lambda_1 + 2\lambda_2$, $\dfrac{\mu(4\lambda_1 + 2\lambda_2)}{4\lambda_1 + 3\lambda_2}$

7.38 *(ii)* $\dfrac{1}{2\lambda}, \dfrac{1}{\mu + \lambda}, \dfrac{1}{\gamma}, \dfrac{1}{2\mu}$

(iii) Transition rates: $\dfrac{2\lambda\mu + 2\lambda^2}{\mu + 2\lambda}, \dfrac{2\lambda\mu^2\gamma + 2\lambda^2\mu\gamma}{\lambda^2\gamma + 2\lambda\mu^2}$

(iv) Frequency of failure $= \dfrac{2\lambda + \dfrac{2\lambda^2}{\mu}}{1 + \dfrac{\lambda^2}{\mu^2} + \dfrac{2\lambda}{\mu} + \dfrac{2\lambda}{\gamma}}$

7.39 *(ii)* $\dfrac{1}{2\lambda}, \dfrac{1}{\lambda + \mu}, \dfrac{1}{\gamma}, \dfrac{1}{\mu}$

(iii) Transition rates: $\dfrac{2\lambda\mu + 2\lambda^2}{\mu + 2\lambda}, \dfrac{2\lambda\mu^2\gamma + 2\lambda^2\mu\gamma}{2\lambda^2\gamma + 2\lambda\mu^2}$

(iv) Frequency of failure $= \dfrac{2\lambda + \dfrac{2\lambda^2}{\mu}}{1 + \dfrac{2\lambda}{\mu} + \dfrac{2\lambda}{\gamma} + \dfrac{2\lambda^2}{\mu^2}}$

7.40 Ternary model transition rates:

[0–1: 2λ]; [1–0: μ]; [1–2: λ]; [2–1: μ]

7.41 $P_1(t) = e^{-3\lambda t}$

$P_2(t) = 3e^{-2\lambda t} - 3e^{-3\lambda t}$

$P_3(t) = 3e^{-3\lambda t} - 6e^{-2\lambda t} + 3e^{-\lambda t}$

$P_4(t) = 1 - 3e^{-\lambda t} + 3e^{-2\lambda t} - e^{-3\lambda t}$

CHAPTER 8

8.1 (*i*) $-4.9997 \times 10^{-3}\%$ (*ii*) -0.4975%

8.2 176.969 hr

8.3 $(\lambda_1 + \lambda_2 + \lambda_3)^{-1}[\lambda_1 r_1 + \lambda_2 r_2 + \lambda_3 r_3 + \lambda_1\lambda_2 r_1 r_2$

$$+ \lambda_1\lambda_3 r_1 r_3 + \lambda_2\lambda_3 r_2 r_3 + \lambda_1\lambda_2\lambda_3 r_1 r_2 r_3]$$

8.4 0.00357 failure/yr

8.5 0.5 yr^{-1}

8.6 0.02%; 1.999%

8.7 0.00597 yr^{-1}; 0.0006667 yr or 5.84 hr

8.8 (*i*) From 1 per year to 3×10^{-6} per year

(*ii*) From 8.76 hr to 2.913 hr

8.9 (*i*) 1.7995×10^{-5} failure/yr   (*ii*) 0.00125 yr   (*iii*) 2.2494×10^{-8} (*iv*) 0.0278%

8.10 0.121212 yr^{-1}; 198 yr^{-1}; 8.25 yr; 198^{-1} yr

8.11 3.0609×10^{-3} hr^{-1} and 297 hr^{-1} as compared to 2.9109×10^{-3} hr^{-1} and 295.50454 hr^{-1}

8.12 (*i*) 5×10^{-12} failure/yr; 2×10^{-4} yr   (*ii*) 2×10^{-8} failure/yr; 2.5×10^{-4} yr

(*iii*) 3×10^{-5} failure/yr; 3.333×10^{-4} yr   (*iv*) 0.02 failure/yr; 5×10^{-4} yr

(*v*) 5 failures/yr; 10×10^{-4} yr

8.13 1.125×10^{-10} failure/yr; 1.667×10^{-3} yr; 1.875375×10^{-13}

8.14 (**b**) $P_1(t) = 0.990283\, e^{-0.001088t} + 0.0097172\, e^{-1.0199t}$

$P_2(t) = 0.9102095 - 0.9102\, e^{-0.001088t} - 9.5276 \times 10^{-6}\, e^{-1.0199t}$

$P_3(t) = (0.00981535)[e^{-0.001088t} - e^{-1.0199t}]$

$P_4(t) = 0.0901188 - 0.090215\, e^{-0.001088t} + (9.62384 \times 10^{-5})e^{-1.0199t}$

(**c**) 10.1 times

8.15 (*i*) 918.92 hr   (*ii*) 910.811 hr   (*iii*) 10.1 times   (*iv*) 9.09 times

8.16 $\lambda = \dfrac{(1 - x)f}{\alpha}$; $\lambda' = \dfrac{xf}{1 - \alpha}$

8.18 (*i*) 0.202   (*ii*) 181.8   (*iii*) 2   (*iv*) 0.001 yr   (*v*) 0.002 yr

8.19 (*i*) 1.0976   (*ii*) 0.7   (*iii*) 100.5   (*iv*) 12.7506 hr   (*v*) 13.994976 hr/yr

8.20 (*i*) 1.0377398   (*ii*) 9.81816 $\times$ 10^{-4} yr   (*iii*) 1.0188699 $\times$ 10^{-3} yr

8.21 3.95153 $\times$ 10^{-3} yr^{-1}; 7.23353 $\times$ 10^{-4} yr

8.22 3.18926 $\times$ 10^{-3} yr^{-1}; 6.25486 $\times$ 10^{-4} yr

8.23 1.27585 $\times$ 10^{-4} yr^{-1}; 5.0795 $\times$ 10^{-4} yr; 6.4807 $\times$ 10^{-8}

8.24 $\rho = 3{,}893$ yr^{-1}; $n = 7$

8.26 $P_{\text{UP}} \quad = \dfrac{\alpha(\alpha^2 + \beta^2)}{(\alpha^2 + \beta^2)\,(\alpha + \lambda) + 2\alpha^2\lambda}$

$\quad\;\; P_{\text{DOWN}} = \dfrac{\lambda\{(\alpha^2 + \beta^2) + 2\alpha^2\}}{(\alpha^2 + \beta^2)\,(\alpha + \lambda) + 2\alpha^2\lambda}$

Transition rates: $\lambda;\ \dfrac{\alpha(\alpha^2 + \beta^2)}{(\alpha^2 + \beta^2) + 2\alpha^2}$

8.27 $P_{\text{UP}} = \dfrac{\mu}{\lambda + \mu};\ P_{\text{DOWN}} = \dfrac{\lambda}{\lambda + \mu}$

8.28 (0.0297, 0.0106, 0.0197, 0.00151) occurrences per year; (32.29, 1.92, 0.99, 0.66) yr

8.31 790.485 hr

8.32 121.795 hr

8.33 $\dfrac{\lambda_1^2 + \lambda_2^2 + \lambda_1\lambda_2}{\lambda_1\lambda_2(\lambda_1 + \lambda_2 + \lambda_{12})}$

8.34 $\dfrac{8\lambda^2 + \lambda_c(\lambda + \mu) + 3\lambda\mu + \mu^2}{4\lambda^3 + \lambda_c^2(\lambda + \mu) + \lambda_c(4\lambda^2 + 3\lambda\mu + \mu^2)}$

8.36 $\beta = 1$; not realistic

8.37 0.1513

8.39 0.14391

8.40 0.999899, 0.9999; 0.98901, 0.99009

8.41 (*i*) 10^{-5} (*ii*) (0.98506)10^{-5}

8.42 $R(t) \cong 1 - \dfrac{kt^2}{2}$

8.43 (**a**) $1 - n\left(\dfrac{kt^2}{2}\right);\ 1 - \left(\dfrac{kt^2}{2}\right)^n$

$\quad\;\;$ (**b**) $1 - {}_nC_{n-r+1}\left(\dfrac{kt^2}{2}\right)^{n-r+1}$

CHAPTER 9

9.1 $[\ln(1 - R^{1/n})]/[\ln(1 - p)];\ R/p^n$

9.2 $\displaystyle\sum_{i=1}^{n} \dfrac{c_i \ln(1 - R^{c_i/c_o})}{\ln(1 - p)}$

9.3 $\left[\{\ln(1 - p_i)\} \sum_{j=1}^{n} \left\{ \frac{1}{\ln(1 - p_j)} \right\} \right]^{-1}$

9.4 $(12, 9, 6)$; $C_u = 48$; $R_u = 0.9918$

9.5 $(9, 8, 8, 7)$

9.6 $C_s = 854$ units; 7.835

9.7 $(30, 14, 8, 4)$; $C_u = 122C_0$; $C_s = 135C_0$

9.8 $n = 2$

9.9 1.145 years

9.10 3.735 years

9.11 Choose option 3 with a (worth/cost) ratio of 1.46

9.12 $w = 2$

9.13 $w = 3$

CHAPTER 10

10.1 $f_n(t) = (1/\mathscr{A}) f_a[(t/\mathscr{A}) - t_0]$
$F_n(t) = F_a[(t/\mathscr{A}) - t_0]$
$\lambda_n(t) = (1/\mathscr{A}) \lambda_a[(t/\mathscr{A}) - t_0]$

10.2 $(MTTF)_n = \mathscr{A}(MTTF)_a$

10.3 $(MTTF)_n = \mathscr{A}\{(MTTF)_a + t_0\}$

10.4 $T_{50n} = \mathscr{A}T_{50a}$

10.5 $T_{50n} = \mathscr{A}T_{50a}$

10.6 (i) 2.154, (ii) 1.585, (iii) 1.29

10.7 41.5; 29.34

10.8 (i) 10^{-5} hr^{-1}, (ii) 69,315 hr, (iii) 16.07%

10.9 193.7

10.10 70

10.11 54.1%

10.12 60.65%

10.13 37.84; 8.95

10.14 $\ln \mathscr{A} = B[(1/T_1) - (1/T_2)]$

10.15 (i) 16.4, (ii) 8.123, (iii) 2.78

10.16 (i) 269, (ii) 65.98, (iii) 7.71

10.17 26.82

10.18 18,078; 1.558

Index